T0261488

SNAKES

SNAKES

Ecology and Conservation

EDITED BY

STEPHEN J. MULLIN
DEPARTMENT OF BIOLOGICAL SCIENCES
EASTERN ILLINOIS UNIVERSITY

RICHARD A. SEIGEL
DEPARTMENT OF BIOLOGICAL SCIENCES
TOWSON UNIVERSITY

COMSTOCK PUBLISHING ASSOCIATES
A DIVISION OF CORNELL UNIVERSITY PRESS
ITHACA AND LONDON

First published 2009 by Cornell University Press

Library of Congress Cataloging-in-Publication Data

Snakes : ecology and conservation / edited by Stephen J. Mullin
and Richard A. Seigel.
   p. cm.
  Includes bibliographical references and index.
  ISBN 978-0-8014-4565-1 (cloth : alk. paper)
  1. Snakes—Ecology. 2. Snakes—Conservation. I. Mullin,
Stephen J., 1967– II. Seigel, Richard A. III. Title.

  QL666.O6S655 2009
  597.96'17—dc22

2008046823

Cornell University Press strives to use environmentally responsible
suppliers and materials to the fullest extent possible in the publishing of its books. Such materials include vegetable-based, low-VOC
inks and acid-free papers that are recycled, totally chlorine-free, or
partly composed of nonwood fibers. For further information, visit
our website at www.cornellpress.cornell.edu.

Cloth printing   10 9 8 7 6 5 4 3 2 1

Contents

Preface

This book follows in the footsteps of two previous efforts—*Snakes: Ecology and Evolutionary Biology* (1987) and *Snakes: Ecology and Behavior* (1993)—to provide established and new researchers with a current synopsis of snake ecology. In the preface to each of these earlier works, one of us (R. A. S.) admitted that he had erred in assuming that another "Biology of the Serpentes" book was not worth tackling. And after the first two books, we thought that perhaps yet another book was not needed—we were wrong again. Because our understanding of snake ecology continues to evolve, this field of study provides a seemingly inexhaustible source of research topics to pursue. Furthermore, even more time has now elapsed between this book and its predecessor than between the publications of the first and second books. As such, the need to enlighten our audience about recent advances in methodology and analysis is obvious. Like the two previous volumes, we developed the concept for this book with three goals in mind: (1) to summarize what is known about the major aspects of snake ecology and conservation, (2) to provide a compilation of the primary literature on this topic that is equally valuable to experienced and developing researchers, and (3) to stimulate new and innovative research on snakes by drawing attention to those areas in which there is a paucity of effort.

Given the ever-increasing number of quantified declines in both population size and species diversity among a variety of taxa, this book has an urgent fourth purpose that almost overshadows the previous three—to provide an awareness of the threats to snake populations and examine the strategies available to protect these unique organisms from further population declines or extinction. Indeed, if the reader is familiar with the contents of the previous snake ecology books, you are already aware that conservation is a topic that carries over from both of them. It is clear to us that the exponential growth of the world human population has already exacted a toll, both directly and indirectly, on snake populations. Furthermore, because of their typical role in most trophic webs, it is not a great leap for us to suggest that the health of snake populations is indicative of overall environmental health—in much the same way that amphibians, over the past two decades, have been viewed as the canaries in the environmental coal mine.

Other significant events that have occurred since the publication of the second *Snakes* book include the second through fourth meetings of the Snake Ecology Group, a loosely organized collection of biologists who are united in their enthusiasm for snakes. The attendance and level of participation have increased steadily with each successive conference, and we have observed that they are especially conducive to promoting collaborative efforts among several, sometimes disparate, subdisciplines. It is from the presenters at the 2004 meeting that we solicited many of the contributions to this book. Because the field of snake ecology has continued to evolve, it should come as no surprise that the authors of these chapters include many individuals who did not contribute to either of the earlier *Snakes* books. We encouraged these authors to interact frequently when writing their chapters and to cross-reference one another's work.

As was the case for the two previous books, our primary audience is the professional scientist; we are hopeful that curatorial staff in zoological parks and nongame wildlife managers will also find this information of interest. When the previous volumes were published, one of us (S. J. M.) was a student who was further encouraged by them; similarly, we trust this book will stimulate creative research and be an invaluable reference for today's developing snake ecologists. If nothing else, we hope that our efforts will continue to foster both interest in and scholarship about snake populations with objectives that include their conservation.

STEPHEN J. MULLIN
RICHARD A. SEIGEL

Acknowledgments

Even though its meetings are irregular, whenever the Snake Ecology Group gets together, one recurring theme is that studying the natural history of snakes is *really* fun but also potentially challenging because funding is scarce. So, it is only natural for us not only to recognize the excellent work of our authors (and their patience with our requesting multiple revisions) but also to acknowledge support provided to all snake ecologists, especially from the ever-decreasing pool of funding agencies that still support research in basic natural history. We also thank all the snake researchers whose work provided the foundations for many of the ideas presented in these chapters.

We are grateful for the rewarding interactions with our team at Cornell University Press: Candace Akins, Scott Levine, Heidi Lovette, Susan Specter, and Emily Zoss. In addition to our own internal reviewing process, several colleagues provided critical feedback at various stages of this project, including G. Blouin-Demers, G. Brown, C. Dodd, H. Greene, J. Mitchell, C. Phillips, H. Reinert, G. Rodda, and J. Rodríguez-Robles.

S. J. M. thanks the administration and staff of Department of Biological Sciences at Eastern Illinois University (EIU) for support of this project, and he thanks the students in the EIU Herpetology Lab for their feedback and tolerance of his extended spells of absent-mindedness during its completion. Portions of this book were completed while S. J. M. was on a sabbatical leave granted by Mary Anne Hanner, dean of the College of Sciences, EIU. Previous guidance from mentors during his training (R. Cooper, H. Greene, W. Gutzke, and H. Mushinsky) is also appreciated. R. A. S. thanks Towson

University for funding and logistical support during the writing of this book, with special thanks to Dean Intemann, Dean David Vanko, and Provost James Clements. Support for R. A. S. was also provided by the Dynamac Corporation, with special thanks to Ross Hinkle for this long-term support.

Contributors

OMAR ATTUM, Department of Biology, Indiana University Southeast

STEVEN J. BEAUPRE, Department of Biological Sciences, University of Arkansas

XAVIER BONNET, Centre d'Etudes Biologiques de Chizé, Centre National de la Recherche Scientifique (France)

FRANK T. BURBRINK, Department of Biology, College of Staten Island–The City University of New York

GORDON M. BURGHARDT, Departments of Psychology and Ecology & Evolutionary Biology, University of Tennessee

TODD A. CASTOE, Department of Biochemistry & Molecular Genetics, University of Colorado–School of Medicine

DAVID CHISZAR, Department of Psychology, University of Colorado

MICHAEL E. DORCAS, Department of Biology, Davidson College

LARA E. DOUGLAS, Department of Biological Sciences, University of Arkansas

CHRISTOPHER L. JENKINS, Project Orianne, Ltd.

GLENN JOHNSON, Department of Biology, State University of New York at Potsdam

MICHAEL HUTCHINS, The Wildlife Society

RICHARD B. KING, Department of Biological Sciences, Northern Illinois University

BRUCE A. KINGSBURY, Department of Biology, Indiana University–Purdue University Fort Wayne

THOMAS MADSEN, School of Biological Sciences, University of Wollongong (Australia)

STEPHEN J. MULLIN, Department of Biological Sciences, Eastern Illinois University

JAMES B. MURPHY, National Zoological Park

CHARLES R. PETERSON, Department of Biological Sciences, Idaho State University

KENT A. PRIOR, Critical Habitat, Parks Canada

RICHARD A. SEIGEL, Department of Biological Sciences, Towson University

RICHARD SHINE, School of Biological Sciences, University of Sydney (Australia)

KEVIN T. SHOEMAKER, Department of Environmental and Forest Biology, College of Environmental Science and Forestry–State University of New York

PATRICK J. WEATHERHEAD, Program in Ecology & Evolutionary Biology, University of Illinois

JOHN D. WILLSON, Savannah River Ecology Laboratory, University of Georgia

SNAKES

Introduction

Opening Doors for Snake Conservation

STEPHEN J. MULLIN AND RICHARD A. SEIGEL

An unfortunate certainty associated with the ever-growing human population is the loss or alteration of habitat. Coupled with this population increase, technological advances have allowed humans to become more mobile, and with that mobility comes the increased likelihood that other organisms will—intentionally or not—move with them. These are just a few of the reasons why many species of nonhuman organisms are experiencing population declines. Although many people are willing to extend some effort for conservation when endearing animals like pandas or parrots are concerned, the sympathy extended to the marvelous variety of snake species is rather limited. This book provides an examination of current research concerning the ecology of snakes, with an emphasis on how this research has been, or has the potential to be, applied to their conservation.

Snakes have intrigued humans for centuries, and were incorporated into several mythologies (e.g., the staff of Aesculapius) and cultures (e.g., Irula snake-catchers; Whitaker 1989). Among the biologists who study snakes, there is little question of their fascination about the natural history of snakes. In spite of a limbless ectothermic body, snake species have radiated to inhabit all of the Earth biomes except the polar regions—even then, species can be found within the Artic circle (e.g., *Vipera berus*; Carlsson and Tegelström 2002). The variety of locomotory modes observed in snakes has garnered much interest (see Gans 1986 and references therein), perhaps exceeded only by that allocated to snake size–prey size relationships (reviewed in Arnold 1993). There is also considerable enthusiasm for snakes in a rapidly growing and dedicated sector of the commercial pet trade.

Sadly, the considerable amount of effort by researchers and enthusiasts has not translated into public support for snakes. Declines in the sizes of snake populations do not receive the same level of attention as has recently been the case for sea turtles (Meylan and Ehrenfeld 2000) or any number of amphibian species (Miller 2000; Norris 2007). The same enthusiasm for snakes observed among commercial breeders might be exacting a negative, but poorly quantified, impact on wild populations (Nilson et al. 1990; Schlaepfer et al. 2005). Other human activities are known sources of declines in wild snake populations (Gibbons et al. 2000), even among venomous species (Whitaker and Shine 2000). In the United States, the continued sanctioning of rattlesnake round-ups clearly does not provide any benefits for the populations of these species (mostly *Crotalus adamanteus, C. atrox,* and *C. horridus;* Fitzgerald and Painter 2000). The troubling nature of this treatment of snakes is compounded by the fact that many of these species represent the highest levels in their respective trophic webs. As such, continued declines in snake populations are likely to leave their prey populations (several of which are commonly construed as pests) unchecked.

The field of snake ecology has advanced considerably over the last 15 years—conceptual frameworks have been revised in light of new findings, and improvements in technology have afforded opportunities for new avenues of research. The contributors to this book represent a healthy mix of the seasoned developers of some of these frameworks and techniques, and the up-and-coming pioneers who have built on these advances to lead conservation efforts in new directions. In addition to describing some of the challenges associated with studying snake ecology in Chapter 1, Michael Dorcas and J.D. Willson discuss several of the recent applications of marking and modeling snake populations. In Chapter 2, Frank Burbrink and Todd Castoe not only describe the latest and most appropriate techniques used in phylogeographic studies, but also tackle the monumental task of summarizing the recently published research on snake phylogeography. In Chapter 3, Richard King summarizes the latest work in population genetics and illustrates the processes that generate population structure on fine geographic and temporal scales. In Chapter 4, Christopher Jenkins, Charles Peterson, and Bruce Kingsbury couple their expertise with geographical information systems and spatial modeling to answer questions associated with the ecology of snakes at the landscape level.

The next couple of chapters encompass our attempt to update reviews of areas within snake ecology that have received a fair amount of attention over the past two decades. In Chapter 5, Patrick Weatherhead and Thomas Madsen summarize the central concepts in behavioral ecology, with particular emphasis on thermal ecology and predator-prey interactions. And because successful reproduction is critical to population viability, we asked Richard Shine and Xavier Bonnet to interpret the latest research examining snake reproductive biology (Chapter 6). In spite of these authors' efforts to combine benchmark studies in these areas with the volume of recent literature, we are still left with the impression that there is much to learn about the behavioral and reproductive ecology of snakes, particularly as it pertains to their conservation.

We have also included contributions designed to address a few relatively new, and sometimes controversial, fields that focus specifically on the importance of conserving snake populations in the field. In Chapter 7, Bruce Kingsbury and Omar Attum discuss the efficacy of management strategies such as repatriation, translocation, and captive propagation. In Chapter 8, Kevin Shoemaker, Glenn Johnson, and Kent Prior describe how various techniques of habitat manipulation can be used to promote the stability of snake populations or minimize the impacts of human alteration of habitat. The impacts of snakes in various ecosystems are further illustrated by Steven Beaupre and Lara Douglas, who describe in Chapter 9 the methodology associated with using snakes as biological monitors of environmental quality.

An enduring mystery to most snake biologists is that the curiosity aroused in the general public by various aspects of snake biology does not also generate sympathy for the plight of many of these species. It is for this, and other reasons, that we have asked Gordon Burghardt, James Murphy, David Chiszar, and Michael Hutchins to contribute Chapter 10, which examines human perceptions of, and interactions with, snakes in natural and educational settings and what can be done to improve the image that snakes have with the general public. Although the emphasis on conservation might be perceived as being greater in this chapter, we hope that this theme can be easily detected in all the contributions to this book.

We expect that this book will be of interest to ecologists, conservation biologists, and curatorial staff at zoological parks and to be of particular value to herpetologists and wildlife and resource managers. We especially dedicate this book to new workers in the field, and we hope that our audience will share our enthusiasm for snakes and the ecological insights that have been generated by studying them. Given the amount of information that is yet to be discovered, we are confident that this book will motivate future generations of researchers to pursue additional avenues of research as well as encourage them to advocate the conservation of snakes.

Readers familiar with the first two volumes in this series of books on snakes might find this one to be lacking in the number of tables that summarize data from the primary literature. Our explanation is that in this book, to a certain extent, we are navigating in uncharted waters with the coverage of conservation measures that are specific to snakes. Simply put, studies addressing the conservation of snakes are relatively few in number, and many conservation tools that have been applied to other taxa remain to be tested with snake species. The advances in molecular techniques used to better understand evolutionary relationships among snake species mandated another change in this book. The taxonomy used throughout reflects this improved understanding and follows the Integrated Taxonomy Information System catalogue (ITIS 2006) and Crother et al. (2008).

In the second of the *Snakes* books, Dodd (1993b) suggested that some snake species might not persist into the twenty-first century. Although Dodd's prediction has not been borne out (we hope!), a number of snakes are still critically endangered (e.g., *Alsophis;* Sajdak and Henderson 1991) or continue to have serious implications for the conservation of other species (e.g., *Boiga;* Rodda et al. 1999d). Environmental threats to other taxa are also generating negative impacts on snakes (e.g., amphibian populations becoming extinct following a chytrid fungus infestation; Lips et al. 2006) because of trophic cascades, competitive displacement, or other ecological relationships. The lack of public interest in how these phenomena are affecting snake populations is juxtaposed with the continued public fascination with snakes (Greene 1997) and the people who study them (Montgomery 2001). Our appreciation for snakes and our continued puzzlement over their maligned reputation are shared by the contributors to this book. Together, we hope the following chapters provide an examination of current research concerning the ecology of snakes, with an emphasis on how this research can, or has, been applied to their conservation. Because conservation goals can benefit from increased public outreach, we also hope that this book inspires our colleagues to expand their sphere of influence and render extinct the ill-deserved reputation suffered by these marvelous animals.

1

Innovative Methods for Studies of Snake Ecology and Conservation

MICHAEL E. DORCAS AND JOHN D. WILLSON

Snakes are fascinating to many laypeople and scientists alike, and numerous studies of snake ecology and natural history have been conducted. For nearly all snake species, however, a comprehensive understanding of their ecology, and especially population biology, is lacking. Such gaps in our knowledge limit our ability to develop effective conservation and management strategies or, more often, prohibit arguments that conservation is needed at all. We argue that snakes, although often challenging to study, offer many opportunities for ecological study unparalleled by other taxa.

One of the main reasons ecologists often shy away from snakes as study animals is the perception that their secretive natures make them difficult to study. Developing a more complete understanding of snake ecology and its application to conservation has been hampered by this perception (warranted or not). Unfortunately, because of their apparent rarity we often know least about the species that are most in need of conservation. Efforts to study snakes can sometimes be hindered by an enthusiasm for the animals that actually inhibits the development of meaningful questions and study designs. Many researchers who begin snake studies either (1) do not have a question at all, (2) have a question but do not know why that question is important, (3) do not match their question with appropriate methodology, or (4) select a species or group of species that are not particularly amenable to addressing the question(s) of interest (Seigel 1993). For example, many herpetologists have embarked on radiotelemetric studies of a species of snake with no clear question or hypothesis (i.e., the goal becomes the study in itself). Such herpetologists sometimes have a question (e.g., What is the home range of my study species?), but do not know whether or why that question is important. Although we have historically learned much about snake ecology through basic studies of snake natural history, the information required for the effective conservation of snakes nearly always requires answers to specific questions relating to such things as diet, habitat requirements, and population status.

Despite the lack of comprehensive information on many snake species and the perception that they are difficult to study, snakes have been proposed as model organisms (Beaupre and Duvall 1998b; Secor and Diamond 1998; Shine and Bonnet 2000). In fact, snakes are particularly amenable to numerous techniques used in ecology and conservation biology. For example, some snakes are particularly good subjects for mark-recapture studies because they occur at high densities and are easily trapped and marked. Many species are particularly amenable to focal animal studies such as radiotelemetry, allowing a detailed examination of habitat use, movement, and physiological ecology. Although snakes pose significant challenges for effective ecological study in some situations, snakes also offer many ideal opportunities for in-depth investigation of ecological phenomena, especially if the correct questions are matched with appropriate capture techniques, study design, and analyses (see also Seigel and Mullin, Chapter 11).

Our goal here is to discuss innovations in methodology for the design and implementation of ecological and conservation-oriented studies of snakes. We take the approach that the reader can find information on details of the basic techniques elsewhere in this book and in other sources; here we focus instead on the development and use of newer techniques and question-oriented approaches to studying snake ecology.

In this chapter, we discuss techniques related to (1) the capture and marking of snakes in the field, (2) focal studies of individual snakes, and (3) studies of snake populations. In each section, we discuss which types of questions can be addressed and which methodological and analytical techniques are best for addressing those questions. Our hope is that, during the course of a well-designed snake ecology study, researchers will seize the opportunity develop and investigate new and exciting questions (Greene 2005; Blomquist et al. 2008). In this chapter, we also make the reader aware of biases associated with certain techniques and how those biases can affect the interpretations of data. The reader should note that we present information on techniques that we have used or with which we are most familiar. Thus, unlike good snake ecology studies, this review is biased toward techniques used by us and our colleagues.

Capturing and Marking Snakes

In the first volume of the *Snakes* series, an entire chapter is dedicated to describing techniques for capturing and marking snakes (Fitch 1987a).

Although these techniques remain the standards among snake ecologists, numerous refinements have been proposed, along with novel methods employing recent technological advances. In addition, studies have elucidated sampling biases that can hamper the interpretation of capture data. Next, we review advances in methods for capturing and marking snakes, with particular emphasis on how the choice of capture methods can influence the analytical tractability of data and interpretation of results.

Active Capture Methods

Active capture methods involve the observer's searching out free-ranging snakes. These methods take advantage of an a priori understanding of snake behavior and can be among the most effective methods for capturing large numbers of snakes. Because such methods rely on the competence of the observer, they are sensitive to observer bias (Table 1.1). For example, interobserver variability was one of the strongest sources of variation in visual counts of Brown Treesnakes (*Boiga irregularis*) on Guam (Rodda and Fritts 1992b). In addition, visual searches often target snakes only in specific habitats or involved in specific behaviors (e.g., basking, foraging, or hiding beneath cover). Because snake activity is highly dependent on environmental conditions (Peterson et al. 1993), active capture methods may suffer from low repeatability as a result of a variation in capture rates caused by environmental variation (Table 1.1).

TABLE 1.1
Strengths and weaknesses of frequently used capture methods for snake population studies

Capture Method	Capture Type	Effort	Capture Rate[a]	Repeatability[b]	Observer Bias
Opportunistic search	Active	− −	++	− −	++
Transect/quadrat survey	Active	++	− −	−	+
Coverboard	Active	−	+	−	−
Road survey	Active	− −	+	−	−
Drift fence/trap	Passive	++	+	+	− −
Funnel trap[c]	Passive	+	+	+	−

Note: Plus and minus signs represent high (+), very high (++), low (−), or very low (− −) values within a category. Thus, pluses are strengths for capture rate and repeatability, but are weaknesses for effort and observer bias.

[a] Efficacy of capture methods varies by snake species (e.g., stand-alone funnel traps are effective for many aquatic and arboreal species but not for many terrestrial species). Thus, capture rate is considered here for species for which the given method is effective.

[b] *Repeatability* refers to comparability of sampling events, independent of differences among observers (observer bias) and, particularly, of sensitivity to environmental stocasticity, changes in snake behavior, or other factors that cause short-term variation among samples. For example, even within a single day, captures under coverboards can vary greatly depending on the environmental conditions at the time of the survey.

[c] *Funnel traps* here refers to stand-alone funnel traps, including aquatic minnow traps and arboreal snake traps.

Visual encounter surveys (VES), the simplest active capture method, are effective for surface-active species or for those that use specific habitat types or bask conspicuously. Although the basics of VES have not changed, increasing standardization by constraining time, effort, or the spatial pattern of sampling (e.g., transects or area-constrained searches) has increased the utility of VES for analytical techniques that rely on standardized sampling (e.g., relative abundance indices). Moreover, several authors have addressed potential sources of bias in VES, improving our ability to interpret results. For example, biotic and abiotic factors that influence census counts have been examined in Shedao Pit Vipers (*Gloydius shedaoensis;* Sun et al. 2001).

Two other active capture methods commonly applied to snakes are the turning of natural or artificial cover objects (coverboards; Fitch 1992; Grant et al. 1992) and road surveys (Fitch 1987a). Although these techniques are essentially variants of VES and suffer from similar repeatability issues, they are less prone to observer bias than VES (Table 1.1). Both methods are highly effective for collecting many snake species, some of which are not sampled effectively using other methods (e.g., traps). However, both coverboards and road surveys have been used relatively infrequently for snake population monitoring (but see Mendelson and Jennings 1992; Sullivan 2000).

Passive Capture Methods

Passive capture methods generally involve trapping animals. Although passive capture methods often yield a lower catch per unit effort than active methods, they are usually preferable for population studies because they are insensitive to observer bias and maximize repeatability by integrating captures over time (Table 1.1; Willson and Dorcas 2003; Willson et al. 2005). Most snake traps are variants of funnel traps (Fitch 1951) that have been used to sample snakes in both aquatic (e.g., minnow traps; Keck 1994a; Willson et al. 2005) and arboreal (Rodda et al. 1999a) habitats. Several new terrestrial funnel trap designs have been developed, most of which are wooden and are used in conjunction with drift fences (e.g., Burgdorf et al. 2005; Todd et al. 2007). Although unbaited funnel traps can be effective, baiting increased capture rates in both aquatic (Keck 1994a; Winne 2005) and arboreal (Rodda and Fritts 1992b; Rodda et al. 1999a) habitats. Escape rates from traps can be high for both arboreal (Rodda et al. 1999a) and aquatic (Willson et al. 2005) traps. Although flaps covering the funnel openings have been shown to reduce rates of entry to the traps, they increase snake retention rates by 170% (Rodda et al. 1999a). As with VES, quantifying biases is crucial to the interpretation of capture data because nearly any trap will not representatively sample all species or demographics within species (see examples in Enge 2001; Willson et al. 2005; Rodda et al. 2007b; Todd et al. 2007; Willson et al. 2008).

Marking Snakes

Individually marking snakes is necessary for mark-recapture studies and allows the researcher to assess movement and changes in body size, condition, or reproductive status. Scale-clipping (Weary 1969; Brown and Parker 1976b; Fitch 1987a) remains one of the most effective and inexpensive methods for marking snakes; even small species can be scale-clipped by using a large-gauge needle to excise a portion of scale (Mao et al. 2006). Clipped scales, however, can regenerate rapidly and, after long periods, marks may be difficult to recognize (Conant 1948; Fitch 1987a).

An alternate method for marking snakes involves the implantation of passive integrated transponders (PIT tags; Camper and Dixon 1988; Gibbons and Andrews 2004). PIT tags are typically injected into the body cavity using a large-bore needle and provide a presumably permanent and unambiguous unique identification number when a reader passes within a short distance (usually <7 cm). Disadvantages of PIT tags include cost (US$6–8 per tag) and size—most snake ecologists agree that they should not be used in very small snakes. Some companies (e.g., BioMark) are now making smaller PIT tags that may be amenable to smaller snakes. Studies have documented no detrimental effects of PIT tags on the growth and movement of Pigmy Rattlesnakes (*Sistrurus miliarius;* Jemison et al. 1995) or on the growth and crawling speed of neonatal Checkered Gartersnakes (*Thamnophis marcianus;* Keck 1994b). PIT tag loss can occur either through the skin (Germano and Williams 1993) or via expulsion through the gut (Roark and Dorcas 2000).

An effective and inexpensive method has been described for branding snakes using field-portable cautery units designed for ophthalmic surgery (Winne et al. 2006a). Cautery units can be used to brand the ventral scutes and adjacent dorsal scales (Fig. 1.1) and have been shown to be effective over several years and useful even on small individuals or species (Winne et al. 2006a).

Focal Animal Studies

Focal animal studies are ecological studies that rely on the in-depth examination of individual animals. Although the focus of many conservation-oriented studies is assessing population status (size or trends), measuring only population status often does not provide information about the mechanisms underlying population dynamics, which are critical for effective management (Beaupre 2002). The secretiveness of some species makes evaluation of population status impractical and thus, focal animal studies provide the most feasible way to obtain the information necessary to make reasoned conservation or management decisions (Seigel et al. 1998).

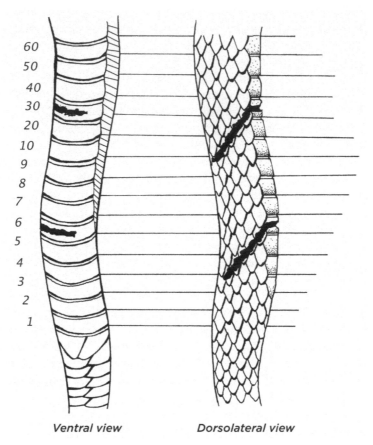

Ventral view   **Dorsolateral view**

Fig. 1.1. Illustration of a snake heat-branded with ID #36 using a medical cautery unit. For each mark, the researchers branded the anterior portion of the ventral scale and extended the mark diagonally onto the adjoining dorsal scales. (Illustration drawn by R. Taylor; used with permission of Society for the Study of Amphibians and Reptiles from Winne et al. 2006a)

Focal animal studies can be used to address questions about spatial ecology, habitat use, diet, energy acquisition and allocation, reproductive ecology, behavioral ecology, and predator-prey relationships. These studies can also provide information useful for the control of invasive snake species such as *B. irregularis* on Guam (Rodda et al. 1999d) or Burmese Pythons (*Python molurus bivittatus*) in Everglades National Park (Snow et al. 2007). Although the basic techniques used in focal animal studies have not changed, refining these techniques, combining them with other methodologies, considering study design, and using advanced analytical methods allow increasingly insightful perspectives on the ecology and conservation of snakes.

For focal animal studies to be effective and meaningful, investigators must (1) develop thoroughly the question(s) of interest and understand how their results can be applied to our understanding of ecology and/or effective conservation efforts; (2) consider carefully what technique(s) are most appropriate to address their question(s); (3) consider how their study will be designed to maximize inferential capability; and (4) consider the inherent limitations of their study, such as sample size, expenses, and required time and effort.

Collection and Selection of Animals

Because the results of focal animal studies are often extrapolated from a small number of individuals to the entire population or even species, the means by which animals are collected and selected for study are extremely important. When the study species is secretive, researchers often have no alternative than to use any and all animals that become available through trapping or incidental captures. In such cases, researchers should be aware of, and attempt to correct for, any biases inherent in the animal selection and how those biases affect their results. For example, if all animals were collected on roads, investigators might infer a far greater use of roadside habitats than if they had a sample truly representative of the population.

Radiotelemetric Studies

The miniaturization of radiotransmitters and the development of surgical techniques to implant radiotransmitters (Reinert and Cundall 1982) have allowed insights into the details of snake ecology unimagined 25 years ago. The basic techniques of radiotelemetry in snakes have been described elsewhere (Reinert and Cundall 1982; Reinert 1992; Ujvári and Korsós 2000; Millspaugh and Marzluff 2001). Here we discuss novel or often-overlooked issues that should be considered when conducting radiotelemetric studies. We recommend that anyone wishing to use radiotelemetry seek hands-on assistance from a snake ecologist experienced in the technique before and during the initial stages of his or her study.

A Few Considerations

The intensive nature and cost of radiotelemetric studies often limit the number of animals that can be sampled. Within the constraints of the study, however, as many snakes as possible should be studied because, in nearly all analyses, each snake represents a single data value. Moreover, in comparisons among groups (e.g., sexes, species, or treatments), the number of animals is divided among groups, thus limiting the ability to discern effects. Combined with the large interindividual variability often observed in radiotelemetric studies (Millspaugh and Marzluff 2001), statistical power is often limited.

In many cases, snake researchers miss the opportunity to gain insights into the ecology of their animals because they do not take time for careful observation. When snake ecologists radio-track an animal, they often just record the geographic coordinates and other information and then move as quickly as possible to tracking the next animal. Often, researchers radio-track their animals at the same time each day, further limiting their ability to observe the full spectrum of activity and behaviors afforded by radiotelemetric studies. Relocating animals at different times of day (e.g., at night) may be less convenient, but it may provide unique insights into the ecology of the study species.

Surgical Considerations

Radiotelemetric studies of snakes have been conducted using transmitters that were force-fed (Fitch and Shirer 1971; Lutterschmidt and Reinert 1990), implanted subcutaneously (Anderka and Weatherhead 1983), or attached externally (Ciofi and Chelazzi 1991). Intraperitoneal implantation of radiotransmitters (Reinert and Cundall 1982), however, allows for the long-term monitoring of individual snakes with minimal disruption of normal physiological processes (e.g., digestion) and behaviors and is currently the method used by most snake ecologists.

Most snake ecologists use gas anesthesia and have found that isofluorane generally works more quickly than others (e.g., halothane) and causes less liver damage (at least in humans; Goldfarb et al. 1989). Generally, inhalation of anesthesia is induced passively by placing the snake's head in a chamber or tube (Hardy and Greene 2000). We have found that using a refurbished anesthesia machine connected to an endotracheal tube and intubating (i.e., placing the tube directly into the glottis) the snakes allows oxygen to be administered during anesthesia and facilitates direct inhalation, resulting in shorter induction times. We have successfully used this technique with ratsnakes (*Pantherophis [Elaphe]*), kingsnakes (*Lampropeltis*), Timber Rattlesnakes (*Crotalus horridus*), and *Python molurus bivittatus*. Propofol has been used by some veterinarians as a form of short-term anesthesia in reptiles. Propofol can be injected directly into the heart (or caudal vein) in snakes and causes rapid and complete anesthesia in many species (Anderson et al. 1999). We have used propofol to anesthetize ratsnakes before the application of gas anesthesia, and it appeared to reduce the stress associated with intubation, allowing the immediate initiation of surgery.

Transmitter Expulsion

It is not uncommon for researchers to find a radiotransmitter but no snake in the field when locating their animals and to assume that the snake died or was depredated. During a radiotelemetric study of pythons, radiotransmitters were found, often within snake fecal material, suggesting that the snakes expelled radiotransmitters implanted intraperioneally (Pearson and

Shine 2002). Dissection of a dead subject revealed a radiotransmitter that was partially incorporated into the stomach. Such expulsion of radiotransmitters from the peritoneal cavity through the gut wall is apparently accomplished through the same physiological mechanism as seen in fish (Chisholm and Hubert 1985) and in PIT tag expulsion in snakes (Roark and Dorcas 2000). We concur with Pearson and Shine (2002) that investigators finding a radiotransmitter but no snake remains should exercise caution in assuming the death of their study animal.

Automated Radiotelemetry

The majority of snake radiotelemetric studies have been conducted in a similar manner—the investigator determines the position of the snake at specified intervals by manually tracking the animal. Today, the use of radiotransmitters outfitted with global positioning systems (GPSs) allows for the real-time automated tracking of animals ranging from whales to turtles (Rogers 2001). Unfortunately, the small size of most snakes and the need to implant radiotransmitters currently prohibits the use of automated GPS in snake studies. Systems have been developed, however, that allow the tracking of animals automatically using a series of directional antennas. Such a system has been used in Panama to follow the movements of various avian and mammalian species (Wikelski et al. 2007), and we see no reason why such a system could not be used for snakes.

Automated monitoring of body temperature (T_b), especially when combined with simultaneous measurements of environmental temperatures, can provide substantial insight into the habitat use and activity patterns of snakes (Peterson et al. 1993). Automated systems (Fast-Data System, Telonics, Mesa, Ariz.) have been used to continually monitor the T_b values of Rubber Boas (*Charina bottae*) in southeastern Idaho and have documented nocturnal activity at low temperatures (Dorcas and Peterson 1998). One system (Lotek—SRX-400 with W21 event logging) has been used for several years to automatically monitor the T_b values of *Crotalus horridus* in Arkansas (S. Beaupre, pers. comm.). This system uses directional antennas that record signal strength as well as temperature and in certain circumstances (e.g., flat, relatively uniform terrain) might be used to estimate the locations of snakes.

Analysis of Radiotelemetry Data

Numerous methods for the analysis of spatial data collected via radiotelemetry have been developed (White and Garrott 1990; Reinert 1992, 1993). Snake researchers frequently evaluate habitat use and home range size of snakes using geographical information systems (GISs; e.g., ArcGIS from ESRI, Redlands, Calif.). Several publications on the analysis of radiotelemetric data (e.g., Millspaugh and Marzluff 2001) and software applications allow relatively easy calculation of spatial parameters (Hooge and

Eichenlaub 2000), but we remind researchers that their question(s) should drive the choice of analytical methods. All too often, snake researchers measure the home ranges of snakes without a thorough understanding of how the analytical technique used (e.g., minimum convex polygon, MCP, or kernel) might influence their conclusions. For example, snake ecologists might calculate a home range for an animal that migrates annually from one area to another. The calculation of a MCP home range for that animal might show a much larger area than is actually used by the animal and include large areas of unsuitable habitat.

Snake ecologists often evaluate habitat use or habitat selection using data generated from radiotelemetric studies. It is important to understand that the analytical methods for the determination of habitat selection must involve the determination of habitats available to the snake (Reinert 1992, 1993).

Automated Cameras

The use of automated photography can provide insights into the ecology of many secretive animals, including snakes. Automated 35-mm film cameras, triggered by the removal of a rat carcass resting on a mechanical switch, have been used to film scavenging *Crotalus horridus* (DeVault and Rhodes 2002). Digital cameras used in this manner increase the image capacity of these systems and, because film developing is not required, greatly reduces costs of operation (Guyer et al. 1997).

Automatically controlled still and video cameras can document predation by various predators, including snakes (e.g., Renfrew and Ribic 2003; Peterson et al. 2004). Most researchers use time-lapse video (2–5 frames/s) that allows recording for a relatively long time (Weatherhead and Blouin-Demers 2004b; Clark 2006). Setting video cameras, positioned at places of high snake activity (e.g., hibernacula), to record based on triggering stimuli such as a switch or the breaking of a light beam should be possible and may allow the deployment of a system without maintenance for longer periods of time.

Automated Monitoring of PIT-tagged Snakes

Automated systems for monitoring animals implanted with PIT tags have been used in studies of fish (Prentice et al. 1990), voles (Harper and Batzli 1996), and bats (Kunz 2001). In some situations, automated monitoring of PIT-tagged snakes could provide considerable insights into snake activity patterns. To monitor PIT-tagged animals, a reader must be placed in an opening or area through which the animal is expected to move. An automated system that reads PIT tags was used to monitor the movements of Desert Tortoises (*Gopherus agassizii*) when they were diverted under highways through culverts (Boarman et al. 1998). Each time a tortoise passed over the reader's detecting coil, the system recorded the PIT-tag number, time of day, date, and duration of time the PIT tag was within reading distance of the coil. Similar systems could be used to monitor snake movements at communal hibernacula (e.g., Prior and Weatherhead 1996) or snakes passing through openings in drift fences (Gibbons and Semlitsch 1982).

Snake Thermal Ecology

Because temperature affects nearly every aspect of their biology, understanding thermal biology allows us to achieve a more complete understanding of snake ecology (Peterson et al. 1993; Weatherhead and Madsen, Chapter 5). When combined with studies of the effects of temperature, measurements of snake temperatures can be used to estimate the effectiveness of locomotion, prey capture, or digestion and can provide insight into the energetics limitations of snakes in various environments (Beaupre 1995b; Dorcas et al. 1997). For snake thermal ecology studies, proper measurement of snake T_b values and the thermal environment is essential.

When conducting field studies, measuring only air and/or substrate temperatures provides an inaccurate representation of the thermal environments available to snakes (Peterson et al. 1993). Fortunately, it is relatively easy to construct biophysical models for most species of snakes from copper tubing (Peterson 1982). Automated monitoring of these "snake models" using a datalogger provides an integrated and more accurate measurement of the thermal environment (i.e., operative temperature) available to snakes (Bakken and Gates 1975; Peterson et al. 1993).

Traditionally, the temperatures of snakes and other reptiles were measured by capturing an animal and inserting a quick-reading thermometer into its cloaca. In addition, measurements of the thermal environment usually consisted of air and possibly substrate temperatures (Dorcas and Peterson 1997). We now know that cloacal temperature measurements result in a biased sampling of snake T_b values and that measuring air or substrate temperatures provides an inadequate characterization of snakes' thermal environments (Peterson et al. 1993). Automated monitoring of both snake and environmental temperatures provides detailed and unbiased measurements that allow a more accurate understanding of snake thermal ecology and can provide insights into the activity and habitat use of snakes (Peterson and Dorcas 1992, 1994).

Automated monitoring of snake T_b values has primarily been conducted using temperature-sensitive radiotransmitters in conjunction with an automated receiving system (Peterson et al. 1993; Beaupre and Beaupre 1994). Such systems are costly and often require considerable maintenance. In addition, when snakes move out of the range of the system, no data are collected. The recent miniaturization of single-channel temperature dataloggers allows the automated collection of T_b values without a receiving station. We have used miniature dataloggers (Tidbits; Onset, Bourne, Mass.) to automatically monitor the T_b values of Eastern Diamondback Rattlesnakes (*Crotalus adamanteus*) and *Python molurus bivittatus* while tracking their movements using radiotelemetry. Dataloggers were programmed and coated with plastic tool dip (PlastiDip) before implantation into the snake's body cavity. After a period of time (e.g., months), the dataloggers were removed surgically and the data downloaded. It may be possible for snakes to expel implanted dataloggers through their gut in a manner similar to that described for radiotransmitters and PIT tags (Roark and Dorcas 2000; Pearson and Shine 2002).

An examination of T_b plots for *P. molurus bivittatus* in Everglades National Park allowed us to determine that, in November, snakes apparently remained in the water to stay warm at night and then emerged to bask when environmental temperatures were favorable (Fig. 1.2). This information would have been difficult to obtain without automated data acquisition because the snakes were in remote areas of the park, reachable only by helicopter.

Considerably smaller temperature dataloggers (iButton Thermochron; Dallas Semiconductor, Dallas, Tex.) that are inexpensive (approximately US$28) and hold more than 8000 date/time-stamped readings have been used successfully in Eastern Racers (*Coluber constrictor*; Green 2005). Other researchers have used these dataloggers successfully on various species of turtles (Grayson and Dorcas 2004; Harden et al. 2007).

Temperature-sensitive PIT tags have been used in the laboratory to examine thermoregulation in Corn Snakes (Roark and Dorcas 2000) and Rubber Boas (Zhang et al. 2008). The development of PIT readers (Blomquist et al. 2008) that can collect data from distances of greater than 25 cm might allow investigations into the thermal ecology and movements of small snakes in natural or semi-natural conditions.

Energetics

Understanding snake energetics can reveal aspects of snake ecology critical for conservation and management (Beaupre 2002). Because metabolic rate is dependent on temperature, incorporating T_b often allows more realistic models to be developed. For example, we modeled the energetics of *Crotalus adamanteus* using data on thermal dependency of metabolic rate and, thus, could predict the food required to meet the resting energetic demands of various-size snakes (Dorcas et al. 2004). When we incorporated the T_b values from free-ranging snakes (collected using implanted microdataloggers) into the model, we determined that only two prey items (30% of snake body mass) per year were required to meet resting metabolic demands (Fig. 1.3). Modeling the effects of landscape structure and prey abundances on energy acquisition and allocation in snakes allowed predictions of the effects

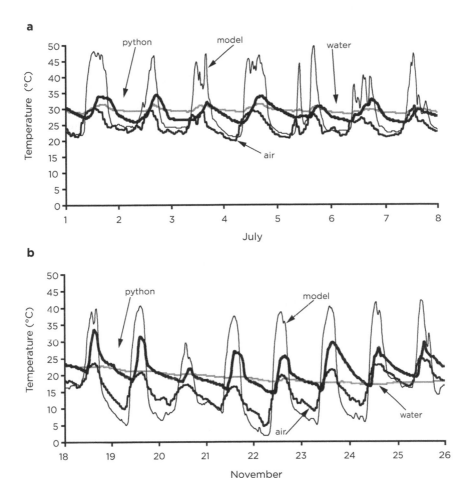

Fig. 1.2. Body temperature variation of an invasive 4.5-m female Burmese Python (*Python molurus bivittatus*) measured using a surgically implanted microdatalogger while being radio-tracked in Everglades National Park. (a) July (b) November. Snake model (operative environmental temperatures), water, and air temperatures were measured using other dataloggers or obtained from a nearby weather station. Note that during July, body temperature was higher and less variable than during November. In contrast, during relatively cold weather in November the snake's body temperature approached or exceeded 30 °C each day, apparently by remaining in the relatively warmer water and by basking when possible.

of landscape manipulation (e.g., forestry practices) on energetics of snakes (Beaupre 2002).

Models using measures of field-metabolic rates using the doubly-labeled water technique have been developed for several snake species. For example, snake T_b data were combined with metabolic rates measured in the field to compare the consequences of foraging mode in ambush (Sidewinder Rattlesnakes, *Crotalus cerastes*), and active foragers (Coachwhips, *Masticophis*

a

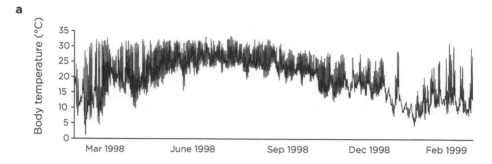

b

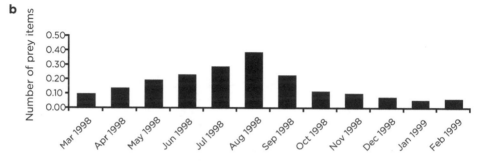

Fig. 1.3. Body temperatures and estimation of number of prey items per month required to sustain resting metabolic rate for an Eastern Diamond-backed Rattlesnake (*Crotalus adamanteus*). (a) Body temperatures obtained from free-ranging snake tracked using radiotelemetry and implanted with a microdatalogger (b) Calculations of the number of prey items required; these assume a 2500-g snake, prey items equivalent to 30% of the snake's body mass with an energy content of 5.9 kJ/g, and 80% assimilation efficiency: $\log_{10}$ SMR = $(0.930 \times \log_{10}$ Mass) + $(0.044 \times$ Temp) -2.589 (equation from Dorcas et al. 2004). Number of prey items required per year (calculated by summing the values across all months) = 1.96. SMR, standard metabolic rate.

flagellum; Secor and Nagy 1994). The doubly-labeled water technique has also been used to examine the field metabolic rates of Rock Rattlesnakes (*C. lepidus;* Beaupre 1995b, 1996).

Diet and Trophic Structure

Traditionally, the diet of snakes has been determined by dissection of museum specimens (Greene 1986) or by forcing captured snakes to regurgitate a recently ingested meal by manual palpation (Mushinsky and Hebrard 1977; Fitch 1987a). Although these techniques are useful, they do have limitations. Diet analyses based on literature or museum specimens often use individuals spanning broad geographic areas and may be inappropriate for determining the diets of specific snake populations (Rodríquez-Robles 1998),

missing subtle but important information, such as spatial or temporal variation in diet (Fitch 1999). Moreover, diet patterns determined by dissection
provide only a snapshot of the diet of that individual snake. For species
that eat infrequently, dissection or palpation of many snakes may be needed
before even a single prey item is recovered. In addition, dissection requires
snakes to be killed and palpation can be stressful to the animal as a result of
the associated physical manipulation and the potential loss of an important
meal. Finally, capture biases may lead to misinterpretations of diet or feeding frequency. For example, after eating a large meal a snake may bask more
conspicuously or be less able to escape, resulting in an overrepresentation
of large prey taxa in the diet analysis. Small diet items may be underrepre-
sented because they are digested more rapidly or are more difficult to detect
by palpation than larger prey items. Using molecular techniques (e.g., DNA
and monoclonal antibodies) to identify prey taxa from gut and fecal material (Sheppard and Harwood 2005) may ameliorate some of these biases.

Recently, stable isotope techniques have been proposed as a method for
assessing diet and trophic relationships without incurring the biases inherent
in traditional gut- or fecal-content analyses (Ehleringer et al. 1986; Gannes
et al. 1997, 1998; Bearhop et al. 2004; Schindler and Lubetkin 2004). Stable isotope techniques use variation in the relative amounts of naturally
occurring stable isotopes of ecologically important elements (e.g., $C^{13}:C^{12}$
and $N^{15}:N^{14}$) as tracers within living systems (Ehleringer and Rundel 1989;
Gannes et al. 1998). Because diet isotopes are often transferred to consumer
tissues in conservative or predictable ways, the isotopic composition of food
sources can be used to draw inferences about consumer trophic relationships. Although stable isotope techniques have been used for a variety of
other taxa, they have only recently been implemented in studies of snake
ecology (Pilgrim 2005, 2007).

Using isotopes to investigate trophic dynamics within or among snake
populations requires the generation of detailed prey isotope profiles across
space and time (e.g., seasons and ontogeny). In addition, prey isotope profiles must be specific to the system in which snake tissues will be sampled.
The greatest inference can be drawn when isotope signatures of prey taxa
or functional groups are distinct and relatively constant through time. For
example, the isotope profiles for amphibian prey taxa available to aquatic
snakes at an isolated wetland in South Carolina clustered into functional
groups based largely on taxonomy (Fig. 1.4a). In contrast, taxonomy was
not a good predictor of isotopic similarity among amphibians at a terrestrial Florida site because the isotopic composition of treefrogs (*Hyla*) en-
compassed the entire range of both the carbon and nitrogen isotope values
observed in the system (Fig. 1.4b). Several authors have reviewed stable
isotope techniques and their use in ecological studies (e.g., Gannes et al.
1997, 1998).

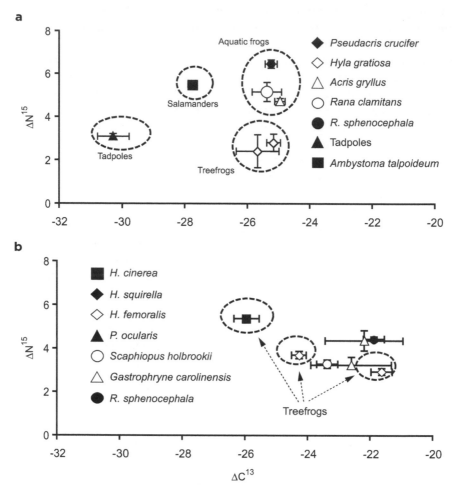

Fig. 1.4. Prey isotope profiles for two ecosystems in the southeastern United States. (a) Amphibian prey taxa available to aquatic snakes (collected in aquatic minnow traps) at Ellenton Bay, Aiken Co., South Carolina, in 2005–2006 (b) Amphibian prey taxa available to terrestrial snakes (collected in terrestrial drift fences) at a site in Volusia Co., Florida, in 2001–2002. Axes represent isotopic composition (carbon and nitrogen) of prey in delta values (proportion of heavy to light isotope in a sample, relative to a standard). Note that in (a) prey functional groups cluster by isotopic composition, whereas in (b) prey taxonomy is a poor predictor of isotope similarity. (Data for [b] adapted from Pilgrim 2005)

Population Studies

The goals of most monitoring efforts, and of many applied ecology studies, lie at the population level and include investigations of demography, site occupancy, population size or density, vital rates (i.e., survival, recruitment, immigration, and emigration), and mechanisms underlying population change. Unfortunately, the accurate estimation of population parameters often requires large investments of time and resources (but see Seigel and Mullin, Chapter 11). Moreover, investigators studying snake population often struggle with low precision in parameter estimates as a result of low recapture rates, high variation in capture rates caused by environmental stocasticity, and unaddressed sources of bias that can cloud results. However, recent advances in efficacy and standardization of collection methodology, in analytical techniques, and in our understanding of snake ecology are paving the way for a new generation of carefully executed, question-driven, snake population studies. In this section, we first discuss conceptual advances in design of snake population studies; then, we detail important analytical advances in methods for studying snake populations at multiple levels of intensity or scale.

Definitions

First, let us define a few terms that are used extensively in this section and discuss how each concept relates to this section.

Species detection probability (p in the presence/absence literature). The probability of encountering any one individual of a given species with a given unit of effort, provided that the species is present. Thus, detection probability is influenced by both abundance and ease of capture/observation. Detection probability is a key parameter in presence/absence monitoring.

Capture probability (p in the mark-recapture literature). The probability of capturing one particular individual of a given species with a given unit of effort. Thus, capture probability is unrelated to population density and is a function only of how easy it is to capture each individual. Capture probability is the primary consideration in mark-recapture studies because it also describes the probability of recapture (although capture and recapture probabilities may differ as a result of trap responses). Generally, mark-recapture analyses lose power when capture probabilities are low.

Heterogeneity Variability in capture probability among individuals, resulting in some individuals being more catchable than others.

Temporary emigration A situation in which a portion of the population is not available for capture during some sampling intervals (Kendall et al. 1997; Bailey et al. 2004a). Individuals may be unavailable for capture due to behavior (e.g., ecdysis, inactivity, or reproduction) or because they are using habitats or geographic areas that are either outside the sampling area or are not sampled effectively by the capture method.

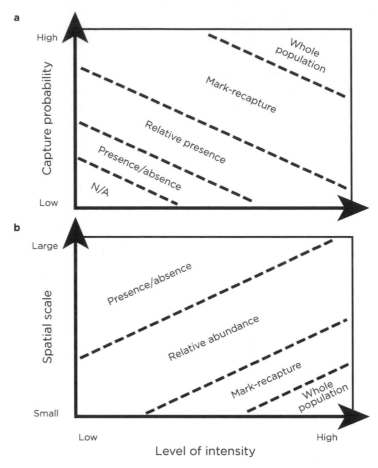

Fig. 1.5. Population-monitoring techniques most appropriate for various combinations of study intensity (time, resources, etc.). (a) Capture probability of the target species (b) Spatial scale of inference. The dashed lines indicate that boundaries between the methods are not rigid and that situations exist in which multiple methods may be applicable. N/A indicates that combinations of detectability and effort that will not yield meaningful results.

Design of Snake Population Studies

Defining a Question

The first step in designing a snake population study is to clearly define the question(s) of interest. All too often snake studies are initiated with little foresight, and ultimately inconsistencies in the sampling methodology preclude useful results. Defining explicit questions will determine the spatial and temporal scale, level of intensity, capture method, and analytical techniques necessary to complete the study given the time and resources available (Fig. 1.5).

Next we introduce several major categories of snake population studies in order of increasing intensity.

1. *Presence/absence, occupancy, or inventory.* Determining occupancy, whether or not a species occurs at a given site or set of sites, is the simplest form of population assessment. Determining occupancy may be an initial step for investigating a snake population or it may be the ultimate goal if the study area is large or the species is particularly intractable (Fig. 1.5). In its simplest form, inventory involves using unstandardized effort and a variety of (often haphazard) techniques with the goal of having the highest probability of documenting the species of interest. Alternatively, and especially for rare or elusive species, standardizing effort at some cost to capture rate is advisable because, with standardized effort, species detection probabilities can be calculated and a likelihood of species absence can be estimated for sites where the species was not found (discussed later in the chapter).

2. *Snapshot population assessment.* Snapshot population assessments seek to understand population characteristics at a single time and are, by definition, short in duration. In fact, lengthening duration can obscure results because of population change over the course of the study. The intensity necessary to complete a snapshot assessment varies depending on the specific question(s) of interest. For example, if the goal is to assess population demography (e.g., size, age, or sex structure) at one time, only enough effort is needed to obtain an adequate sample size using relatively unbiased methods (or methods for which the biases are understood). Alternatively, estimating population size (or density) at a given time requires mark-recapture methods, probably with high-intensity sampling to obtain adequate recapture rates. A snapshot has the potential, however, to miss information if temporary emigration exists. Furthermore, yearly variation in snake populations can be considerable (e.g., Seigel and Fitch 1985; Seigel et al. 1995; Willson et al. 2005), and a short duration study might not be representative of the population in most years.

3. *Monitoring population trends over time.* Studies that monitor trends in populations over time may be conducted with or without in-depth snapshot studies, and the dichotomy between the two is critical in terms of study design. In short, when monitoring a snake population over time, the researcher must ask, is it necessary to determine the population size (or density)? Alternatively, is it sufficient to determine only if the population is growing, declining, or stable? If the goal is simply to assess stability, and especially if the area of interest is large, then an unbiased index of relative abundance may be sufficient. But if the study area is small, funding is sufficient, and the species has a relatively high capture probability, mark-recapture methods may be used to estimate the population size and quantify vital rates. In these cases, however, open or robust design models must be used because the assumption of population closure is violated (discussed later in the chapter).

4. *Monitoring to assess mechanisms for population change.* Both the commitments of time and of resources necessary to complete a study increase when the goals involve assessing mechanisms for changes in population size or structure. In addition to intensive mark-recapture, such studies probably also involve monitoring immigration and emigration rates or focal animal studies that assess movements, reproduction, and sources of mortality. Especially important when designing a study to assess mechanisms for population change is a careful consideration of biotic and abiotic factors (e.g., weather, habitat, prey availability, and predator abundance) that should be monitored in conjunction with animal monitoring.

Level of Intensity and Spatial Scale

Before beginning any snake population study, careful consideration must be given to the scope or scale of the question(s) in light of available resources and tractability of the species (Fig. 1.5). For studies addressing questions on a small spatial scale, mark-recapture is a viable option, provided the species has a relatively high capture probability (Fig. 1.5). For species with low capture probability, however, mark-recapture may not be possible, even on relatively small spatial scales (discussed later in this chapter).

When questions concern large spatial scales, and especially for intractable species, animals may be monitored across the entire area using low-intensity approaches such as occupancy monitoring or indices of relative abundance. Alternatively, one or, ideally, several subpopulations can be studied intensively. Data on subpopulations can be used in conjunction with more limited data on larger populations to address questions on larger scales (see section on Relative Abundance Indices).

Defining the Population of Interest

Explicitly defining a target population for study is a critical step in the design of any population study. We define a population from the biological perspective as a group of individuals in which movement within the group is greater than movement into or out of the group. In some cases, a clear delineation of biological populations is possible and those populations are of a size that can be studied manageably as a unit. Examples of such defined populations include those existing on islands (e.g., King and Lawson 2001; Sun et al. 2001; Bonnet et al. 2002b; Pearson et al. 2002), species with extremely small geographic ranges (e.g., Webb and Shine 1997b; Holycross and Goldberg 2001; Prival et al. 2002), and those centered around naturally patchy habitats such as isolated wetlands (e.g., Winne et al. 2005; Willson et al. 2006; Winne 2006b) or suitable hibernacula (e.g., Diller and Wallace 2002; Weatherhead et al. 2002). In many cases, however, it is impossible to study an entire biological population because the population is too geographically widespread to examine with the desired level of intensity. In such cases, it is necessary to define a study population, which is defined by arbitrary boundaries and thus does not represent a true biological population. Population size, in this case, applies only to the arbitrary study area and represents density per unit area.

Defining the size of the study population must include a consideration of the scale at which results will be interpreted (e.g., unit of land, population, region, or range of species) in light of the resources available and the tractability of the species. When limited resources necessitate the definition of an arbitrary study population, it is important that the study population be as representative of the biological population of interest as possible. In practice, this may mean that the study population should not be situated in the area where snakes are most abundant but, rather, in habitat that is typical of the entire area of interest. In a study of either an entire biological population or an arbitrary study population, knowledge of immigration and emigration rates can aid in the interpretation of results. These rates can be quantified using capture techniques that intercept animals entering or leaving the study area (e.g., drift fences; Dodd 1993a; Willson et al. 2006; Winne et al. 2006b), by following individuals using radiotelemetry, or using genetic techniques (see King, Chapter 3). Regardless of the question of interest, conducting pilot studies to assess individual capture probability will help determine the size of the study population that can reasonably be studied given the available resources.

Exceptional situations exist in which a combination of sampling efficiency and tractability of the snake species allows the researcher to capture nearly all the individuals in a population. Although the communal denning of high-latitude snake populations may be the most familiar example of such a situation, similar opportunities also exist in other habitats. For example, dredging was used to thoroughly sample uniform mats of Water Hyacinth (*Eichhornia crassipes*) at Rainy Slough, Florida (Godley 1980). By assuming that all snakes within the hyacinth mats were captured, densities of Striped Crayfish Snakes (*Regina alleni*) and Black Swampsnakes (*Seminatrix pygaea*) could be directly calculated.

Another promising direction in studies of snake population ecology is the use of closed or experimental snake populations. Other fields have benefited from the use of mesocosms, penned populations, or the experimental manipulation of field populations. For example, numerous studies have used field enclosures (e.g., Wilbur and Collins 1973; Wilbur 1976; Todd and Rothermel 2006) or laboratory mesocosms (e.g., Morin 1981; Semlitsch and Gibbons 1985; Semlitsch 1987a, 1987b) to investigate mechanisms driving population dynamics in amphibians. Few similar studies have been conducted with snakes, however, perhaps as a result of the remarkable escape abilities of many snake species. Studies on *B. irregularis* have shown that even large arboreal species can be confined by relatively simple barriers (Rodda et al. 2007a). Such barriers were used to successfully enclose a 5-ha *B. irregularis* population with no evidence of trespass after 4 years (Rodda et al.

2007b). Many small, sedentary, aquatic, or litter-dwelling snakes are ideal for similar closed population studies. Such studies not only facilitate population monitoring by eliminating immigration and emigration but also allow experimental manipulations (e.g., food availability) that can be replicated at the population level.

Designing a Sampling Scheme

The temporal pattern of data collection largely reflects the analytical technique that will be used. The first concern in this regard is the duration of the study. For a snapshot assessment, a short duration study with high intensity is preferable. The timing of sampling within a study of longer duration also reflects the analytical technique used. For most long-term studies, short-duration, high-intensity sampling is preferred to continuous sampling, and such "pulsed" designs are necessary to meet the assumptions of most mark-recapture analyses. Because the main concern in relative abundance assessment is bias, successive sampling occasions should be timed to minimize differences in behavior and are best conducted in similar seasons and/or under similar environmental conditions.

Likewise, the spatial pattern of sampling will reflect the goals of the study. In mark-recapture analyses, it is preferable to sample as large a subset of population as possible during each interval to reduce heterogeneity. For relative abundance assessments, the entire population need not be sampled, but a sampling scheme that maximizes comparability is preferred. Studies wishing simply to confirm species presence often benefit from focusing on optimal habitat; but a standardized effort is needed to infer species absence with statistical confidence.

Choosing a Capture Method

The first consideration when selecting a capture method is how effective that method is for sampling the target species. It is important to remember, however, that obtaining a high total number of captures is not always the most important goal. In many cases (e.g., relative abundance assessments) the primary concern is repeatability of the samples, even at the expense of total captures. In such cases, passive capture methods are preferable because they minimize bias (Table 1.1).

Understanding potential capture bias is critical because many capture methods underrepresent certain segments of the population. For example, although arboreal funnel traps are highly effective for capturing large *B. irregularis*, small individuals can be detected only in visual surveys (Rodda et al. 2007b). Likewise, aquatic funnel traps differ in their usefulness for sampling aquatic snakes (Willson et al., 2008). Understanding such capture biases is an underappreciated component of studying snake populations. In many cases, comparing samples collected using multiple methods (e.g., Prior et al. 2001; Rodda et al. 2007b; Willson et al., 2008) or using laboratory or enclosed populations to test assumptions of equal catchability can yield invaluable insights for the design and interpretation of snake population studies.

Dealing with Low Capture Rates

Mark-recapture, the most in-depth method for studying populations, requires relatively high individual capture probability (to produce adequate recaptures) to provide useful estimates of population parameters. Although the secretive nature of many snakes leads to low capture probability, this problem can be ameliorated in several ways:

1. *Increase effort or decrease spatial scale.* It is often tempting to try to maximize the number of individual snakes captured by employing low-intensity techniques across a large study area. Increasing effort and/or decreasing the area sampled may reduce the overall capture rate, but it increases individual capture probability, thus improving the precision of parameter estimates (see Fig. 1.5).

2. *Incorporate temporary emigration.* In populations in which individuals exhibit temporary emigration, capture probability may be underestimated because all individuals are not available for capture during all samplings (Kendall et al. 1997; Bailey et al. 2004a). Testing for and/or estimating temporary emigration can improve parameter estimation by ensuring that capture probability estimates include only animals available for capture during each sampling event (Bailey et al. 2004a). Temporary emigration, however, can only be addressed using robust design models (Kendall et al. 1997).

3. *Use a monitoring technique that does not depend on recapturing individuals.* Mark-recapture analyses depend on obtaining adequate numbers of recaptures, but presence/absence designs, indices of relative abundance, and distance sampling do not require recaptures. These methods are sensitive only to species detection probability, which is a rate of capture over an arbitrary sampling unit. By increasing the effort of the sampling unit, sufficiently high detection probabilities can be generated for nearly any species. Note, however, that indices of relative abundance are sensitive to bias and should be interpreted with caution.

4. *Study a surrogate population.* In some cases, individual snakes are sufficiently intractable that mark-recapture is ineffective (Fig. 1.5a). Likewise, the spatial scale of interest might be so small that presence/absence monitoring is inappropriate (Fig. 1.5b). In such cases, the most satisfactory option may be to study a different population of the same species in hopes of gaining insights that could help manage the focal population (see Seigel and Mullin, Chapter 11).

Presence/Absence Monitoring

Presence/absence monitoring uses relatively low-intensity sampling to investigate patterns of distribution, generally on fairly large spatial scales. Recent improvements in analytical techniques have made quantitative presence/absence monitoring a useful tool that can be applied on scales too large for other forms of monitoring (Fig. 1.5b). In addition, presence/absence may be the most appropriate method for monitoring species for which low capture probability makes mark-recapture infeasible (Fig. 1.5a).

Traditional Methods for Presence/Absence Monitoring

Presence/absence data have been the impetus for many snake conservation efforts. For example, declines in the Southern Hog-nosed Snake (*Heterodon simus*) were detected through a comparison of the historical distribution (largely from museum records) and the distribution in recent reports (Tuberville et al. 2000). Traditionally, studies of this type have relied on haphazard observations made over long time scales. Such anecdotal reports are difficult to interpret, however, because the lack of standardized effort makes it impossible to calculate detection probability, thus making it impossible to confirm species absence with statistical confidence from nondetection data (Kery 2002; Bailey et al. 2004b).

Advances in Presence/Absence Monitoring

Analytical software.   Recently, the applicability of presence/absence monitoring has been enhanced by advances in analytical techniques and software that makes those techniques accessible to the public. The software program PRESENCE (http://www.mbr-pwrc.usgs.gov/software/) uses likelihood-based techniques (developed by MacKenzie et al. 2002) to estimate site occupancy and species detection probability. PRESENCE requires presence/absence data from repeated sampling occasions (consisting of any standardized unit of effort) and allows for the inclusion of site (e.g., habitat characteristics) and sampling (e.g., climatic conditions, season) covariates that can improve precision of parameter estimates. PRESENCE has recently been used to model factors influencing the distribution and species detection probability of amphibians (Bailey et al. 2004b; Gooch et al. 2006) and at least one group of snakes (Luiselli 2006).

Inferring species absence.   A related, but alternative question involves determining the confidence of species absence at sites where the species has not been detected. One study used repeated visits to 87 sites to calculate the species detection probability for three common European snakes, the Aspic Viper (*Vipera aspis*), Smooth Snake (*Coronella austriaca*), and Grass Snake (*Natrix natrix;* Kery 2002). By calculating the detection probability at sites where each species was known to occur, the researcher determined the number of unsuccessful visits necessary to declare the absence of each species with statistical confidence.

Relative Abundance Indices

Traditional Methods for Assessment of Relative Abundance

Relative abundance indices are generally rates of capture standardized for effort or time (e.g., captures per trap-night or sightings per kilometer; see sources in Parker and Plummer 1987). Relative abundance data, although often easy to collect, must be interpreted with caution. The key assumption in a comparison of relative abundance is that individual capture probability is relatively constant and thus that overall capture rate reflects population density. In other words, any comparison of relative abundance assumes a consistent (generally assumed to be positive and linear) relationship between density and capture rate. In reality, however, capture rate reflects a combination of factors, including population density, behavior (activity levels, habitat use, etc.), and individual capture probability. Thus, when using indices of relative abundance, care must be taken to ensure that the sampling method used is as repeatable and unbiased as possible, even at the expense of increased overall captures. In addition, because individual capture probability often varies among species and among populations, relative abundance indices generally provide poor indicators of community composition or differences in density between sites, unless they are used in conjunction with detection probability estimates obtained using mark-recapture.

Advances in Relative Abundance Assessment

Methods for assessing relative abundance are straightforward (generally consisting of standardized sampling events repeated across the spatial or temporal scale of interest; Parker and Plummer 1987), and recent progress has been made in the assessment of bias and testing of the critical assumption that capture rate correlates directly with population density. The efficacy of visual surveys as abundance indicators has been tested for introduced *Boiga irregularis* on Guam (Rodda et al. 2005). When the sighting rate was correlated with population density (estimated using mark-recapture), no correlation existed between snake relative abundance (sighting rate) and population density (Fig. 1.6a), demonstrating that not all capture methods yield useful indices of relative abundance. Conversely, capture rate (hand captures) correlated strongly with population density across 11 insular populations of the Lake Erie Watersnake, *Nerodia sipedon insularum* (Fig. 1.6b; King et al. 2006a). This correlation allowed for the estimation of population density at 19 additional sites where there were insufficient data to provide mark-recapture population estimates.

Distance Sampling

Distance sampling uses data on species detection probability to estimate density from VES without relying on mark-recapture (Buckland et al. 2001,

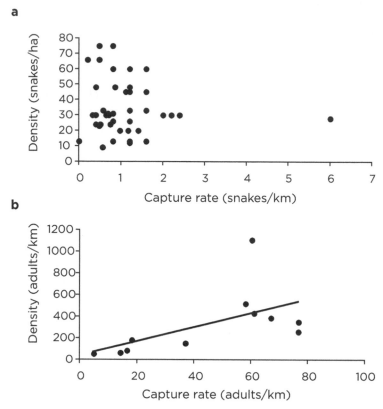

Fig. 1.6. Tests of the efficacy of relative abundance indices (sighting rate) for predicting population density (estimated via mark-recapture). (a) Brown Treesnakes (*Bioga irregularis*) on Guam (b) Lake Erie Watersnakes (*Nerodia sipedon insularum*) on islands in Lake Erie, United States. For *B. irregularis*, there is no relationship between sighting rate and population density (R^2 = .0005; not significant), whereas for *N. sipedon* there is a significant relationship between ln(Sighting rate) and ln(Population density) (R^2 = .851; p = .001). Removing outliers from either analysis does not change the results. (Data for [a] adapted from Rodda et al. 2005, with permission; data for [b] adapted from King et al. 2006a, with permission)

2004). Essentially, distance sampling works by measuring the distance at which animals are observed from a transect line or observation point. The distribution of captures around the line is then used to calculate a detection function assuming total (100%) detection along the transect and declining detection probability at increasing distances from the transect. One limitation of distance sampling is the need for a relatively large number of observations (a minimum of 60–80 observations from 10–20 transects; Buckland et al. 2001). to generate meaningful density estimates. In addition, many snake species may violate the assumption of 100% detection along the transect, although distance sampling models that relax this assumption have been proposed (Buckland et al. 2004). Distance sampling could be used in some snakes that are easily observed using VES and has been used to investigate habitat associations of sympatric vipers (*Bitis*) in West Africa (Luiselli 2006). However, a study designed to validate the efficacy of distance sampling for reptiles found that distance sampling underestimated densities of *B. irregularis* by 700% (Rodda and Campbell 2002).

Mark-Recapture Studies

Since Henry Fitch's pioneering snake population studies in Kansas (reviewed in Fitch 1999), mark-recapture has been the favored approach for high-intensity monitoring of snake populations. Executing a snake mark-recapture study, however, demands more than simply capturing and marking as many snakes as possible. Both the analytical technique used and the required sampling design vary depending on the question(s) of interest, and a consideration of study design is necessary to ensure these question(s) can be evaluated. Moreover, the combination of sampling intensity and spatial scale must produce recapture rates sufficient to make population estimation possible. To some extent, as a result of low recapture rates, many previous studies used population demography (e.g., size distribution, proportion reproductive, and body condition) or relative abundance indices rather than direct population estimates when assessing population status over time (e.g., Shine and Madsen 1997; Madsen and Shine 2000a Lacki et al. 2005; Madsen et al. 2006; Willson et al. 2006; Winne et al. 2007). Given careful design consideration, however, meaningful mark-recapture studies are possible for many snake species.

Traditional Methods for Mark-Recapture
Closed population models. Closed population models are the simplest form of mark-recapture analysis and most are derivations of the Lincoln-Peterson estimator (Lincoln 1930; discussed in detail in Pollock et al. 1990). Closed models use two or more samples collected over a short period to estimate population size (Fig. 1.7a) and assume population closure. For this reason, they do not provide estimates of vital rates (e.g., survivorship and population growth rate). Moreover, because closed population studies are necessarily of short duration, they may underestimate the population size if a portion of the population is unavailable for capture during the study.

A major advantage of closed population models is that they do not necessarily assume equal capture probability (Pollock et al. 1990). Closed models are available that account for time-varying capture probability, trap responses (i.e., "trap-happy" or "trap-shy" responses), and heterogeneity in capture probability. Perhaps because of their simplicity, closed models have been used to estimate population size in several snake species including Rough Greensnakes (*Opheodrys aestivus;* Plummer 1997), Tiger Snakes

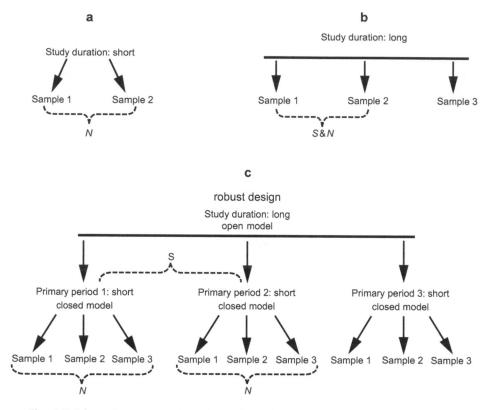

Fig. 1.7. Schematic representation of sampling schemes used in mark-recapture studies. (a) Closed population models (b) Open population models (c) Robust design (mixed population) models. For each design, the intervals over which population size (*N*), survivorship (*S*), or both are estimated are indicated.

(*Notechis scutatus;* Bonnet et al. 2002b) and *V. aspis* (Lourdais et al. 2002), among others (reviewed in Parker and Plummer 1987).

Open population models.   Open population models allow the estimation of vital rates (e.g., survivorship and population growth) when populations are open to births, deaths, immigration, and emigration. The most popular open models are based on the Jolly-Seber group of estimators (Seber 1982; Pollock et al. 1990; Lebreton et al. 1992) and require a sampling design comprising at least three samples separated by relatively long intervals, across which population parameters are estimated (see Fig. 1.7b). The major drawback of open models is that they assume constant capture probability and thus cannot account for capture probabilities that vary due to temporal effects, behavioral effects, or heterogeneity in capture probability. For this reason, they generally do not provide particularly robust estimates of population size (Pollock et al. 1990). Despite this, open and closed models produced similar population size estimates for *N. sipedon insularum,* although standard errors were large due to low recapture rates (King et al. 2006a. Open models have been used successfully to estimate survivorship and population size over long time scales in Water Pythons (*Liasis fuscus;* Madsen et al. 2006), and ratsnakes (*Pantherophis;* Weatherhead et al. 2002).

Advances in Mark-Recapture Studies
Analytical software.   Recently, mark-recapture analyses have become more accessible and user-friendly through advances in publicly available software packages such as the programs CAPTURE (http://www.mbr-pwrc.usgs.gov/software/; Otis et al. 1978) and MARK (http://www.warnercnr.colostate.edu/~gwhite/mark/mark.htm; White and Burnham 1999). These programs allow the analysis of large mark-recapture data sets using a variety of models (including open, closed, and robust designs) and use maximum likelihood selection procedures to compare multiple competing models using Akiake's information criterion (AIC) or similar methods. Moreover, MARK allows for partitioning of data sets into demographic groups (e.g., sexes and cohorts) and inclusion of both individual (e.g., body length, mass, and age) and sampling (e.g., environmental conditions and effort) covariates, all of which can help partition variance in data sets, thus maximizing the precision of parameter estimates.

Robust design models.   Robust designs (Pollock 1982) require a sampling scheme in which the population is sampled several times over short secondary sampling intervals (often successive days) separated by longer primary sampling intervals (Fig. 1.7c). Population size and capture probability are estimated within secondary intervals using closed population models and survivorship is estimated over the longer, open, primary intervals. Thus, the design is robust in the sense that it permits the estimation of both survivorship and population size without violating the assumptions of either open or closed models (Pollock 1982). When using robust design models, all individuals must be available for capture at each sampling event, so if secondary sampling intervals are successive days, animals must be marked and released on the day of capture.

One advantage of robust design models is that they allow the investigator to test for the presence of temporary emigration. Standard open and closed mark-recapture analyses assume that all animals are available for capture during all sampling events and thus may provide imprecise parameter estimates if temporary emigration exists but is not accounted for (Kendall et al. 1997; Bailey et al. 2004a). Although temporary emigration probably exists in many snake populations, to our knowledge no studies have tested for temporary emigration in snake populations.

Because robust designs use closed analyses to estimate population size, they can account for time-varying capture probability, trap responses, and heterogeneity. There is strong evidence that these factors are important in some snake species, necessitating the use of robust design models for long-term monitoring of population size and vital rates. For example, models including heterogeneity were favored within closed population estimates of population size for *S. pygaea* in an isolated wetland in South Carolina (Fig. 1.8a). This heterogeneity was probably due in part to differences in capture probability among seasons and between sexes (Fig. 1.8b). Unfortunately, heterogeneity is difficult to account for analytically. Indeed, although robust design models that use Pledger's finite mixtures (Norris and Pollock 1996; Pledger 2000) to model heterogeneity are available in the software program MARK, their properties and precision have not been examined in the literature. However, heterogeneity can often be mitigated in several other ways, including:

1. *Design of sampling methodology.* Heterogeneity can result directly from sampling methodology. For example, uneven sampling within the study area or variation in the effectiveness of capture method across habitats can lead to some individuals being captured more often than others. Thus, homogeneous sampling across the entire study area and the use of multiple capture methods can minimize heterogeneity.

2. *Inclusion of individual and sampling covariates.* Heterogeneity can result if demographic subsets of the population have different capture probabilities. For example, nonreproductive female *V. aspis* have lower capture probabilities than reproductive females (Bonnet and Naulleau 1996). In such cases, heterogeneity can be reduced by dividing the population into subgroups (e.g., sexes or cohorts) that are analyzed separately and thus may differ in capture probability. Analyzing the capture probabilities of *S. pygaea* separately by sex reduced the support for models favoring heterogeneity in favor of null models (constant capture probability; Fig. 1.8). Likewise, the inclusion of both individual and sampling covariates can reduce heterogeneity if there are relationships between those covariates and capture probability (Pollock 2002).

3. *Investigation of temporary emigration.* The presence of temporary emigration can lead to apparent heterogeneity in capture probability. Robust design analyses can address temporary emigration, ameliorating the effects of heterogeneity on population estimation.

Although robust design models are not particularly new, they are poorly represented in current herpetological literature. However, several recent studies have used robust design sampling to examine aspects of the biology of woodland salamanders in the Appalachian Mountains of the eastern United States (Bailey et al. 2004a, 2004c, 2004d). To our knowledge, no published studies have yet used robust design analyses to investigate snake population dynamics.

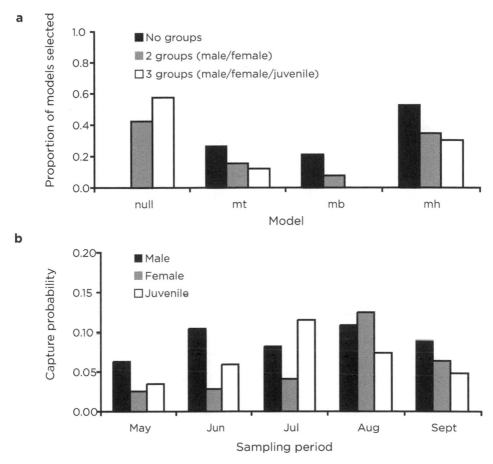

Fig. 1.8. Capture probability (P_c) of Black Swampsnakes (*Seminatrix pygaea*) at Ellenton Bay, Aiken Co., South Carolina, in 2005. (a) Factors influencing P_c in *S. pygaea* as indicated by the proportion of total selected models that included a constant P_c (null model), time-varying P_c (mt), behavioral effects (mb), and heterogeneity in P_c (mh), with data divided in to one, two, or three groups by sex and life stage (b) Variation in P_c across seasons and sexes. Figures based on a total of 1192 captures of 462 individual snakes during monthly 10-day sampling periods between May and September 2005. Model selection was performed within 10-day periods using closed population models in program CAPTURE.

Population Viability Analyses and Population Modeling

With data on population size and vital rates comes the ability to model population trends over time. Indeed, population viability analyses (PVAs), which use various population models to project population trajectories, have been used in the conservation and management of variety of animal taxa (reviewed in Reed et al. 2002). Moreover, many PVA software packages are now available, allowing researchers to conduct a variety of analyses including the calculation of extinction risk, minimum viable population size, and sensitivity analysis (weighing of factors that drive population change). Unfortunately, PVAs and other forms of population modeling have seldom been applied to snakes (but see Ferriere et al. 1996; Altwegg et al. 2005; Row et al. 2007; Shine and Bonnet, Chapter 6), perhaps due to the lack of vital rate estimates for many species.

Another approach to population modeling is to project individual data to populations using individual-based population models (IBMs; DeAngelis et al. 1991). For example, physiological data from laboratory experiments was used to generate a mechanistic IBM of time-energy allocation for the lizard *Sceloporus merriami* (Dunham 1993). This model was combined with environmental data to predict that a rise in mean temperature as small as 2–5 °C would constrain lizard activity enough to drive populations to extinction. IBMs may be particularly amenable to snakes because of the wealth of laboratory and field studies that have investigated snake physiological ecology. For example, detailed physiological data were used to model individual time-energy allocation decisions in *C. horridus* (Beaupre 2002). Individual allocation decisions predicted by the model were extrapolated to the population level, predicting the relative influences of changes in temperature and prey availability on population persistence.

Future Research

In this chapter, we have provided information designed to improve the effectiveness of snake studies. Our hope is that future snake researchers will use this information to address the myriad of ecological questions about snakes that remain. A more complete understanding of snake ecology at the various scales (individual, population, and landscape) will then aid in the development of more effective conservation programs. We hope that the recent methodological advances we describe will both prompt meaningful question-oriented field studies of snakes in the future and also encourage theoretical investigations that seek to understand the unifying factors common to many snake species.

Acknowledgments

For advice and assistance, we thank L. Bailey, S. Beaupre, D. Cundall, V. Cobb, S. Gary, J. Gibbons, M. Pilgrim, S. Price, R. Reed, G. Rodda, B. Todd, and C. Winne. We thank S. Bennett, M. Cherkiss, W. Hopkins, W. Kalinowski, F. Mazzotti, R. Snow, and C. Winne for allowing our use of unpublished data collected collaboratively with M. E. D. or J. D. W. We thank R. King and G. Rodda for supplying raw data for the figures. Support was provided by Duke Power; the Department of Biology at Davidson College; the Savannah River Ecology Laboratory; National Science Foundation grants REU DBI-0139153 and DEB-0347326 to M. E. D.; a National Science Foundation Graduate Research Fellowship to J. D. W.; and the Environmental Remediation Sciences Division of the Office of Biological and Environmental Research, U.S. Department of Energy, through Financial Assistance Award number DE-FC09-96SR18546 to the University of Georgia Research Foundation.

2

Molecular Phylogeography of Snakes

FRANK T. BURBRINK AND TODD A. CASTOE

Phylogeography is a relatively young field that investigates the historical and contemporary processes that affect the geographic distribution of genealogical lineages, particularly those at the intraspecific level (Graves et al. 1984; Avise et al. 1987; Avise 1998). Phylogeography occupies a place between microevolutionary (demography, population genetics, and ethology) and macroevolutionary (systematics, historical biogeography, and paleoecology) fields (Avise 2000). Ironically, since the term was coined, the lines that demarcate phylogeography from phylogenetic and population genetic studies has substantially blurred, and it may be more reasonable to consider this subdiscipline to be research that incorporates both macro- and microevolutionary processes rather than occupying a discrete space between these two scales.

The main benefit of phylogeographic studies is that they reveal patterns that are too difficult to discover using other less integrative approaches. For instance, phylogeographic studies can detect cryptic genetic diversity in the geographic range of a taxon, which may be an early step in the recognition of new species (e.g., Zamudio and Greene 1997; Burbrink et al. 2000; Parkinson et al. 2000; Burbrink 2001; Rodríguez-Robles et al. 2001; Feldman and Spicer 2002; Castoe et al. 2003, 2005). In this sense, phylogeography may provide the initial information about the geographic range of a newly defined lineage; this in turn may supply critical information used to prioritize conservation efforts aimed at maintaining viable populations of newly discovered lineages (Fig. 2.1). Because assessing species boundaries and recognizing the true biodiversity of a region is a primary goal for conservationists, phylogeographic research is tied intimately to conservation biology. Phylogeographic methods also allow the examination of hypotheses concerning the effects of dispersal on population structure and the temporal and geographic origins of lineage diversity (Kolbe et al. 2004; Driscoll and Hardy 2005; Holland and Cowie 2007; Rodríguez-Robles et al. 2007; see Fig. 2.1). Comparative phylogeography has also emerged as a method for inferring the role of historical events and demographic processes in shaping genetic diversity in ecological communities. This approach asks, are species within a particular area affected by similar historical events and if so, how? Compared with single-species studies, this multispecies approach enables broader inferences to be made about the importance of particular geographic areas, the diversity these areas harbor, and the common historical processes that have generated biodiversity (Moritz and Faith 1998; Feldman and Spicer 2006; Rowe et al. 2006; Huhndorf et al. 2007).

In comparison with other vertebrates, particularly mammals and birds, few snake species have been the subject of substantial and well-sampled phylogeographic research. For example, only 3 of the 148 species (Lawson 1987; Burbrink et al. 2000; Burbrink 2002) included in a comparative phylogeographic study of taxa occurring in the southeastern United States were snakes (Soltis et al. 2006), despite the fact that more than 30 snake species occur between the Mississippi River and the Florida panhandle (Crother et al. 2000, 2003; Gibbons and Dorcas 2005). A small percentage of snakes found in the United States have actually been examined phylogeographically and the species that have been the subject of phylogeographic studies worldwide is shockingly low, particularly in the most biologically diverse areas, the New and Old World tropical regions (Greene 1997). Of the approximately 3000 described species of snakes, we estimate less than 3% of them have been examined phylogeographically. Consequently, studies on snakes have contributed little to the methodological development of the field of phylogeography. This is unfortunate because phylogeographic research on snakes has the potential to yield valuable information on snake biodiversity, taxonomy, and evolution, and may further contribute strongly as a model system for elucidating and validating broader phylogeographic patterns that may have shaped many components of a region's biota.

Data needed for phylogeographic studies of animals are most often derived from DNA sequences. Other genetic markers, such as allozymes, tandomly repeated microsatellite markers (sequences made up of a single, short, repeated motifs), randomly amplified polymorphic DNA (RAPDs; Harris 1999; Ali et al. 2004), amplified fragment length polymorphisms (AFLPs; Bensch and Åkesson 2005), and short and long interspersed elements (SINEs and LINEs, respectively; Shedlock et al. 2004; Ray 2007) can also be used in phylogeographic studies. Given that most studies currently infer phylogeographic estimates (i.e., trees) from DNA sequences, particularly mitochondrial DNA (mtDNA), our focus in this chapter includes methods of analysis, experimental design, and other considerations predominantly for mtDNA-based phylogeographic data.

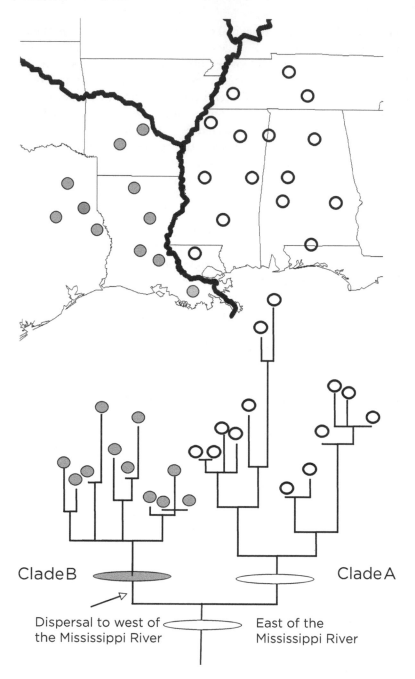

Fig. 2.1. Hypothetical scenarios demonstrating processes responsible for the phylogenetic and associated spatial (geographic) relationships at a common genetic barrier for snakes—the Mississippi River. In examples (a–e), the shading of the circles represents clade designation, whereas in (f) they represent sampling location relative to the barrier.
(a) Reciprocal monophyly of lineages distributed east and west of the barrier, indicating the function of the river in limiting migration and promoting lineage diversification.

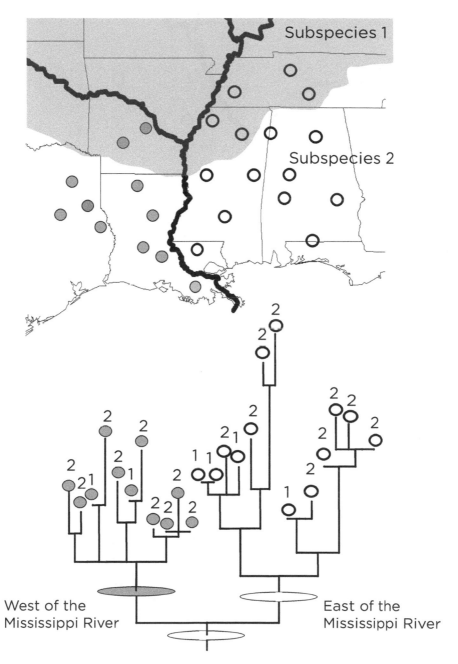

Fig. 2.1 continued. (b) Reciprocal monophyly of lineages with respect to spatial orientation at the barrier in (a); members of each traditionally recognized subspecies do not share a most recent common ancestor.

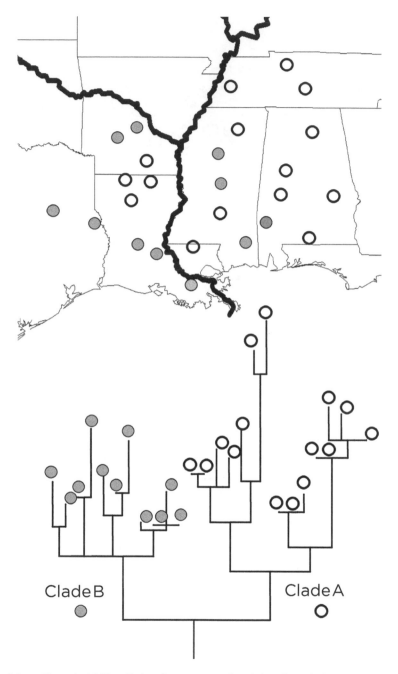

Fig. 2.1 continued. (c) Two distinct lineages are inferred, but these clades are not restricted to specific geographic areas with respect to the barrier. Therefore, it must be assumed that the barrier was not integral to the formation of these lineages or subsequent dispersal has obscured the impact of the barrier.

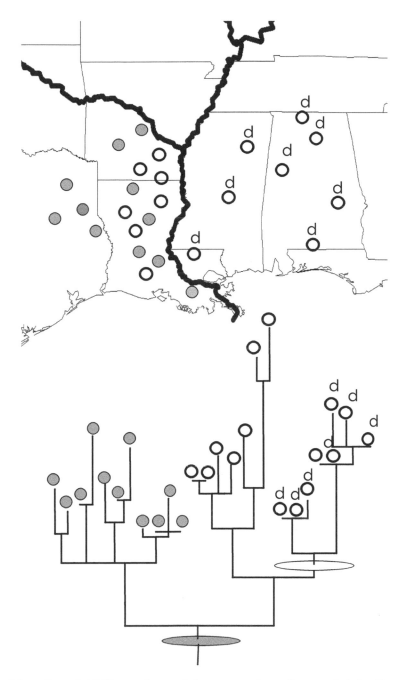

Fig. 2.1 continued. (d) The two deepest clades are not reciprocally monophyletic with respect to geographic region. The initial divergence was not caused by the putative barrier. Due to the dispersal and divergence of the d lineage across the Mississippi River, members of the western (open-circle) clade are more closely related to the lineage east of the barrier (the d clade) than to the other (shaded-circle) lineage west of the barrier.

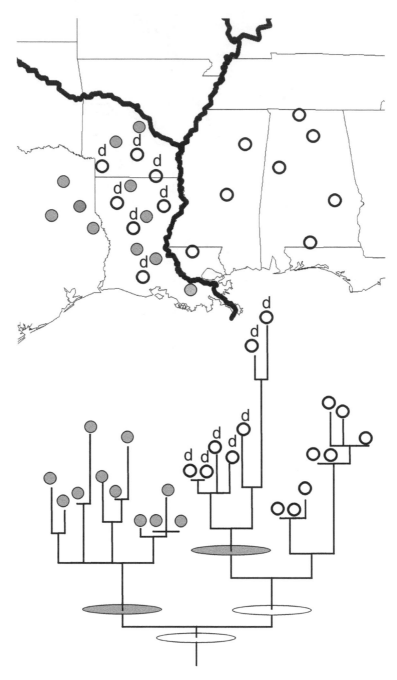

Fig. 2.1 continued. (e) The earliest and deepest divergence was caused by the barrier. Subsequently, dispersal of the d lineage occurred west of the Mississippi River. This indicates that some individuals found west of the barrier are more closely related to individuals east of the barrier (open-circle clade) than they are to neighboring individuals (shaded-circle clades).

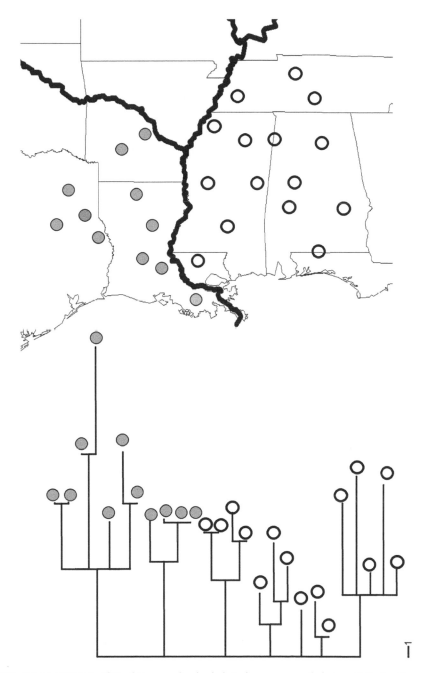

Fig. 2.1 continued. (f) In this example, shaded circles were sampled west of the barrier and open circles were sampled east of the barrier. However, poor resolution in the phylogenetic tree impairs our ability to make a connection between geography and evolutionary history.

In this chapter we provide researchers with a broad spectrum of guidelines and suggestions and a general overview of methods for conducting phylogeographic studies of snakes. Our intention is to provide the fundamental information that will allow the novice phylogeographer to design and implement studies that may reveal spatial patterns in the distribution of genetic variation in conspecific populations and uncover the historical processes responsible for producing these patterns. We also address the expanding statistical and computational methods that have permitted researchers to answer questions relating to lineage diversity, divergence dating, demographics, species boundaries, and comparative phylogeography. Last, we provide some examples and summarize current findings and future directions for the field of snake molecular phylogeography.

Data Collection

Tissue Acquisition

Acquiring tissues for phylogeographic research from most species, particularly those with wide ranges, is a time-consuming task that can require several years. Ideally, a phylogeographic study includes numerous individuals from as many populations as possible for the researcher to attain the goal of adequately characterizing the major phylogeographic patterns and elucidating the detailed genetic structure and historical demographics of the study species. As with any quantitative study, the number of samples required depends on the power needed to address specific hypotheses of interest. The hard reality is that the sampling scope for any phylogeography project relies on the availability of tissues already collected and the feasibility of legally collecting further tissues in the field.

Tissues can be acquired directly from field-caught animals or indirectly through colleagues or museum collections. The first method can be very time consuming and costly, involving completing permit requests, traveling, and conducting fieldwork. The direct acquisition of tissues, however, is most rewarding because it allows researchers to familiarize themselves with the natural history of their study species (e.g., habitat preferences and behavioral patterns). This knowledge is inherently valuable (Futuyma 1998; Wilcove and Eisner 2000) and provides a more holistic understanding of the ecology of the target organisms, which may ultimately play a vital role in interpreting the fundamental aspects that have played key roles in determining phylogeographic patterns. Field collectors must also keep detailed field notes and georeferenced locality records of their samples; these data contain the critical documentation that gives validity to scientific specimens (Simmons 2002). When tissues are acquired indirectly, researchers should make an effort to verify the identity and locality information of the voucher specimens before publishing their results.

Tissues can also be obtained from colleagues or museum research collections, and many institutions have searchable tissue databases on their websites. Because acquiring tissues is a time-consuming and expensive endeavor, it is inappropriate for researchers to ask colleagues or museums for their samples as the sole source for a phylogeographic study. Workers should petition institutions and individuals for only a small number of samples to supplement their own material collected in the field (or obtained from captive animals with reliable locality data). The contributions of field collectors to a genetic study should not be undervalued. Considering the costs, time, and effort associated with obtaining, preparing, and maintaining tissue collections, researchers who provide a significant fraction of the tissues included in a phylogeographic (or systematic) study should be invited to coauthor the article that reports the findings of the research, depending on the relative contribution of their samples.

Nearly all countries and most states in the United States require researchers to obtain a government-issued permit or license *before* attempting to collect tissues from live or dead animals (Duellman 1999; Simmons 2002). Collecting specimens without appropriate official authorization can lead to a substantial fine, a felony conviction, and severe restrictions on future research. Investigators who wish to collect specimens or tissues in the United States should consult *A Field Guide to Reptiles and the Law* (Levell 1997). The institutions issuing the permit often request a considerable amount of ancillary documentation, including a research proposal, a description of the numbers of specimens to be collected, and the areas where and dates when the collecting will take place. For wide-ranging taxa, the permit process may require a considerable investment of time and money. Therefore, investigators should submit permit applications several months before their scheduled field expeditions to avoid bureaucratic delays that could force a change or cancellation of collecting plans. Consulting the list of species defined as endangered, threatened, or of special concern prior to applying for permits or visiting targeted areas is recommended. A species with few collecting restrictions in one region may be listed as threatened in another (which drastically complicates the permitting process). Researchers must always keep on hand copies of all collecting permits issued to them or their collaborators when in the field, and some journals require these permit numbers in the acknowledgments section of the resulting article. A government agency can also examine the list of specimens included in a study and ask the author(s) to produce copies of all permits under which the specimens and tissues were collected.

Sources of DNA

Many types of tissues, given the various tissue preservation strategies, produce sufficient sources of DNA. Although it is possible to obtain sufficient DNA from nearly any tissue, the higher the quality of the source (and resulting DNA), the faster and easier it will be to produce results. Poor-quality template DNA can easily increase the effort required to obtain data. The most commonly used tissues are liver and skeletal muscle; these also tend to produce the highest quantity and quality of DNA. If nonlethal collection of tissues is desired (or required), ventral scale clips that include a thin layer of connective tissue at the base of the scales or tail tips of live or road-killed specimens often yield usable DNA. Road-killed specimens in a relatively advanced state of decomposition can yield enough material for genetic studies, provided that the tissue is not taken from the decaying internal organs. Obtaining blood from larger specimens and shed skins are also viable nonlethal means of obtaining quality DNA. The highest DNA yields from shed skins are often obtained from the harder, threadlike, opaque base of the ventral scales. Ultraviolet radiation damages DNA, and very dry and brittle shed skins (or other tissues), substantially exposed to sunlight, are poor sources of DNA.

All tissues should be placed in sterile (typically 1.5–2 ml) plastic vials and immediately stored in 95% ethanol, in lysis buffer (see later in the chapter), on dry ice (solid carbon dioxide), in liquid nitrogen, or an ultracold (–70 to –150 °C) freezer. Samples stored in alcohol (or lysis buffer) should be minced or cut in strips (and not overfilled) to allow the preservative to permeate the entire tissue; otherwise parts of the sample may decay with time, yielding unusable DNA. Samples preserved in alcohol should always be kept in dark, dry, and cold places, preferably a –10 °C (or colder) freezer. Storing tissues in ethanol is convenient and inexpensive, but it has its shortcomings; this type of long-term storage precludes samples from being used to obtain non-DNA molecular data (from protein or RNA molecules). Shed skins should be placed in plastic bags and stored in an ultracold freezer. Freezing tissues immediately in liquid nitrogen and storing these either in liquid nitrogen or in a –80 °C freezer is the best way to preserve all molecules (protein, DNA, and RNA) for later use. Qiagen and other companies sell preservative buffers for the long-term preservation and storage of RNA. See Dessauer et al. (1996) for a thorough discussion of issues pertaining to the collection and storage of tissue samples for genetic studies.

An alternative to alcohol for preserving DNA from tissues in the field is a detergent-based lysis buffer, which is fairly inexpensive, is easy to make, and produces higher-quality DNA than alcohol storage even after long-term (> 5 years) storage. Unlike alcohol, this type of buffer digests the tissue within several days at room temperature and releases DNA into the buffer. This method also appears to work equally well with fresh tissue, sheds, scale clips, and blood. There are numerous recipes, and we provide one that we have used extensively (with excellent results). The concentrations of reagents for this lysis buffer are 0.5 M tris (hydroxymethyl) aminomethane (Tris), 0.25% ethylenediaminetetraacetic acid (EDTA), and 2.5% sodium dodecyl sulfate (SDS), all in purified distilled water. The recipe for 1 l is as follows. Combine 60 g Tris, 2.5 g EDTA, 25 g SDS, and water to a volume of 1 l. Autoclave the buffer, and aliquote it into sterile vials prior to use. Samples in lysis buffer can be kept at or below room temperature for months (at least) and can also be frozen for long-term storage. This preservation method is also convenient because a small volume of frozen sample (mostly buffer) can be easily scraped off and used to extract DNA without the need for sample separation or thawing.

DNA Extraction

Extraction and purification of high-quality (substantial amounts of relatively pure and high–molecular weight) DNA from tissues can be accomplished relatively easily with simple laboratory equipment. Two basic options are available: traditional and (commercial) kit methods. Traditional methods include long-established procedures such as SDS-proteinase K/phenol/RNAase and phenol or chloroform extraction protocols (Sambrook et al. 2001). These procedures are cost-effective and generally produce good results, but are time consuming and inefficient for processing numerous samples simultaneously. Furthermore, phenol and chloroform are toxic chemicals and severe environmental pollutants. The kit methods consist of commercial extraction protocols and regents (other than phenol and chloroform); they include the DNeasy Blood & Tissue Kit (Qiagen) and the AquaPure Genomic DNA Tissue Kit and Chelex 100 Molecular Biology Grade Resin (Bio-Rad Laboratories). These kits typically yield sufficient amounts of quality DNA, are fast, and are far more practical for large numbers of samples, but are more expensive.

The extraction of DNA with traditional or kit methods works well for most tissues, but there are special considerations for shed skins. To perform DNA extractions from shed skins, we normally use approximately 2 cm of shed from the base of the ventral scales. Larger amounts tend to require more digestion buffer than can fit into a 1.7-ml tube, resulting in poorly digested samples. Suitably fragmented shed skin allowed to digest at 55 °C for 2 or more days in cetyl-trimethylammonium bromide (CTAB) with proteinase K on a rocker platform usually produces enough high-quality DNA for polymerase chain reaction (PCR). Alternatively, grinding sheds into a powder using liquid nitrogen and a mortar and pestle, prior to digestion, appears to dramatically increase the concentration of high-quality DNA (even when standard traditional or kit extraction protocols are used). Thus, high-quality DNA can be extracted from sheds but with slightly more work and a higher failure rate than from standard tissues.

Amplification of Molecular Markers

After DNA has been extracted from the source tissue, the next step is to select the appropriate segment of DNA that will be amplified with PCR—this is accomplished using gene- or region-specific oligonucleotide primers. Primers anneal to a complementary sequence in a single-stranded DNA or RNA template, and the DNA polymerase then extends the complementary sequence from the primer. We recommend that researchers not familiar with PCR and DNA sequencing consult a basic text that describes these methods. After gene- or region-specific primers have been either designed or obtained (e.g., from the literature), a PCR thermal cycling reaction (i.e., denaturation, annealing, and extension times and temperatures) must be constructed that will efficiently amplify the gene of interest. Thermal cycling conditions are quite variable due to the different annealing temperatures of primers and the types of PCR kits used. Researchers should consult the literature for PCR reaction chemistry and thermal cycling conditions as a starting point. We have also found that colleagues have been very helpful and have readily shared advice, primer sequences, and PCR protocols.

Mitochondrial Gene Sequences: Pros, Cons, and Considerations

Despite some potential shortcomings of mitochondrial markers, these genes have been the workhorse of phylogeographic and phylogenetic studies for several reasons, including the relatively high rate of nucleotide evolution (~5–10 times greater than nuclear protein-coding genes), a general lack of recombination, single-copy status, the large number of mitochondrial genomes per cell that facilitates easy amplification of these genes, and the availability of published primers (and complete mitochondrial genome sequences) that simplifies cross-species primer design. The rapid rate of mitochondrial gene evolution is critical for discerning relatively recent evolutionary events and demographic changes required for addressing many questions in intraspecific phylogeographic studies (Birky 1991; Moore 1995; Broughton and Harrison 2003). The high rate of evolution and smaller effective population size (because mitrochondrial DNA, mtDNA, is inherited maternally as a single allele) lead to a relatively rapid coalescence process that should increase the probability of an mtDNA gene correctly tracking the species tree compared to a nuclear gene (Moore 1995).

There are also drawbacks to using exclusively mtDNA data, and we stress that mitochondrial genes are not a panacea. One constraint is that all mitochondrial genes are from single linked genome that is inherited together as a single haplotype or allele only along the maternal lineage. Therefore, phylogenetic patterns from different mitochondrial genes do not provide independent evolutionary information, and they represent only the matrilineal perspective of genealogies. In addition, mitochondrial genes represent a single genetic coalescent event and may not completely characterize the phylogeographic history of the species (Rosenberg and Nordborg 2002; Knowles and Carstons 2007; Edwards et al. 2007). In general, any single-locus estimate may confound the "true" phylogeographic pattern due to hybridization, horizontal transfer, lineage sorting (deep coalescence), gene duplication, and allelic extinction (Zhang and Hewitt 1996; Avise and Wollenberg 1997; Maddison 1997; Hare 2001; Nichols 2001; Rosenberg and Nordborg 2002; Edwards et al. 2007). Given these considerations, we encourage the corroboration of the conclusions from any phylogeographic study that uses only mtDNA genes with estimates from the nuclear genome wherever possible.

In snake phylogeographic studies, the most commonly used mitochondrial genes include cytochrome *b*, nicotinamide adenine dinucleotide (NADH) dehydrogenase subunit 4, the control region, adenosine triphosphotase (ATPase) 6, ATPase subunit 8, 12S ribosomal RNA (rRNA), and 16S rRNA. Primers suitable for the amplification of mtDNA segments across most species of Colubroidea (i.e., the Atractaspididae, "Colubridae," Elapidae, and Viperidae) are described or referenced in several sources (e.g., Rodríguez-Robles and de Jesús-Escobar 1999; Burbrink et al. 2000; Burbrink 2001; de Queiroz et al. 2002; Douglas et al. 2002, 2006; Utiger et al. 2002; Vidal and Hedges 2002, 2004; Nagy et al. 2004; Lawson et al. 2005; Wüster et al. 2005a, 2005b; Burbrink and Lawson 2007). Primers to amplify regions of any of the 22 transfer RNAs (tRNAs), two rRNA genes, 13 proteins, and two control regions (in the alethinophidia) could be easily constructed for all species of snakes due to the expanding diversity of complete mitochondrial genome sequences of snakes available from the National Center for Biotechnology Information (NCBI; GenBank). With alignments of these genomes, conserved regions can be identified and primers with similar melting temperatures and appropriate GC nucleotide contents can be developed and synthesized cheaply.

There are several important points to consider before choosing a mitochondrial gene or region to use in a phylogeographic study. It is possible that there are already many sequences available on the GenBank website for one particular gene; therefore, it may be most fruitful to use that gene so that a larger data set with more individuals can be constructed. Given the highly variable rates of evolution among mitochondrial genes and taxa (e.g., Mueller 2006) and the unevenness in population structuring, all mtDNA regions will not necessarily produce sufficient variation to facilitate an interesting phylogeographic estimate. For example, the rRNA genes (12S and 16S) and, to a lesser extent, cytochrome oxidase genes tend to evolve fairly slowly compared with other commonly used genes (such as cytochrome *b*) and the NADH dehydrogenase subunits (Pesole et al. 1999; Jiang et al. 2007). Protein-coding genes have some advantages because many tests of neutrality and population expansion require a protein-coding sequence (to compare synonymous and nonsynonymous rates of evolution). Moreover, they are also easy to align and rarely contain gaps in alignment in phylogeographic studies. Most of the mtDNA protein-coding genes evolve rapidly, particularly at the third codon position, and should produce phylogeographic structure, if it exists. Furthermore, models of evolution, particularly complex partitioned models (see later in the chapter) seem to fit protein-coding genes well, whereas tRNA and rRNA genes tend to evolve under more complex patterns that are difficult to model due to the impact of secondary structure that leads to compensatory changes.

We recommend against using the control region in alethinophidian snakes as a phylogeographic marker, despite the fact that this region is often used in studies of other vertebrates. All alethinophidian snakes sampled contain two mitochondrial control regions that have identical (or extremely similar) sequences that evolve in synchrony through a poorly understood mechanism of concerted evolution (Kumazawa et al. 1998; Kumazawa 2004; Dong and Kumazawa 2005; Jiang et al. 2007). There is also some indication that control regions in alethinophidians may evolve fairly slowly within some species (Ashton and de Queiroz 2001; Jiang et al. 2007). Avoiding this region may be the best strategy for phylogeographic studies due to the complexity, heterogeneity, and duplication associated with the control regions of alethinophidian snake mtDNA.

Nuclear Gene Sequences: Pros, Cons, and Considerations

The use of nuclear genes, with their comparatively reduced rate of evolution, for phylogeographic studies in many vertebrates is becoming more common, but it has lagged behind mtDNA studies. The slow evolutionary rate provides little or no variation within species, and it ultimately yields little phylogeographic information. There is currently insufficient nuclear genomic data on snakes to design primers for new nuclear genes that are single copy and evolve rapidly enough for phylogeographic studies. In addition, when single-copy nuclear genes are used for phylogeographic studies, they generally provide much less resolution than mtDNA (Heckman et al. 2007) due to many factors, including recombination, gene conversion, large effective population sizes, and incomplete lineage sorting (see Johnson and Clayton 2000; deBry and Seshadri 2001; Palumbi et al. 2001; Allen and Omland 2003; Whittall et al. 2006). Collectively, we have sequenced 14 independent, single-copy nuclear genes in snakes, all of which either provided little or no information at phylogeographic scales or actually turned out to be multicopy. The lack of known nuclear genes suitable for phylogeographic studies represents a substantial hindrance for obtaining non-mtDNA-based phylogeographic inferences. Until nuclear gene regions are identified in snakes that are sufficient to infer phylogenetic structure below the species level, other nuclear markers (not based on DNA-sequence determination) appear to be the most effective resource for assessing nuclear-based population structure. Microsatellites, RAPDs, allozymes, and AFLPs have been used successfully in this role in snakes (see King, Chapter 3). The development of these markers is slow and time consuming, although data collection after initial development is often more rapid and inexpensive. Developing these types of nuclear markers and combining them, especially microsatellites, with mtDNA sequence information are areas that should be major priorities for future research.

DNA Sequence Alignment

Phylogeographic inferences rely critically on accurate DNA sequence alignment because this alignment is an explicit inference of the homology of DNA characters across sequences. Numerous computational methods, included in programs such as ClustalW (Chenna et al. 2003) and Praline (Simossis and Heringa 2005), have been developed to automate this procedure. The alignment of phylogeographic-scale data sets is, however, often trivial and straightforward because insertions or deletions are rare at such shallow divergences, particularly in protein-coding genes. Moreover, gaps in any protein-coding genes among individuals should occur only at the level of a complete codon (in multiples of 3 bp) for the gene to remain in the correct reading frame required to yield a functional protein. A researcher should be suspicious of any alignments of protein-coding genes that have gaps not placed in multiples of three or that contain internal stop codons. These suggest that an error was made in amplification (possibly due to a pseudogene copy of a mtDNA gene found in the nuclear genome), DNA sequence determination, alignment, or application of the correct genetic code to translate the DNA sequence. Even if automated alignment methods are used, it is always best to visually inspect the alignment for apparent problems. If regions of the alignment appear tenuous to the extent that the homology of the positions is not obvious, then these regions should be excluded from later analyses; this is often relevant only for alignments of non-protein-coding genes (e.g., tRNAs and rRNAs) with insertions or deletions of nucleotides.

Phylogenetic Inference

Phylogenetic relationships among DNA sequences can be estimated and represented in a number of ways. First, we must consider the most appropriate way that sequences should be related: in a network fashion or in a tree-based fashion. Phylogenetic trees represent a subset of phylogenetic networks that are constrained to produce only bifurcating relationships; in trees, only two descendent lineages may stem from a single ancestor and no reticulations (back mutation, hybridization, horizontal gene transfer, recombination, or gene duplication) are allowed in the graphic structure. Phylogenetic trees also assume that ancestral and descendent DNA sequences (i.e., haplotypes) do not coexist in time. The assumptions about bifurcating relationships and the extinction of ancestral sequences are not always satisfied at the intraspecific level (Templeton et al. 1995; Page and Holmes 1998). It may be useful to construct both phylogenetic networks and phylogenetic trees and to compare the two because trees are most appropriate at the higher levels of sequence divergence and networks are more realistic portrayals of fine-scale relationships (e.g., within a population). Alternatively, phylogenetic trees can be used to portray relationships among major evolutionary lineages and networks can be used to represent relationships within each major lineage.

Traditional Tree-Building Methods

Producing the best phylogenetic estimate from aligned DNA sequences is essential. An incorrect estimate of the phylogeny will ultimately result in erroneous interpretations of the geographic distribution of lineages and historical demographic processes. A detailed explanation of all methods used to infer trees is beyond the scope of this chapter. We do, however, provide a brief overview of the most useful and reliable methods. Phylogenetic methods can be divided into distance-based methods, such as neighbor-joining (NJ) or unweighted pair group method with arithmetic mean (UPGMA), and discrete-character methods, such as maximum parsimony (MP), maximum likelihood (ML), and Bayesian inference (BI). Distance methods rely on the assumption that evolutionary distances can be estimated accurately between all sequences in the matrix and that these can be used to infer phylogenetic relationships, essentially by minimizing the overall distance across the tree (Page and Holmes 1998; Felsenstein 2004). The major objections to using distance-based methods are (1) loss of information from converting character data into distance data, (2) inaccurate estimation of evolutionary distances and branch lengths, and (3) the simplified assumption that overall similarity among individuals is equivalent to evolutionary relationship (Page and Holmes 1998; Felsenstein 2004). Because distance methods yield trees in a matter of seconds, however, they are often very useful for obtaining rapid estimates of phylogeny that can be used to check progress during a study or to verify the accurate labeling of samples or concatenation of data sets. Distance measures can also be used as a starting point for model-based discrete-character methods (BI and ML).

MP relies on the assumption that a tree that connects sequences with the fewest number of changes best represents the evolutionary relationships of these individuals. In MP, support for common ancestry is derived only for characters that are presumed to represent shared derived characters (synapomorphies). The method works well when rates of evolution are not highly variable among the terminal taxa, sequence divergence is low, all sites evolve at similar rates, and different types of change occur at similar rates. Many phylogenetic software programs have been created that implement MP; one of the most common is PAUP* (Swofford 2000). The pathological behavior of the method when evolutionary rates vary substantially across a tree (or when some very long branches are included) is referred to as long-branch attraction, or the Felsenstein Zone. This results in individuals with long-branch lengths being incorrectly placed as sister taxa due to homoplasy or convergence being erroneously inferred to be due to common ancestry (Felsenstein 1978; Huelsenbeck and Hillis 1993; Siddall 1998; Swofford et al. 2001). Variation in rates of evolution across sites, and variation in the rates of different types of substitutions (e.g., transitions vs. transversions), may also contribute to the failure of parsimony, especially when more divergent sequences are analyzed. These shortcomings are probably not major issues for inferences using extremely closely related sequences. Certain studies of "single" snake species, however, have revealed remarkably evolutionary distant separate geographic lineages, as much as 13% sequence divergence, in mtDNA genes (Burbrink et al. 2000). Accurately estimating branch lengths (i.e., estimated number of substitutions per site), often used to infer rates of evolution and various demographic parameters, is an impossible task under the assumption of MP. Therefore, any subsequent analyses using an MP tree, such as those for demographic estimates, divergence dating, or statistical character mapping, will require more accurate branch-length estimates, disqualifying MP as a particularly useful method. (See Swofford et al. 2001 and Felsenstein 2004 for a comprehensive review of the methods and problems associated with MP estimation.) It is unlikely that most journals in the field would accept results solely from MP estimates (or distance-based methods); they would minimally require additional ML or BI phylogeny estimates.

Modern Model-Based Tree Methods

Maximum Likelihood

Currently, the most commonly used phylogenetic methods are based on likelihood criteria (Cavalli-Sforza and Edwards 1964; Felsenstein 1973). Likelihood-based methods (both ML and BI) require some understanding of model building and statistics because these probabilistic methods generally attempt to take into account the stochastic rates and patterns of DNA evolution. Numerous studies have shown that ML is particularly robust to many of the potential problems that lead to errors in traditional distance- or parsimony-based methods. In their most basic form, ML methods aim to maximize the likelihood of the observed data (the DNA matrix), given a tree (with branch lengths) and a model of evolution. Essentially, ML methods attempt to find the single most likely estimate (MLE) of the tree that produced the observed DNA data set by evaluating different topologies and branch lengths using credible stochastic models of evolution.

Because likelihood-based methods rely on models of DNA evolution, it is critical to carefully select which model is appropriate for each data set.

These nucleotide models typically comprise three groups of parameters: (1) the nucleotide frequencies, (2) the relative rates (i.e., instantaneous rates or stochastic probabilities) of change among different nucleotide states, and (3) the variation of evolutionary rates across sites. In terms of the rates of change among nucleotide states, the least complex model that implies equal probability of change among all nucleotides is the Jukes-Cantor (JC) model (Jukes and Cantor 1969): the most complex of the typical models, is the general time reversible (GTR) model, which allows for different probabilities of change between all possible nucleotides but equal rates for forward and reverse substitutions (e.g., the rate of A → C = C → A). These models may also account for variation among the frequencies of the four nucleotides (Lanave et al. 1984; Tavaré 1986; Rodríguez et al. 1990). Many other models of complexities intermediate between JC and GTR exist and are also commonly used. Two other important parameters are often included in typical models to account for the variation of evolutionary rates across sites: the gamma parameter (Γ), which permits rates of evolution to vary in a predefined number of classes across all sites, and the invariable sites parameter (I), which aids Γ by allowing a certain percentage of sites to be classified as invariable (Hasegawa et al. 1987; Jin and Nei 1990; Yang 1996).

Identifying the most appropriate substitution model is crucial to finding the MLE of the tree and associated branch lengths. The best-fit model will vary given the data set, and there is no consensus about a single model that is appropriate for all snake phylogeographic projects. In general, the size of a data set and the sequence variation present determine how complex of a model should be used because there must be sufficient variation to accurately estimate all the parameters of a model. Thus, it is not necessarily a matter of applying the most realistic model; it is often more of an issue of determining how complex a model can be accurately inferred based on the data being analyzed. The DNA data matrix will almost always fit the more complex model with a higher likelihood but at the possible price of overparameterization (Rannala 2002). If more model parameters are included than can be reliably estimated, the resulting inferences may be highly inaccurate or otherwise unreliable. Therefore, several statistical methods including the likelihood ratio test (LRT), Akaike information criteria (AIC), Bayesian information criteria (BIC), or Bayes factors (BF; used for Bayesian inference) are applied to choose the most appropriate model prior to phylogenetic estimation (Posada and Crandall 2001; Bollback 2002; Huelsenbeck et al. 2002; Nylander 2004; Nylander et al. 2004; Posada and Buckley 2004). This phase of model testing can be automated using software programs Modeltest (Posada and Crandall 1998) and MrModeltest (http://www.abc.se/~nylander/).

Several models can be used simultaneously in a single analysis to accommodate different patterns and rates of evolution that may characterize various parts of a single data set; this is commonly referred to as model partitioning. For instance, a complex set of models could characterize a DNA data set composed of a protein-coding gene, intron, tRNA, and rRNA. In this example involving a single phylogenetic estimate, one model could account for each of the three codon positions of the protein-coding gene, a second model for the intron, a third for the tRNA, and yet a fourth for the rRNA gene (Nylander et al. 2004; Brandley et al. 2005; Castoe and Parkinson 2006; Burbrink and Lawson 2007). The most difficult part of choosing the appropriate partitioned model centers on determining the number of distinct models and the groups of genes or sites that should be included in the various partitions. At present, manual estimation is necessary to determine which scheme is best. (Detailed examples and suggestions for these partitioned model approaches and model selection can be found in Nylander et al. 2004; Castoe et al. 2004, 2005; Brandley et al. 2005; Castoe and Parkinson 2006; Castoe et al. 2007a). Generally, low (evolutionarily shallow) divergence often characterizes phylogeographic data, and such low divergence and sequence variation may not justify extremely complex models. Finally, quantitative estimates of branch lengths, and even tree topologies, may be quite different depending on the evolutionary models used, particularly for deeper, more ancient divergences (Castoe et al. 2004, 2005; Castoe and Parkinson 2006), even in moderate-scale phylogeographic studies (Castoe et al. 2005).

Competently searching or exploring the enormous number of possible phylogenetic trees to identify the most likely topology is a difficult problem for any phylogenetic method. Fortunately for MP and ML, heuristic methods reduce the set of all possible trees to be searched, although heuristic searches do not necessarily guarantee that the best tree will be found (Felsenstein 2004). These methods function by first producing a relatively reasonable tree that joins all individuals together using a method of low computational intensity, such as NJ or stepwise addition. Subsections of the tree are then moved throughout the topology and reconnected to find more likely trees, and topologies with high likelihood scores (ML) or fewer changes (MP) are retained. These hill-climbing algorithms accept only trees with higher likelihoods (or fewer changes, in MP) than those previously visited, and it is possible that the overall (global) best tree may never be reached (see Page and Holmes 1998; Felsenstein 2004). Because a single heuristic search does not guarantee the identification of the most optimal tree, these searches are often repeated many (e.g., 10–1000) times and the best estimate from this set is taken, under the assumption that one of the searches should have reached the global optimum.

Support for trees inferred using distance-based, MP, and ML methods are often derived from jacknifing or nonparametric bootstrapping. The former resamples the data set without replacement, whereas the latter resamples with replacement. In nonparametric bootstrapping, the most common approach, pseudo-replicated data sets are produced by resampling the original DNA data sequence with replacement (Felsenstein 1985). Trees are then estimated for each pseudo-replicate, and the frequency of observance for any relationship is summed across all pseudo-replicates—this frequency is used to represent the bootstrap support (bootstrap percentage) for relationships. For instance, 1000 pseudo-replicated data sets will yield 1000 trees, from which support for any node is assessed by determining the frequency that any node is found among all pseudo-replicated trees. If there are 995 trees that contain a relationship where snake A is sister to snake B, then we can use this as measure of confidence to indicate that 99.5% of the bootstrap trees contain this relationship (Hedges 1992; Hillis and Bull 1993). Generally, relationships (or nodes) found in at least 70 to 80% of the pseudo-replicated trees are considered credible, depending on the data and models used to infer the tree, although warnings against the assumption of a standard measure of support by bootstrapping have been argued (Felsenstein 2004).

Bayesian Inference

Bayesian inference of phylogeny is becoming increasingly common and is gradually displacing the use of ML methods. For the most part, these two probabilistic methods appear to produce the most reliable tree topologies with the most accurate support values and branch-length estimates. Like ML, BI is also a likelihood model–based method of tree inference, but it has some very key differences. The two most important differences are that BI relies critically on the prior expectations of inferred parameters (including trees) and that the results of BIs represent a distribution of optimal estimates (the posterior distribution) rather than a single-point estimate of the "best" hypothesis (as in ML). (See Holder and Lewis 2003 for an excellent philosophical and practical contrast between these approaches; see also Felsenstein 2004). One major desirable property of the BI approach is that the result of a Bayesian analysis (the posterior) represents a distribution of all the highly optimal estimates; when this distribution is summarized, the resulting estimate is integrated across all these very highly likely estimates. For example, the tree topology and support are averaged over all the highly optimal values of model parameters in the posterior distribution. This approach also allows an enormous increase in computational efficiency over ML while maintaining much of the same positive qualities.

Ultimately, the goal for the phylogeographer is the posterior probability distribution of trees ($P(ti|X)$ = the probability of the tree, given the data), which not only yields a final tree estimate but also support for that tree (Fig. 2.2). BI examines the posterior probability by inferring the likelihood of a tree given the data multiplied times prior information about that tree and scaled over all possible arrangements given the data. Unfortunately, the integration (or summation) over all possible trees and parameters is impossible due to computational complexity, or the sheer number of possibilities. To compensate for this and to obtain a posterior probability distribution of trees, the Markov chain Monte Carlo (MCMC) method is used

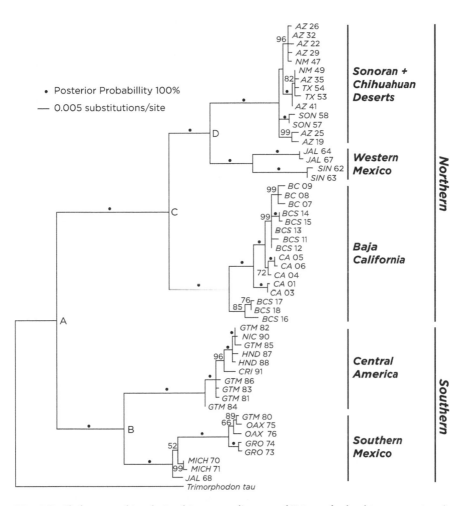

Fig. 2.2. Phylogeographic relationships among lineages of *Trimorphodon biscutatus* using the ND4 gene and flanking tRNA sequences (Ser and Leu).

(a) Tree produced using Bayesian inference with the model GTR + *Γ* + *I* with posterior probability support values placed above branches. *Γ, gamma parameter;* GTR, general time reversible model; I, invariable sites parameter. (By permission of Thomas J. Devitt from Devitt 2006)

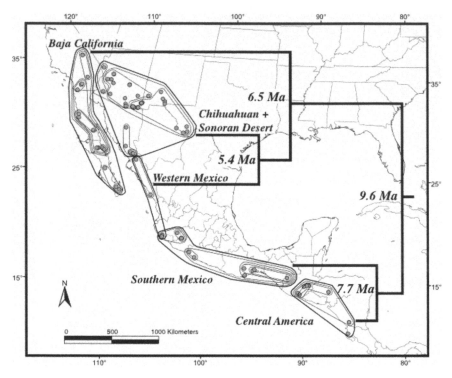

Fig. 2.2 continued. (b) Geographic distribution of lineages and dates of divergence using the Bayesian relaxed clock method in MultiDivTime. Ma, millions of years ago. (From Thorne and Kishino 2002)

(Metropolis et al. 1953; Hastings 1970; Griffith and Tavaré 1994; Kuhner et al. 1995; Larget 2006). This method implements a series of links to form a chain, in which each link in the chain represents a newly sampled tree with substitution and branch-length parameter states. A new parameter state is proposed and forms the next connected link in this chain. Each adjacent link in the chain is similar to the previous one, but slight changes to parameters have been made in some cases, and proposal mechanisms (such as those described in the Metropolis-Hastings method; see Nielsen 2006) determine whether a new set of parameters will be accepted in the new link. Of importance here is that this method does not necessarily always climb hills (or directly optimize) because not all proposals (even if more optimal) are necessarily accepted. Generations are the number of links in a chain, and the MCMC chain may run for many millions of generations (Gilks et al. 1996; Huelsenbeck et al. 2002; Larget 2006). The chain generally moves into areas of high posterior probability, and the amount of time spent in these regions of tree space is equivalent to the support for any topology. The tree samples taken before the chain moves into the region of high probability is referred to as burn-in and is discarded (Gilks et al. 1996; Huelsenbeck et al. 2002; Nylander et al. 2004; Larget 2006).

A well-constructed chain, or multiple chains, will move through tree space and sample many different topologies and model parameters. At the end of a run, after burn-in generations are removed, the researcher is presented with the posterior probability distribution that can be summarized in the form of a consensus of topologies and branch lengths, with support values (posterior probabilities) for various branches or clades (Huelsenbeck et al. 2002; Holder and Lewis 2003; Larget 2006).

Conducting a Bayesian analysis requires knowledge of some very detailed statistical issues. For instance, the proposal mechanism for the MCMC is quite crucial for adequately searching tree space (i.e., to avoid local optima), and the length of the chain is also important to determine that burn-in has occurred (Huelsenbeck et al. 2002; Archibold et al. 2003; Larget 2006). The Bayes theorem also requires the researcher to specify priors for all parameters, which ultimately may affect the posterior probability distribution.

Methods of BI are becoming necessary for phylogeographic studies due to the massive sizes of data sets and complexity of models. The most common software programs used to implement BI are the freely available and easy-to-use MrBayes (Ronquist and Huelsenbeck 2003) and BEAST (Drummond and Rambaut 2006). These programs run in a reasonable amount of time and allow the incorporation of some very flexible models, including complex partitioned models and Bayesian relaxed molecular clocks for divergence time estimation. A third program, BEST (Edwards et al. 2007; Liu and Pearl 2007), uses a hierarchical Bayesian method to infer a species (or population) tree from the joint estimate of gene trees while incorporating information from the coalescent.

Comparative Phylogeography

Examining the phylogeographic patterns of independent species with overlapping or partially overlapping ranges may reveal common events that have affected the evolutionary patterns of many taxa in similar ways. This broader field of study—comparative phylogeography—examines codistributed taxa and infers historical, geological, and climatic events that have shaped biogeographic patterns in communities of species (Bermingham and Moritz 1998; Schneider et al. 1998; Avise 2000; Arbogast and Kenagy 2001; Zink 2002; Steele and Storfer 2006). It is expected that, if codistributed species share similar reciprocally monophyletic phylogeographic topologies with genetic discontinuities occurring at the same geographic barriers, they also share a similar relatively stable and long-term history in these areas (Zink 2002). Even if different taxa share similar phylogeographic breaks, however, it is possible that the origins of these lineages in different species occurred at different times or in slightly different geographic areas, thus producing only pseudo-congruent phylogeographic patterns.

As an example, the Mississippi River embayment (forming at the confluence of the Ohio and the Mississippi rivers in southern Illinois, and distinct from the upper Mississippi) has been implicated as a barrier to gene flow for several snake species, including *Coluber constrictor, Pantherophis guttatus,* and *Pantherophis obsoletus* (Burbrink et al. 2000, 2008; Burbrink 2002), although the timing of the divergence at this barrier is not known. The pattern of genetic discordance at this river has been found for many unrelated reptiles and amphibians (Burbrink et al. 2000, 2008; Leaché and Reeder 2002; Moriarity and Cannatella 2004; Howes et al. 2006; Soltis et al. 2006). Even if the discordance occurred at a variety of times among the taxa, their phylogeographic structures still demonstrate the power that the Mississippi River embayment has, or once had, in separating formerly connected populations. To determine whether phylogeographic lineages sharing the same geographic range also share similar dates of divergence, several methods should be used to infer lineage age. Therefore, comparative temporal phylogeography assesses the degree of overlap in the dates of origin for codistributed lineages using various nonclocklike methods, including penalized likelihood with error estimation (Sanderson 2002, 2003) and Bayesian relaxed molecular clocks (Drummond et al. 2006; Thorne and Kishino 2005).

In addition, incongruence among codistributed taxa may occur due to lineage sorting, variation in effective population size, extinction, dispersal, sympatric speciation, or a lack of response to vicariant events (Mason-Gamer and Kellog 1996; van Veller et al. 1999; Crisci et al. 2003). The method of approximate Bayesian computation (ABC), implemented in the software program MsBayes (Hickerson et al. 2006, 2007) is designed to test for simultaneous divergence (vicariance) across various population pairs that span the same barrier. Simultaneous separation at a barrier in MsBayes is tested on all population pairs from all taxa of interest by estimating three hyperparameters that characterize the degree of variability (the mean, variability, and number of splitting events) in divergence times across codistributed population pairs while allowing for variation in several within-population-pair demographic parameters (subparameters) that affect the coalescent.

Properly conducting a comparative phylogeographic study not only requires a good knowledge of phylogeographic methods but also a detailed understanding of geology and other relevant historical events (e.g., climate change and glacial cycles). Several other barriers have been implicated in the formation of distinct lineages within species having overlapping ranges in the United States (Bermingham and Moritz 1998; Soltis et al. 2006). Many phylogeographic studies on terrestrial vertebrates, including snakes, in western North America have shown that common barriers to gene flow occur at the Rocky Mountains, the Great Basin, the division between the Chihuahuan and Sonoran deserts and the associated Cochise filter barrier/Continental Divide, and the Transverse Mountains in southern California (Zamudio et al. 1997; Pook et al. 2000; Avise 2000; Devitt 2006; Feldman and Spicer 2006; Castoe et al. 2007b). In eastern North America, major barriers to gene flow have been identified at the Mississippi River, the Tombigbee River and Mobile Bay, the Appalachian Mountains, the Apalachicola River, the Teays/Ohio River, and the river systems situated at either side of the eastern Continental Divide (Burbrink et al. 2000, 2008; Moriarty and Cannatella 2004; Howes et al. 2006; Kozak et al. 2006; Soltis et al. 2006; Lemmon et al. 2007). These ancient geological barriers may have had complex and non-uniform effects in separating populations among unrelated taxa (Soltis et al. 2006). The identification or corroboration of mutual genetic breaks is usually conducted in organisms only in particular areas of North America (e.g., east or west of the Continental Divide, the southwestern United States, and southeastern United States) and not across the entire continent. Peninsular Florida also provides another example in which distinct and endemic lineages of snakes are found, including *Coluber constrictor, Thamnophis sirtalis,* and *Agkistrodon piscivorus* (Burbrink et al. 2008; Guiher and Burbrink 2008). These distinct lineages highlight a former barrier to gene flow that is no longer as evident as the Mississippi River. The rise in sea level during interglacial periods throughout the Pliocene and Pleistocene most likely separated continental populations from those occurring on isolated highland islands in central Florida (Webb 1990; Wiens and Graham 2005). Today, the sea levels are 35 m lower, and there is a broad land connection between southern-central and northern Florida and Georgia, in contrast to the earlier periods when marine incursions separated these areas.

Feldman and Spicer (2006) examined comparative phylogeographic and demographic patterns in several lizards and snakes in California: *Contia tenuis, Diadophis punctatus, Elgaria multicarinata, Charina bottae,* and *Lampropeltis zonata.* They concluded that the basic deep genealogical divisions are the same spatially and temporally for all taxa at the Transverse Ranges, the Monterey Bay and Sacramento–San Joaquin Delta regions, and southern Sierra Nevada in California. Interestingly, demographic methods assessing population growth (discussed later in the chapter) suggested that lineages of these species in the north have all experienced rapid population growth due to the increase of woodland habitat in Holocene.

Historical Demography

The Coalescent

Inferring demographic change is another major area of study often associated with phylogeography. Although many of the statistics used for these analyses are connected with population genetics (King, Chapter 3), we include a brief discussion of the important methods to encourage their use in snake phylogeographic studies. The strict divisions between population genetics, phylogeography, and phylogenetics are, deservedly, becoming blurred. For the most part, we focus here on the expansion or contraction of populations of phylogeographic lineages.

Modern descriptions of historical population demographics usually begin with a discussion of coalescent theory. This theory models genealogical relationships backward in time to common ancestors. This is an extension of the classic population genetics concept of neutral evolution and is an approximation of the Fisher-Wright model for large populations (Fisher 1930; Wright 1931; Kingman 1982; Emerson et al. 2001). Although coalescent theory fits well within the field of population genetics, the benefits for phylogeographers in understanding historical demographic changes in a lineage are substantial.

Modeling lineage sorting in reverse time permits the researcher to examine questions relevant to populations and phylogeographic lineages, including estimates of population size, structure, selection, mutation rate, and recombination (Wakely 2007). Effective population size predicts the probability that two gene sequences will coalesce. Compared with a large population, two individual sequences drawn from a smaller population have a higher probability of sharing a more recent common ancestor (i.e., they coalesce more quickly or there are fewer substitutional differences that exist between them; Kingman 1982; Emerson et al. 2001). Therefore, the dynamics of population size over time leaves an imprint in the sequence differences among individuals and ultimately in the trees inferred from these sequences. The shapes of the trees and differences among sequences can then be used to estimate recent or ancient population growth and declines (Emerson et al. 2001; Drummond et al. 2005).

When conducting coalescent or demographic analysis, it is critical to consider how limited sampling may bias the experimental results. Because these methods are dependent on the tree used and the distribution of variation across the data set, nonexhaustive sampling or uneven sampling in certain areas may substantially bias inferences.

The Mismatch Distribution and Other Measures of Population of Expansion

Under an infinite-sites model (i.e., each mutation occurs at a new site), coalescent theory yields an understanding that demographic changes in a population may be evident in the amount and type of genetic variation retained in individuals (Hudson 1990; Donnelly and Tavaré 1997). The earliest studies using coalescent theory to assess population size changes relied on the effect of pairwise differences (i.e., the number of nucleotide sites for which alleles differ in their nucleotide state) in DNA sequences among haplotypes and the number of segregating sites (i.e., the number of sites with polymorphisms) in a population (Tajima 1989; Slatkin and Hudson 1991; Rogers and Harpending 1992; Emerson et al. 2001). These measures of pairwise differences can predict the sudden expansion of lineages using the theoretical expectations of a Poisson distribution. Population expansion should yield an unresolved phylogeny, a reduction of segregating sites, a large proportion of low-frequency mutations, or a unimodal distribution of differences (Slatkin and Hudson 1991; Rogers and Harpending 1992; Fu and Li 1993; Bertorelle and Slatkin 1995; Aris-Brosou and Excoffier 1996; Tajima 1996; Fu 1997). Effective population sizes through time directly impact coalescent times and thus influence the shapes of phylogeographic trees (Avise 2000). Therefore, different topologies with variable branch lengths (expected numbers of substitutions) are indicative of alternative population demographics (Tajima 1989; Harpending et al. 1993; Eller and Harpending 1996). For instance, clustering of older nodes through time may indicate that the phylogeographic lineage of interest grew rapidly in the past and slowed toward the present (Avise 2000). Rapid population growth often follows genetic bottlenecks and produces an unresolved star phylogeny with most of the lineage diversification occurring directly at the time of population expansion (Slatkin and Hudson 1991). One of the most widely used statistics to incorporate these measures is the mismatch distribution.

The mismatch distribution examines the number of site differences among all pairs of haplotypes and provides information about spatial and historical population expansion. A histogram of these differences is plotted against an observed distribution of differences, and inferences based on the differences between the theoretical and experimental distributions yield insights into past population demographics. A unimodal mismatch distribution indicates a recent range expansion, a multimodal (including bimodal) distribution suggests diminishing or structured population sizes, and a ragged distribution reveals that the lineage was widespread (see Fig. 2.3) (Rogers and Harpending 1992; Rogers et al. 1996; Excoffier and Schneider 1999). The multimodal distribution may also indicate that the population is influenced by migration, is subdivided, or has undergone historical contraction (Marjoram and Donnelly 1994; Bertorelle and Slatkin 1995; Ray et al. 2003).

The statistical significance of these distributions can be tested using the sum of squares distances (SSD) and Harpending's raggedness index (r_g; Harpending 1994) against a null distribution of recent population expansion using bootstrap replicates, as in the program Arlequin v3.0 (Excoffier et al. 2005), or Monte Carlo simulations, as in the program DnaSP 4.10.8 (Rozas et al. 2006). The R_2 statistic of Ramos-Onsins and Rozas (2002) can also be used to examine population expansion and is particularly powerful when population sizes are small. Although the error estimate is generally

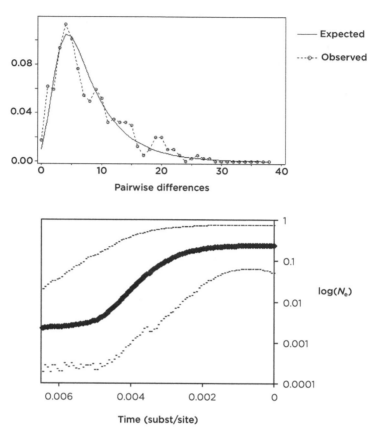

Fig. 2.3. Estimation of population expansion through time using the mismatch distribution with tests against the expectation of growth (upper graphs) and Bayesian skyline plots (BSPs; lower graphs) for three lineages of snakes occupying similar ranges in the eastern United States from the cytochrome *b* gene. N_e, expected population; subst, substitutions.
(a) *Coluber constrictor*—the unimodal mismatch distribution is not significantly different from the expectation of recent population growth; similarly, the increasing BSP toward time zero indicates that this lineage has undergone recent expansion in time.

high, the dates of population expansion can examined using the formula $T = \tau/2\mu$, where T is time since expansion, τ is the expansion time produced in Arlequin v 3.0, and μ is the mutation rate generation time sequence length. It is important to note that the mutation rate cannot be assumed to be identical across all snake species or across genes. It is best to estimate the rates of substitution for lineages using either penalized likelihood in the software program 8S v 1.07 (Sanderson 2003) or uncorrelated relaxed Bayesian clocks (Drummond et al. 2006) in BEAST using fossils or geological calibration points to remove the effect of time from the rate. In addition, the formula for T indicates that knowledge of the generation time is known

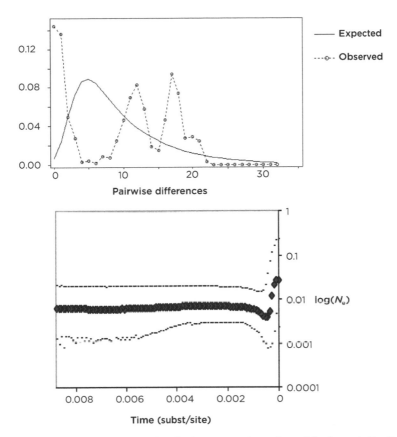

Fig. 2.3 continued. (b) *Pantherophis alleghaniensis*—the multimodal mismatch distribution indicates this lineage does not exhibit the genetic signature of recent population expansion, but the more sensitive coalescent method using BSP indicates a constant population size through time, with a rapid expansion in nearly modern times.

and that generations do not overlap. This information may or may not be known and may also differ among populations of snakes in different environments (Fitch 1999).

Population growth in each lineage can also be examined using Tajima's (1989) D^* and Fu and Li's (1993) F^* in DnaSP 4.10.8 (Rozas et al. 2006) or Arlequin (Excoffier et al. 2005). Because the results of both statistics may not separate the effects of population expansion from purifying selection (Braverman et al. 1995; Simonsen et al. 1995; Fu 1997; Fu and Li 1999; Hahn et al. 2002), it is recommended that each test for both π_S (within lineage synonymous sites) and π_N (within lineage nonsynonymous sites) be conducted separately (Hahn et al. 2002). If population expansion has occurred, then statistics for both π_S and π_N should be significantly negative. In contrast to the homogeneous effects on both types of substitutions testing

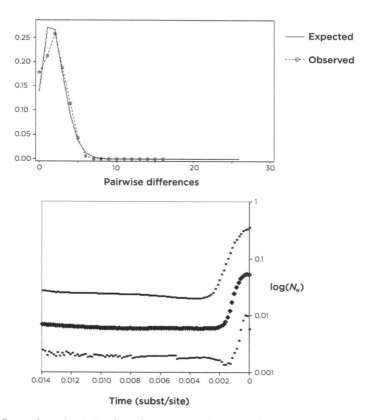

Fig. 2.3 continued. (c) *Pantherophis guttatus*—the unimodal mismatch distribution indicates that this lineage has undergone recent population expansion; the BSP indicates that the species exhibits constant population growth since the origin of the lineage, with a sharp population increase before time zero.

expansion, purifying selection is expected to yield significantly negative test statistics for π_N only (Rand and Kann 1996; Hahn et al. 2002). Population expansion should be evident in most or all unlinked genes with rapid rates of evolution, whereas purifying selection should be evident only in one gene or closely linked genes.

Bayesian Skyline Plots

Methods using pairwise differences do not easily consider populations under constant growth, nor do they provide a clear picture of growth patterns through time (Felsenstein 1992). These methods also suffer from a lack of independence of sites, which can be corrected by using a genealogical estimate (see Emerson et al. 2001 for the assumptions made by different methods). There are several methods that may incorporate genealogical (or phylogeographic) information to examine effective population sizes through time, including lineage through time plots (LTT) and skyline plots (Nee et al. 1995; Pybus et al. 2000); there is also the isolation with migration model (Hey and Nielsen 2004; see the example in Castoe et al. 2007b). Therefore, given a phylogeographic estimate and different slices of coalescent times though the tree, it is possible to model population sizes through time. Recent modifications to these coalescent models by Drummond et al. (2002) and Pybus et al. (2003) have provided a powerful and flexible method using MCMC methods for the joint estimation of genealogy, demographic patterns, and substitution parameters. Drummond et al. (2005) introduced a visual modification of these methods called the Bayesian skyline plot, which eliminates the prespecification of population demographic models. These models assess demographic parameters that determine whether the lineages have experienced a reduction in population size, remained constant, or undergone logistic or exponential growth. This flexibility allows us to view changes in effective population sizes through time without specifying any of the possible growth curves a priori. Currently implemented in BEAST (Drummond and Rambaut 2006), the method assesses various population demographic patterns through time using the coalescent while estimating the probability and uncertainty of tree topology. Combining the relaxed Bayesian clocks method with Bayesian skyline plots in BEAST also permits the researcher to attach an estimation of time to demographic events.

Example of Population Demographic Inferences in Co-Occurring Snake Lineages

Comparative demographics is another burgeoning field of study, but is not discussed as often as comparative phylogeography. This type of research examines the population dynamics of codistributed lineages found in a similar area. These studies are capable of addressing questions relating to population growth or decline of codistributed lineages in different species. For instance, we can ask, do the codistributed lineages currently occupying formerly glaciated habitats in the United States and Canada show similar patterns of population growth due to the recent re-opening of habitable areas following glacial retreat? If they do not, then we ask, what is it about these organisms (e.g., niche) that prevents them from having similar demographic responses to the same climatic event?

Although the methods used to examine population demographics are commonly used in other vertebrates, we have found few instances in which phylogeographers have applied them to snakes (Douglas et al. 2006; Castoe et al. 2007b; Burbrink et al. 2008). Here, we present two different measures of effective population size changes for geographically defined lineages of *Coluber constrictor, Pantherophis guttatus,* and *P. alleghaniensis,* all occupying similar geographic areas east of the Appalachian Mountains

(Burbrink et al. 2000, 2008; Burbrink 2002). Two of these species, *P. alle-ghaniensis* and *C. constrictor,* occupy areas in formerly glaciated regions of the northeastern United States and we expect that they will show evidence of population expansion. After choosing the appropriate phylogenetic model (GTR + Γ + I) and using a relaxed exponential clock, we examined Bayesian skyline plots in BEAST v1.3 (Drummond et al. 2005; Drummond and Rambaut 2006) for each lineage using sequence data from the mtDNA gene Cytochrome *b*. Population growth patterns are very different for these three snakes in the eastern United States (Fig. 2.3). *Coluber constrictor* reveals logistic population expansion over a long period of time, whereas *P. guttatus* has experienced only recent and rapid expansion. In contrast, it appears that *P. alleghaniensis* shows a population expansion after a population crash, possibly caused by a reduction in effective population size following glacial advances. Mismatch distributions predict growth in *C. constrictor* and *P. guttatus,* but did not estimate the population crash and recovery in *P. alleghaniensis*. Statistical tests of the null mismatch distributions (Fig. 2.3) and Fu and Li's *F** and Tajima's *D** with significantly negative values ($P < .001$) also confirm this; however, they do not indicate the shape or nature of the growth curve. This example suggests that although lineages of snakes living in similar areas might show evidence of population expansion, there may be notable variation in the rate and timing of growth and possible population contraction in different lineages or species.

Nested Clade Analysis

The geographic structure of haplotypes that researchers observe may be due to (1) current or historical restricted gene flow, (2) past population fragmentation, (3) range expansion, or (4) colonization. These hypotheses can be addressed simultaneously using nested clade analysis (NCA). This methodology attempts to reduce the errors that occur in inferring processes that produced phylogeographic structure by simply overlaying trees on a map (Templeton 2004). Grounded in coalescent theory, this method uses haplotype networks (rather than phylogenetic trees) as the basis to test these hypotheses. Haplotype networks are not constrained to bifurcations; instead, multifurcations are permissible and possibly better represent the actual dynamics and relationships of haplotypes than do purely bifurcating trees, especially for fine-scale phylogeographic studies (Fig. 2.4) (Panchal 2007). The estimation of minimum spanning haplotype networks using the principle of statistical parsimony with the software program TCS (Clement et al. 2000) is the first step in addressing these hypotheses; note that this statistical parsimony approach is very distinct from standard MP (see Templeton et al. 1995). Second, these networks must be nested by grouping haplotypes into clades by the number of mutational differences (or steps) from the lowest (haplotypes or 0 step clades at the tips of the network) to highest (internal

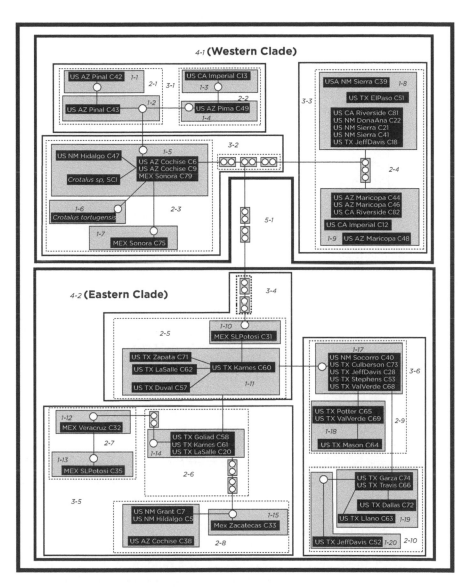

Fig. 2.4. Phylogeographic analysis of *Crotalus atrox* in the southwestern United States and Mexico using nested clade analysis. (Nested clade analysis adapted from Castoe et al. 2007b, with permission of Molecular Phylogenetics and Evolution [Elsevier])

(a) Grouped haplotype network showing the geographic distribution of clades (see Castoe et al. 2007 for more demographic and dichotomous key interpretations of this network)

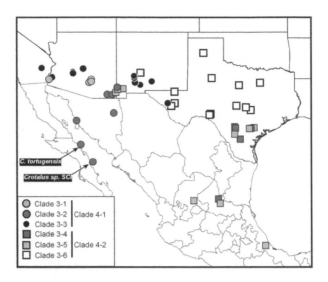

Fig. 2.4 continued. (b) Geographic range of grouped haplotype networks shown in (a).

clades). This process is continued until all clades have been hierarchically joined and the highest nesting levels represent those with the greatest genetic distances, which will include all haplotypes in the form of an entire network. The clades or haplotypes at the tips of the network are assumed to be younger than those in the interior (Castelloe and Templeton 1994); thus, NCA relies on a relative temporal relationship between tips and interiors. The spatial distribution of the haplotypes and clades are quantified using two measures of distance: the geographic distance and the nested distance (Templeton et al. 1995; Templeton 2004).

The geographic distance (D_c) measures how far a single individual with a given haplotype is located from the geographic center of all individuals with this haplotype. The nested measure of the distance (D_n) indicates how far away a haplotype or clade is located from those in which it is nested into the next higher hierarchical level. That is, this essentially measures how far away a single haplotype sampled from the clade of interest is from the center of the next hierarchically nested clade. Inferences based on the statistical significance of these distance measures against random distance distributions permit inferences about restricted gene flow, past fragmentation, range expansion, or colonization in reference to any particular clade. Contrasting D_c and D_n between tip and interior clades provides evidence for population or lineage structuring and gene flow (Templeton 1998). Posada et al. (2000; http://darwin.uvigo.es/software/geodis.html) provided a key to guide the interpretation of these phylogeographical inferences based on statistical tests comparing different distance measures estimated in the program GEODIS.

This updated key also determines where haplotype information is lacking and when inferences cannot be satisfactorily made.

The use of NCA in snake phylogeography is uncommon (Creer et al. 2001; Castoe et al. 2007b), possibly because NCA was, until recently, quite laborious and required substantial manual annotation of the haplotype network structure. Thankfully, Panchal (2007) provided a fully automated program to perform NCA and interpret the phylogeographic and demographic patterns. Criticisms of NCA suggest that the method cannot effectively distinguish between historical and current gene-flow processes responsible for the creation of simulated data (Knowles and Maddison 2002). High false-positive rates with respect to concluding isolation by distance or restricted gene flow for a clade are, however, possible (Panchal 2007). The soundest approach to using NCA is to cross-validate inferences obtained using the other coalescent and demographic approaches already described (Knowles and Maddison 2002; Castoe et al. 2007b).

For an example of NCA used to examine phylogeographic structure in snakes, we turn to the phylogeographic research on *Crotalus atrox* (Castoe et al. 2007b) (see Fig. 2.4). Their results demonstrated that this species is composed of two major lineages (eastern and western) separated in the southwestern United States by the Continental Divide. These primary divisions apparently occurred during the mid to late Pliocene, as inferred from an approximation of the mutation rate at 1.4% divergence per million years for the gene *ND4*. This major split in the southwestern U.S. deserts is geographically concordant with divisions found in other squamates (Ashton and de Queiroz 2001; Leaché and Reeder 2002; Leaché and McGuire 2006). Within the western clade, the authors discovered two well-resolved lineages, 3-1 + 3-2 and 3-3. The former is predominantly distributed in the central and western Sonoran Desert and the latter is found in California, New Mexico, and western Texas. The western clade most likely existed in Pleistocene refugia in the Sonoran Desert, with NCA inferring restricted gene flow and isolation by distance for this clade. However, clade 3-2 in the western lineage may have undergone recent population expansion from a refugium, as predicted by NCA. The NCA results also appear to be validated by the unimodal mismatch distribution and significantly negative Fu's F^* value, and by coalescent analyses using the isolation with migration model (Hey and Nielsen 2004). The clade east of the continental divide occupies an area approximately five times the size of the western clade and is structured into three smaller geographic clades (3-4, 3-5, 3-6; see Fig. 2.4). The NCA results indicate that one of the smaller clades (3-4) may have undergone a recent range expansion, as confirmed by the mismatch distribution and negative Fu's F^* value. Clades 3-4 and 3-5 may also have shared a common refugium in the Mapimian subregion of the Chihuahuan Desert (Castoe et al. 2007b).

The Present and Future of Snake Phylogeography

One prevalent theme that has emerged from snake phylogeographic research is that species- and, especially, subspecies-level taxonomies can be
poor indicators of phylogeographic and phylogenetic groups. In many cases,
in both temperate species (Rodríguez-Robles and de Jesús-Escobar 1999;
Burbrink et al. 2000, 2008; Burbrink 2002; Fontanella et al. 2008) and
tropical species (Wüster et al. 2002, 2005a; Castoe et al. 2003, 2005, 2008),
phylogeographic studies have often discovered substantial genetic diversity
and structure below the level of the recognized species, much of which is
not concordant with subspecific taxonomy. The genetic divergence among
lineages in a single species is extremely large (e.g., 10–13% uncorrected in
lineages of *C. constrictor* and the *P. obsoletus* complex) and possibly greater
than the divergence of many recognized species of other vertebrate clades.

Many authors have suggested that most recognized species examined
phylogeographically comprise fairly ancient radiations of related evolutionary lineages; thus, most species contain multiple evolutionary lineages that
diverged from one another millions of years ago. For instance, using a Bayesian relaxed clock method (i.e., relaxed phylogenetics) Burbrink et al. (2008)
found that lineage diversification at various geographic boundaries began in
the late Miocene and early Pliocene for *C. constrictor*. These findings beg
the question, are any single-species groups with such large and ancient geographically separated lineages actually a single species comprising shallow
evolutionary diversity? It would also appear that, relative to their genetic
diversity, many single species of snakes are quite conserved morphologically
(i.e., these ancient lineages have no obvious morphological differences).

An obvious general conclusion from previous snake phylogeographic
studies is that the same geographic barriers have effected the diversification
of lineages in multiple species of snakes, in similar ways in some cases. At
this time, there is insufficient information currently to assess (or predict) the
degree to which multiple species may have been historically affected by the
same geographic, tectonic, or physiographic boundaries. As discussed earlier, several barriers (e.g., the Mississippi River) have produced the same
patterns of genetic discordance in unrelated taxa with different habitat requirements (Burbrink et al. 2000, 2008). However, barriers to gene flow
in some snakes may have no effect in other species. Given the diversity of
habitat requirements, life history traits, and population sizes, we should expect that non-uniform responses through time would occur at these barriers
(Fontanella et al. 2008; Guiher and Burbrink 2008), and future research to
test such historical responses to common barriers is needed.

Similarly, historical demographic fluctuations due to changes in glacial
cycles (or other habitat modifications) have been reported in several species.
Major trends observed include population crashes (although not always) at
glacial maxima and population expansions at glacial minima. Some lineages

of *Agkistrodon contortrix*, *A. piscivorus*, and *D. punctatus* in North America have the signature of an increase in effective population size following the last glacial maxima (~21,000 years ago; Fontanella et al. 2008; Guiher and Burbrink 2008). These responses are non-uniform in terms of the timing and intensity of the effective population response to glaciers. In contrast, *C. constrictor* has shown constant growth in populations for all lineages throughout the last half of the Pleistocene (Burbrink et al. 2008). Understanding variation in response to glaciation across diverse snake species may provide significant insights into the historical assembly and glacial impact on genetic diversity across temperate snake communities.

Future Directions of Snake Phylogeography

Several major areas have yet to be explored in snake phylogeography. One major problem is that few species of snakes have been examined phylogeographically, particularly in the tropics. Although most examples used in this chapter focus on taxa found in North America, several other species of snakes have been examined phylogeographically in Europe (e.g., *Malpolon monspessulanus* and *Hemorrhois hippocrepis*, Carranzo et al. 2006; *Natrix maura* and *N. tesselata*, Guicking et al. 2002; *Vipera aspis* and *V. berus*, Ursenbacher et al. 2006), Africa (e.g., *Macroprotodon abubakeri*, *M. brevis*, and *M. mauritanicus*, Carranza et al. 2004; *Naja nigricollis*, Wüster et al. 2007), Asia (e.g., *Cerberus rynchops*, Alfaro et al. 2004; *Deinagkistrodon acutus*, Huang et al. 2007; *Naja kaouthia*, Wüster and Thorpe 1994; *Trimeresurus stejnegeri*, Malhotra and Thorpe 2004), Australia (e.g., *Aipysurus laevis*, Lukoschek et al. 2007; *Hoplocephalus stephensii*, Keogh et al. 2003; *Morelia viridis*, Rawlings and Donnellan 2003; *Notechis ater* and *N. scutatus*, Keogh et al. 2005; *Pseudechis australis*, Kuch et al. 2005), and Central and South America (e.g., *Atropoides* species, Castoe et al. 2003, 2008; *Bothrops jararaca*, Grazziotin et al. 2006; *Bothrops pradoi*, Puorto et al. 2001; *Cerrophidion godmani*, Castoe et al. 2005, 2008; *Crotalus durissus*, Wüster et al. 2005a, 2005b; *Lachesis* species, Zamudio and Greene 1997; *Porthidium nasutum*, Castoe et al. 2005). Many of these wide-ranging taxa are composed of geographically distinct lineages that might represent distinct and unrecognized species under a lineage species concept (de Queiroz 1998). Several major problems may have impeded the process of examining wide-ranging taxa that cross political boundaries, including the difficulty in obtaining tissues of many species due to the cryptic habits of snakes and in acquiring the legal permits and funding to do so. It is critical that more species, especially wide-ranging species, of snakes be examined phylogeographically to provide further evolutionary perspectives on snake taxonomy, conservation, and overall snake biology.

It is not yet clear whether genetic barriers have caused lineages to diverge simultaneously in multiple species of snakes. With further comparative phylogeographic work, it may be possible to address the question, can we summarize what types of geographic, physiographic, or historical processes have repeatedly affected the phylogeographic structure of different snake species? For example, if multiple species diverge simultaneously at the Mississippi River, then this poses the question, what is special about this time for a river that has existed prior to the origin of colubroids? On the other hand, why do certain species fail to diverge at these common barriers? Other areas of exploration related to these concepts are the importance of geographic and physiographic barriers to snake community assemblages and the extent to which phylogeographic patterns in snakes are comparable to other terrestrial animals.

The lack of reliable and practical knowledge regarding the rates of molecular evolution in snake mitochondrial and nuclear genes is currently impeding snake phylogeographic research. This gap in our understanding of snake evolution complicates phylogeographic research because a broad diversity of phylogeographic analyses are probably dependent on these estimates. The extremely wide range of previous estimates of mitochondrial evolutionary rates presently precludes even an approximate understanding of what reasonable rates may be (Zamudio and Greene 1997; Wüster et al. 2002; Castoe et al. 2007b; Jiang et al. 2007). Although new flexible methods of obtaining divergence time and evolutionary rate estimates exist, these require calibration points (e.g., known dated fossils) to derive these estimates, which are typically unavailable for phylogeographic studies. Ultimately, the field will strongly benefit from future studies that clarify the rates of evolution for commonly used snake mitochondrial genes and that quantify the variance of rates across lineages.

It is imperative that non-mtDNA markers be applied in future studies on snakes to corroborate, and also reinterpret, previous mtDNA-based estimates of phylogeographic structure and historical demography. In addition to identifying rapidly evolving nuclear genes for phylogeography, other markers such as microsatellites should be applied to examine population demographic history and assess levels of gene flow among mtDNA-defined phylogeographic lineages. As a demonstration of the importance of such research, Gibbs et al. (2006) found that discrete mitochondrial phylogeographic lineages of *P. obsoletus* appear to be freely exchanging nuclear genes in Canada. The pervasiveness of this type of scenario across different phylogeographic barriers and species is a critically important question for further research. Future studies should include the comparison of phylogeographic structures inferred using both mitochondrial and nuclear-based genetic markers in snakes. It is currently unclear how well mitochondrial phylogenetics represents the entire process of snake phylogeography, demographic history, selection, gene flow, and taxonomy. Given snake life histories that may include sex-biased dispersal in some species, mitochondrial and nuclear marker comparisons can also provide new insights into snake reproductive biology, and its relationship to determining phylogeographic structure and population differentiation.

Along with the estimate of species trees from gene trees, snake phylogeographers should target the assessment of modes of speciation in snakes. Allopatric speciation appears to be a common mode, and it has been suggested by numerous phylogeographic studies that have demonstrated the separation of lineages at physically isolating barriers (Burbrink et al. 2000; Castoe et al. 2007b; Burbrink et al. 2008). Other types of speciation (parapatric, peripatric, or sympatric), however, have not yet been thoroughly examined with phylogeographic data. Such questions addressing speciation are becoming more common in phylogeographic studies of lizards (Morando et al. 2003; Sinclair et al. 2004; Sites and Marshall 2004; Pelligrino et al. 2005). Moreover, examining questions relevant to speciation and lineage formation can readily be aided by assessing differences in current and past niche space for these phylogeographic clades (Wiens and Graham 2005; Carstens and Knowles 2007).

Ultimately, conducting sound phylogeographic research requires a clear understanding of a diverse group of fields: geology, genetics, ecology, statistics, molecular biology, and, of course, herpetology. Snake phylogeographers must simultaneously maintain an awareness of advances in tree inference, population demographics, comparative phylogeography, gene discovery, divergence dating, and niche modeling. Technological advances in these fields occur rapidly and often provide new ways of elucidating the evolutionary history of snakes. Despite these demanding requirements, some of the most intriguing questions in biology may be best addressed by phylogeographic research. By considering the evolutionary and ecological processes that occur at both the microevolutionary and macroevolutionary scales, snake phylogeographic research may provide key insights into the role that physiographic, ecological, evolutionary and genetic processes play in the establishment of biodiversity.

Acknowledgments

Drafts of this chapter benefited from constructive comments from J. Castoe, Juan Daza, A. P. Jason de Koning, S. Mullin, J. Rodríguez-Robles, R. Seigel, and E. Smith. We also thank J. Castoe for copyediting the chapter and formatting the references. We extend our gratitude to A. Pyron, S. Ruane, and T. Guiher for examining drafts of the chapter and discussing topics relevant to snake phylogeography.

3

Population and Conservation Genetics

RICHARD B. KING

Population genetics addresses the effects that microevolutionary processes have on patterns of genetic variation within and among populations (Hedrick 2000; Hartl and Clark 2006). Key processes include natural selection, gene flow, genetic drift, mutation, mating system, and metapopulation dynamics. Historically, discrete traits with simple modes of inheritance, such as visible polymorphisms determined by single autosomal loci, were the focus of population genetic analysis. Over time, however, the field has become increasingly broad with development of molecular techniques. Initially used to assess protein variation (e.g., venom and allozymes), these techniques have subsequently provided direct measures of DNA-based variation. In addition, although traits exhibiting continuous distributions (morphology, behavior, and physiology) are more frequently the subject of quantitative genetic analysis (reviewed for snakes by Brodie and Garland 1993), the distinction between population and quantitative genetics is breaking down.

Conservation genetics is a related discipline that seeks to apply population and quantitative genetic principles to biodiversity management and protection (Frankham et al. 2002; Allendorf and Luikart 2007). Because threats to biodiversity often impact microevolutionary processes, such threats can affect patterns of genetic variation and put species at even greater risk. This connection is well encapsulated in the concepts of the extinction vortex (Gilpin and Soulé 1986) and mutational meltdown (Lynch et al. 1995), in which small population size promotes the loss of genetic diversity through stochastic processes (random genetic drift), resulting in increased homozygosity, expression of deleterious recessive alleles, and inbreeding depression (Crnokrak and Roff 1999; Keller and Waller 2002). The resulting reduction in mean fitness leads to further decreases in population size, promoting an even more rapid loss of genetic diversity and population decline. Habitat fragmentation compounds the problem by reducing effective population size and slowing the rate of gene flow, thus contributing to the extinction vortex. Small population size also reduces the ability of populations to adapt to local environmental conditions because stronger selection is required to overcome the effects of genetic drift. This can be especially problematic when changing environmental conditions (e.g., invasive species or global climate change) impose new selective regimes. Genetic concerns also arise in the design of captive breeding programs and the use of headstarting, reintroduction, repatriation, and translocation as management tools (Kingsbury and Attum, Chapter 7).

The goals of this chapter are to review the empirical knowledge-base of snake population genetics, focusing on both molecular genetic variation and variation in ecologically significant traits; to apply population genetic principles to the problem of snake conservation; and to identify future directions in snake population and conservation genetics.

Molecular Genetic Variation

Much of modern population genetics focuses on patterns of variation in molecular markers. A variety of such markers exist, for example, allozymes; randomly amplified polymorphic DNA (RAPDs); variable number tandem repeats (VNTRs), including minisatellite and microsatellite DNA (also called simple sequence repeats, SSRs); amplified fragment length polymorphisms (AFLPs); inter-simple-sequence-repeats (ISSRs); restriction fragment length polymorphisms (RFLPs); single-nucleotide polymorphisms (SNPs); and DNA sequences (Avise 2004; Lowe et al. 2004). Of these, allozyme and microsatellite DNA loci have been used most extensively in snake population genetic analyses. Laboratory techniques for allozyme analysis are well established (Murphy et al. 1996) and typically require little modification for use in scoring dozens of loci in a wide range of taxa. Other techniques usually involve the extraction and amplification of template DNA using the polymerase chain reaction (PCR) and often require the development of species-specific primer sequences and PCR conditions. Of these, microsatellite DNA loci have been used most widely. At present, published primer sets are available for 188 loci developed in 19 snake species, including members of the Colubrinae, Natricinae, Hydrophiinae, Crotalinae, Viperinae, and Boidae (Table 3.1, part A). Significantly, many of these primers sets have proven useful in amplifying microsatellite DNA loci in other, sometimes distantly related, species (Table 3.1, part B).

Different types of molecular markers provide different insights into population genetic processes (Avise 2004; Lowe et al. 2004). For example,

TABLE 3.1
Microsatellite DNA loci for which primer sequences have been developed in snakes and for which cross-amplification in other snake species has proven successful

Species	Locus Names	Reference
A. Snake Microsatellite Loci		
Colubridae, Colubrinae		
Coronella austriaca	*Ca16, Ca19, Ca20, Ca26, Ca27, Ca30, Ca40, Ca43, Ca45, Ca47, Ca61, Ca62, Ca63, Ca66, Ca78, Ca79*	Bond et al. 2005
Pantherophis obsoletus[a]	*Eobμ1, Eobμ2, Eobμ3, Eobμ4, Eobμ10, Eobμ13, Eobμ16, Eobμ34, Eobμ358, Eobμ366, Eobμ373*	Blouin-Demers and Gibbs 2003
Colubridae, Natricinae		
Natrix tessellata	*μNt1, μNt2, μNt3, μNt5, μNt6, μNt7, μNt8, μNt10*	Gautschi et al. 2000
Nerodia fasciata	*M17, M19*	Jansen 2001
N. sipedon	*Nsμ2, Nsμ3, Nsμ4, Nsμ6, Nsμ7, Nsμ8, Nsμ9 (Nsμ9b), Nsμ10,*	Prosser et al. 1999
Thamnophis elegans	*TelCa2, TelCa3, TelCa18, TelCa29, TelCa50, TE051B*	Garner et al. 2004; Manier and Arnold 2005
T. sirtalis	*2Ts, 3Ts, Ts1, Ts2, Ts3, Ts4, Ts1Ca4, TS010, TS042*	Garner 1998; McCracken et al. 1999; Garner et al. 2002, 2004; Manier and Arnold 2005
Elapidae, Hydrophiinae		
Aipysurus laevis	*AL983, AL28_e1, AL28_f6, AL28_h4, AL29_f6, AL102_c4, AL104_f6,AL105_c4, AL106_d11, AL106_g10, AL107_c2*	Lukoschek et al. 2005
Hoplocephalus bungaroides	*Hb2, Hb30, Hb48, Hb65, Hb70*	Burns and Houlden 1999
Notechis scutatus	*Ns03, Ns05, Ns14, Ns32, Ns40, Ns43, Ns67*	Scott et al. 2001
Rhinocephalus nigrescens	*Rn75, Rn81, Rn94, Rn114, Rn128, Rn32, Rn33, Rn50, Rn78, Rn84, Rn93, Rn111, Rn117, Rn126, Rn54*	Stapley et al. 2005
Viperidae, Crotalinae		
Crotalus horridus	*Ch3-155, Ch5-183, Ch7-144, Ch7-150, Ch7-87, Ch5A*	Villarreal et al. 1996
C. tigris	*Crti05, Crti06, Crti08, Crti09, Crti10, Crti12*	Goldberg et al. 2003
C. viridis	*CvMFRD5, CvMFR12, Cv9, Cv23, Cv15*	Oyler-McCance et al. 2005
C. willardi	*CwA14, CwA29, CwB6, CwB23, CwC24, CwD15*	Holycross et al. 2002
Sistrurus catenatus	*Scμ01, Scμ05, Scμ07, Scμ11, Scμ16, Scμ26*	Gibbs et al. 1998
Viperidae, Viperinae		
Vipera berus	*Vb3, Vb11, Vb21, Vb37, Vb64, Vb71, Vb-A8, Vb-A11, Vb-B1, Vb-B'2, Vb-B'9, Vb-B10, Vb-B'10, Vb-B18, Vb-D6, Vb-D'10, Vb-D12, Vb-D'13, Vb-D17*	Carlsson et al. 2003; Ursenbacher et al. 2008

TABLE 3.1—continued

Species	Locus Names	Reference
Boidae		
Epicrates subflavus	*Esµsat1, Esµsat3, Esµsat10, Esµsat11, Esµsat13, Esµsat16, Esµsat241, Esµsat30, Esµsat36,*	Tzika et al. 2008a
Morelia spilota	*MS1, Ms2, Ms3, Ms4, Ms5 Ms6, Ms7 Ms8, Ms9, Ms10, MS11, Ms12, Ms13, Ms14, Ms15 Ms16, Ms17 Ms18, Ms19, Ms20,MS21, Ms22, Ms23, Ms24, Ms25 Ms26, MS27*	Jordan et al. 2002

B. Cross-species Amplification

Colubridae, Colubrinae

Species	Locus Names	Reference
Coluber constrictor	*Nsµ9, Scµ07, Scµ11, Scµ16, Scµ26*	Gibbs et al. 1998; Prosser et al. 1999
Coronella austriaca	*Nsµ3, Nsµ6, Nsµ9b, Ts1, Ts3, Ts4, Hb2, Hb30, Hb65, Ch5-183*	Hille et al. 2002
Elaphe longissima	*Nsµ2, Nsµ3, Nsµ6, Nsµ7, Nsµ9b, Ts1, Ts3, Ts4, Hb2, Hb30, Hb48, Hb65, Ch5A, Ch7-150, Ch5-183*	Hille et al. 2002
Hemorrhois nummifer	*Nsµ2, Nsµ6, Nsµ7, Nsµ9b, Ts3, Hb30, Hb48, Hb65, Ch5-183*	Hille et al. 2002
Pantherophis gloydi	*Eobµ1, Eobµ2, Eobµ4, Eobµ10, Eobµ13, Eobµ16, Eobµ34, Eobµ358, Eobµ366, Eobµ373, Scµ11, Scµ16, Scµ26*	Gibbs et al. 1998; Blouin-Demers and Gibbs 2003
P. obsoletus[a]	*Scµ11, Scµ16, Scµ26*	Gibbs et al. 1998

Colubridae, Natricinae

Species	Locus Names	Reference
Natrix maura	*Nsµ2, Nsµ3, Ts1, Ts2, Ts3, Ts4*	Hille et al. 2002
N. natrix	*Ca26, Ca78, Nsµ2, Nsµ3, Nsµ6, Ts1, Ts2, Ts3, Ts4, Hb30*	Hille et al. 2002; Bond et al. 2005
N. tessellata	*Nsµ2, Nsµ3, Nsµ6, Ts1, Ts2, Ts3, Ts4, Hb30*	Hille et al. 2002
Nerodia erythrogaster	*Nsµ2, Nsµ3, Nsµ6, Nsµ7, Nsµ9b, Nsµ10,Ts1Ca4*	J. Marshall, pers. comm.
N. fasciata	*Nsµ2, Nsµ3, Nsµ6, Nsµ7, Nsµ8, Ts1, Ts2, Ts3, Ts4*	Jansen 2001; Hille et al. 2002
N. rhombifer	*Nsµ3*	T. L. Wusterbarth, pers. comm.
N. sipedon	*Eobµ1, Eobµ2, Eobµ3, Eobµ10, Eobµ13, Scµ16, Scµ26*	Gibbs et al. 1998; Blouin-Demers and Gibbs 2003
Regina septemvittata	*2Ts, 3Ts*	T. L. Wusterbarth, pers. comm.
Storeria dekayi	*Nsµ2, Nsµ3, Nsµ7, Nsµ8*	Prosser et al. 1999; T. L. Wusterbarth, pers. comm.
S. occipitomaculata	*Nsµ2, 3Ts*	Prosser et al. 1999; T. L. Wusterbarth, pers. comm.
Thamnophis elegans	*Nsµ2, Nsµ3, Nsµ7, Nsµ8, Nsµ10, TS010, TS042, Ts2, Ts3, Ts1Ca4, 2Ts*	Garner et al. 2004; Manier and Arnold 2005
T. gigas	*Nsµ3*	Paquin et al. 2006
T. melanogaster	*3Ts, Nsµ10*	T. L. Wusterbarth, pers. comm.

TABLE 3.1—continued

Species	Locus Names	Reference
T. ordinoides	*Te1Ca2, Te1Ca3, Te1Ca18, Te1Ca29, Te1Ca50, Ts1Ca4, 2Ts*	Garner et al. 2004
T. radix	*Nsμ2, Nsμ3A, Nsμ8, Nsμ9, Ts2, Ts3, Ts4, 3Ts*	G. M. Burghardt, pers. comm.; T. L. Wusterbarth pers. comm.
T. sauritus	*Nsμ2, Nsμ3, Nsμ4, Nsμ7, Nsμ8, Nsμ 9, Nsμ9b*	Prosser et al. 1999; T. L. Wusterbarth, pers. comm.
T. sirtalis	*Nsμ2, Nsμ3, Nsμ4, Nsμ6, Nsμ7, Nsμ8, Nsμ9, Nsμ9b, Nsμ10, TE051B, TelCa2, TelCa3, TelCa18, TelCa29, TelCa50*	Garner 1998; Prosser et al. 1999; Bittner 2000; King et al. 2001; Hille et al. 2002; Garner et al. 2004; Manier and Arnold 2005

Elapidae, Hydrophiinae

Hoplocephalus bungaroides	*Ns03, Ns05, Ns14, Ns32, Ns40, Ns43, Ns67*	Scott et al. 2001
Suta dwyeri	*Rn78, Rn81, Rn138*	Stapley et al. 2005
S. flagellum	*Rn78, Rn81, Rn138*	Stapley et al. 2005
S. gouldii	*Rn78, Rn81, Rn138*	Stapley et al. 2005
S. monachus	*Rn78, Rn81, Rn138*	Stapley et al. 2005
S. nigricepts	*Rn78, Rn138*	Stapley et al. 2005
S. punctata	*Rn78, Rn81*	Stapley et al. 2005
S. spectabalis	*Rn78, Rn81, Rn138*	Stapley et al. 2005
S. suta	*Rn78, Rn81, Rn138*	Stapley et al. 2005

Viperidae, Crotalinae

Agkistrodon contortrix	*Ch7-144, Ch5A, Ch7-87*	Bushar et al. 2001
Bothrops atrox	*Vb11, Vb37*	Carlsson et al. 2003
Crotalus adamanteus	*Ch7-144, Ch5A,Ch 7-87*	Bushar et al. 2001
C. atrox	*Ch7-144, Ch5A, Ch7-87, Ch3-155, CwA14, CwA29, CwB6, CwB23, CwD15*	Bushar et al. 2001; Holycross et al. 2002
C. cerastes	*Ch7-144, Ch5A, Ch7-87*	Bushar et al. 2001
C. durissus	*Ch7-144, Ch5A, Ch7-87, Ch3-155*	Bushar et al. 2001
C. enyo	*Ch7-144, Ch5A, Ch7-87*	Bushar et al. 2001
C. horridus	*CwA29f, CwB6, CwC24, CwD15, Scμ05, Scμ07, Scμ11, Scμ16, Scμ25*	Anderson 2006; Clark et al. 2007
C. molossus	*Ch7-144, Ch5A, Ch7-87, Ch3-155*	Bushar et al. 2001
C. scutulatus	*CwA14, CwA29, CwB6, CwB23, CwC24, CwD15*	Holycross et al. 2002
C. tigris	*CwA14, CwA29, CwB6, CwB23, CwC24, CwD15*	Holycross et al. 2002
C. unicolor	*Ch7-144, Ch5A, Ch7-87, Ch3-155*	Bushar et al. 2001
C. viridis	*Ch7-144, Ch5A, Ch7-87, Ch3-155, CwA14, CwA29, CwB23, CwD15, Nsμ2, Nsμ3, Scμ01, Scμ07, Scμ11, Scμ16, Scμ26, Vb11, Vb37*	Gibbs et al. 1998; Prosser et al. 1999; Bushar et al. 2001; Holycross et al. 2002; Carlsson et al. 2003
C. willardi	*Ch7-144, Ch5A, Ch7-87, Scμ01, Scμ07, Scμ11*	Bushar et al. 2001; Holycross 2002
Sistrurus catenatus	*Ch7-144, Ch5A, Ch7-87*	Bushar et al. 2001
S. miliarus	*Ch7-144, Ch5A, Ch7-87*	Bushar et al. 2001

TABLE 3.1—continued

Species	Locus Names	Reference
Viperidae, Viperinae		
Atheris ceratophora	*Vb11, Vb71*	Carlsson et al. 2003
Echis carinatus	*Vb11, Vb71*	Carlsson et al. 2003
Vipera ammodytes	*Vb3, Vb21, Vb37, Vb71*	Carlsson et al. 2003
V. berus	*Nsμ3, Nsμ6, Nsμ9b, Ts1, Ts3, Ts4, Hb30, Hb65, Ch5-183*	Hille et al. 2002
V. dinniki	*Vb3, Vb11, Vb21, Vb37, Vb64, Vb71*	Carlsson et al. 2003
V. kaznakovi	*Vb3, Vb11, Vb21, Vb37, Vb64, Vb71*	Carlsson et al. 2003
V. ursinii	*Vb37*	Carlsson et al. 2003
Boidae		
Antaresia childreni	*MS1-MS27* (20)	Jordan et al. 2002
A. stimsoni	*MS1-MS27* (19)	Jordan et al. 2002
Apodora papuana	*MS1-MS27* (19)	Jordan et al. 2002
Aspidites melanocephalus	*MS1-MS27* (17)	Jordan et al. 2002
A. ramsayi	*MS1-MS27* (17)	Jordan et al. 2002
Boa constrictor	*Nsμ2, Nsμ3, Nsμ6, Nsμ9b, Ts1, Ts3, Ts4, Hb2, Hb30, Hb48, Hb65, Ch5-183*	Hille et al. 2002
Bothrochilus boa	*MS1-MS27* (9)	Jordan et al. 2002
Eunectes murinus	*Nsμ2, Nsμ3, Nsμ6, Nsμ9b, Ts1, Ts3, Ts4, Hb2, Hb30, Hb65, Ch5A, Ch7-150, Ch5-183*	Hille et al. 2002
E. notaeus	*Nsμ2, Nsμ3, Nsμ6, Nsμ9b, Ts1, Ts3, Ts4, Hb2, Hb30, Hb48, Hb65, Ch5-183*	Hille et al. 2002
Leiopython albertisii	*MS1-MS27* (19)	Jordan et al. 2002
Liasis fuscus	*MS1-MS27* (20)	Jordan et al. 2002
L. olivaceus	*MS1-MS27* (20)	Jordan et al. 2002
Morelia viridis	*MS1-MS27* (20)	Jordan et al. 2002
Python reticulatus	*MS1-MS27* (20)	Jordan et al. 2002
P. timoriensis	*MS1-MS27* (17)	Jordan et al. 2002
Typhlopidae		
Typhlops vermicularis	*Nsμ2, Nsμ3, Nsμ6, Nsμ9b, Ts1, Ts3, Ts4, Hb2, Hb30, Hb48, Hb65, Ch5-183*	Hille et al. 2002

Notes: Locus names come from original references except that the first letter of the genus and species have been added to locus names for *Crotalus horridus, Crotalus viridis, Epicrates subflavus, Hoplocephalus bungaroides,* and *Vipera berus.* Primer sequences and polymerase chain reaction (PCR) conditions can be found in references listed. Species names follow the Integrated Taxonomy Information System Catalogue of Life 2006 Annual Checklist (ITIS 2006). Family- and subfamily-level taxonomy follows Lawson et al. (2005). In part B, the number of loci successfully amplified appears in parentheses.

[a] Includes *Pantherophis obsoleta, P. alleghaniensis,* and *P. spiloides* of Burbrink et al. 2000 (but see Gibbs et al. 2006)

because mitochondrial DNA (mtDNA) is maternally inherited, a comparison of nuclear (e.g., allozymes and microsatellite DNA loci) versus mtDNA-based markers can provide information on sex-biased dispersal patterns. In mammals and other groups with XY sex determination, such comparisons are strengthened by the use of paternally inherited Y-chromosome-linked markers. The ZW sex inheritance of many snakes (Olmo 1986; Beçek et al. 1990), however, means that the development of paternally inherited markers is unlikely. Interestingly, snake W chromosomes are reported to harbor large amounts of repetitive DNA (Jones and Singh 1985), which may prove useful in future population genetic analyses. Rapidly evolving microsatellite DNA loci exhibit high levels of selectively neutral variation and are especially useful in providing detailed information on patterns of variation within populations and on fine geographic scales. Homoplasy (convergence at the molecular level because alleles are identical in state but not identical by descent) may limit the utility of microsatellite DNA loci at large geographic scales, however, in which case allozymes or sequence-based markers may be more useful. Because they allow unambiguous scoring of genotypes at many loci, microsatellite DNA and allozymes markers are especially useful for analytical techniques that require multilocus genotypes (e.g., estimation of relatedness, assignment tests, and estimation of effective population size), whereas the inability to distinguish heterozygotes from dominant homozygotes for RAPDs, AFLPs, and ISSRs limits their utility in this context.

In the sections that follow, emphasis is placed on the utility of allozymes, microsatellite DNA, and mtDNA sequences in assessing patterns of genetic variation within and among populations. Examples involving RAPDs, RFLPs, and ISSRs are also included. In snakes, the use of AFLPs has been limited (Giannasi et al. 2001; Groot et al. 2003), but such markers may prove useful in the future. To date, SNPs and related markers (SNPSTRs and hapSTRs) have not been used to address snake population and evolutionary genetic questions, but promising results have been obtained for other taxa (Mountain et al. 2002; Brumfield et al. 2003; Hey et al. 2004; Morin et al. 2004).

Genetic Variation within Populations

Measures of Genetic Variation within Populations

Biparentally inherited codominant markers such as allozymes and microsatellite DNA can be used to generate multilocus genotype data for large numbers of individuals, providing accurate measures of genetic variation within and among populations. Simple measures of variation within populations include proportion of polymorphic loci (p = the number of polymorphic loci/number of loci assayed) and heterozygosity (either observed frequency of heterozygotes, H_o, or expected frequency under Hardy-Weinberg conditions, H_e). Such markers can also be used to assess mating patterns within populations through the inbreeding coefficient, F, which compares observed and expected heterozygosity as $F = (H_e - H_o)/H_e$ (Hedrick 2000; Hartl and Clark 2006). Inbreeding (the occurrence of matings

among relatives more frequently than expected by chance) results in reduced heterozygote frequencies and F values greater than zero. In contrast,
outbreeding (the occurrence of matings among relatives less frequently than
expected by chance) results in increased heterozygote frequencies and F values less than zero.

Often, F is estimated from a hierarchical sampling design in which individuals (I) are nested within subpopulations (S) nested within the total
population (T). In such cases, deviations from random mating can be described at any level using Wright's F statistics (Wright 1931; Hedrick 2000;
Hartl and Clark 2006). Inbreeding or outbreeding within subpopulations
is characterized as previously described, but is now designated F_{IS}. In addition, the separation of the total population into subpopulations results in a
reduction in expected heterozygosity within subpopulations (H_S) compared
to that in the total population (H_T) and is characterized by $F_{ST} = (H_T - H_S)/$
H_T. Consequently, one caveat to the interpretation of strongly positive values of F (= F_{IS}) is that they may reflect either inbreeding or unrecognized
population subdivision.

Measures of variation in DNA sequences, analogous to those used for allozymes and microsatellite DNA, are also available. Within-population measures include haplotype diversity (H = the probability that any two randomly
chosen individuals differ in haplotype) and nucleotide diversity (π = the
average number of nucleotide differences per site between any two randomly
chosen individuals; Hedrick 2000; Hartl and Clark 2006). At present, most
snake DNA sequence data come from phylogenetic and phylogeographic
analyses (Burbrink and Castoe, Chapter 2), and so within-population measures of variation typically involve small numbers of individuals. This will
probably change as sequencing becomes easier and less expensive. Because
levels of variation can vary dramatically between genes, careful marker selection for population-level analysis is important (Avise 2004; Lowe et al.
2004).

Observed Patterns of Genetic Variation within Populations

Information on genetic variation within snake populations is of interest
for several reasons. (1) Snakes vary dramatically in population ecology
(Parker and Plummer 1987), reproductive ecology (Seigel and Ford 1987;
Duvall et al. 1993), and movement patterns (Gregory et al. 1987), and this
variation can have important genetic consequences. (2) Small population
size and habitat fragmentation result in losses of genetic variation. Hence,
knowledge of typical patterns can provide a baseline for comparison that
may highlight species and populations suffering such losses. (3) Levels of
variation in neutral genetic markers may correlate with that of functional
genes (Merilä and Crnokrak 2001; Leinonen et al. 2008; but see Reed and
Frankham 2001). Consequently, reduced variation at marker loci may

reflect a reduced capacity for adaptive responses to changing environmental conditions.

Patterns of within-population genetic variation have been characterized for a wide variety of snake taxa (Table 3.2). In compiling examples for inclusion here, the following criteria were used:

1. Only studies based on 10 or more individuals per population were included (three or more for studies based on sequence data).
2. Estimates of P were restricted to allozyme-based studies because microsatellite DNA loci are rarely monomorphic.
3. Studies that included only allozyme loci known or expected to be polymorphic (i.e., King and Lawson 1995, 2001; Lawson and King 1996; Rye 2000) were not included among estimates of P.
4. Studies reporting genetic variation of composite samples (consisting of individuals from multiple locations) were included among estimates of P and H_o, but not among estimates of H_e or F_{IS}, because differences in allele frequency among locations bias such estimates.
5. Allozyme-based estimates F_{IS} were rare and so only microsatellite DNA- and RAPD-based estimates were included.
6. Only polymorphic loci were included in generating estimates of H_e and H_o so that results are comparable across marker types and studies.

Given these criteria, estimates of heterozygosity (H_o, H_e, or both) were available from 34 studies (1–13 populations each) using allozymes and from 41 studies (1–20 populations each) using microsatellite DNA. For allozymes, heterozygosity ranges from 0.05 to 0.55 with a median of 0.27; for microsatellite DNA, heterozygosity ranges from 0.35 to 0.87 with a median of 0.60 (Table 3.2 and Fig. 3.1; summary statistics based on H_e unless only H_o was available). The greater heterozygosity seen for microsatellite DNA is expected, reflecting higher mutation rates at these loci (typically on the order of 10^{-3}–10^{-4}) compared to allozyme loci (typically on the order of 10^{-5}–10^{-6}). Unusually low levels of allozyme heterozygosity were observed in *Nerodia fasciata* (H_o = 0.13, based on data in Lawson et al. 1991) and *Ovophis tokarensis* (H_e = 0.05, based on data in Toda et al. 1999; see Table 3.2, part A). Allozyme heterozygosities in another study of *N. fasciata,* however, and of *Protobothrops* (a sister taxon to *Ovophis*) fell within the range typical of other taxa (see Table 3.2, part A). Unusually low levels of microsatellite DNA heterozygosity were observed in two of three studies of *Natrix tessellata* (Gautschi et al. 2002; Guicking et al. 2004) and in *Aipysurus laevis* (Lukoschek et al. 2005) (see Table 3.2, part B). In the case of *N. tessellata,* low heterogeneity may relate to small population size. These populations are considered critically endangered or were (re)introduced using small numbers of founders (Gautschi et al. 2002; Guicking et al. 2004). In *A. laevis,* heterozygosity was especially low (H_o < 0.10)

TABLE 3.2.
Levels of genetic variation within snake populations as measured using allozymes, microsatellite DNA loci, RAPDs, and ISSRs

Species	Loci	N	H_o	$\bar{H}_e$	P	Reference
A. Allozymes						
Colubridae, Colubrinae						
Elaphe dione (12)	17	18–20	0.14 (0.10–0.20)	0.15 (0.11–0.21)	0.32 (0.24–0.41)	Paik and Yang 1987
Pantherophis bairdi (1)	21	15	0.18 (0.07–0.29)	0.29 (0.24–0.34)	0.10	Lawson and Lieb 1990
P. obsoletus (1)	21	14	0.29	0.25	0.05	Lawson and Lieb 1990
P. alleghaniensis (1)	21	13	0.39	0.45	0.05	Lawson and Lieb 1990
Colubridae, Natricinae						
Natrix maura (1)	36	12	0.40		0.28	Busack 1986
Nerodia clarkii (composite)	35	35–69			0.20	Lawson 1987
N. clarkii (4)	33	18–35	0.25 (0.09–0.56)	0.30 (0.14–0.57)	0.14 (0.09–0.27)	Lawson et al. 1991
N. cyclopion (1)	37	10			0.00	Thompson and Crother 1998
N. cyclopion (composite)	35	8–22			0.03	Lawson 1987
N. erythrogaster (1)	12	11			0.00	Rose and Selcer 1989
N. erythrogaster (1)	37	10		0.27 (0.18–0.32)	0.11	Thompson and Crother 1998
N. erythrogaster (composite)	35	23–36			0.11	Lawson 1987
N. fasciata (1)	37	10		0.44 (0.26–0.76)	0.24	Thompson and Crother 1998
N. fasciata (composite)	35	77–181			0.49	Lawson 1987
N. fasciata (2)	33	20–23	0.13 (0.07–0.18)	0.26 (0.16–0.35)	0.19 (0.17–0.20)	Lawson et al. 1991
N. floridana (composite)	37	9			0.00	Thompson and Crother 1998
N. floridana (composite)	35	17–19			0.00	Lawson 1987
N. harteri (1)	12	10			0.00	Rose and Selcer 1989
N. harteri (1)	35	4–13			0.00	Lawson 1987
N. paucimaculata (1)	12	13			0.00	Rose and Selcer 1989

TABLE 3.2—continued

Species	Loci	N	H_o	H_e	P	Reference
N. rhombifer (composite)	35	10–34			0.11	Lawson 1987
N. sipedon (composite)	35	28–83			0.26	Lawson 1987
N. sipedon (7)	7	25–110	0.26 (0.21–0.34)			King and Lawson 1995
N. taxispilota (composite)	35	21–28			0.09	Lawson 1987
Regina alleni (composite)	35	3–5			0.03	Lawson 1987
R. grahamii (composite)	35	4–12			0.03	Lawson 1987
R. rigida (composite)	35	4–15			0.06	Lawson 1987
R. septemvittata (composite)	35	7–16			0.14	Lawson 1987
Storeria dekayi (7)	7	19–70	0.23 (0.19–0.25)			King and Lawson 2001
S. occipitomaculata (1)	12	35		0.18 (0.03–0.49)	0.32	Grudzien and Owens 1991
Thamnophis atratus (2)	31	14–38	0.43 (0.35–0.51)		0.19 (0.15–0.23)	Lawson and Dessauer 1979
T. couchii (composite)	31	13	0.26		0.13	Lawson and Dessauer 1979
T. elegans (2)	31	9–10	0.23 (0.22–0.23)		0.13 (0.13–0.13)	Lawson and Dessauer 1979
T. hammondii (composite)	31	8	0.23		0.06	Lawson and Dessauer 1979
T. proximus (4)	26	12–42	0.32 (0.23–0.36)		0.21 (0.19–0.27)	Gartside et al. 1977
T. sauritus (1)	26	10	0.27		0.27	Gartside et al. 1977
T. sirtalis (4)	14	9–52	0.14 (0.10–0.19)		0.18 (0.07–0.29)	Sattler and Guttman 1976
T. sirtalis (13)	9–12	24–130	0.25 (0.13–0.34)			Lawson and King 1996
T. sirtalis (12)	5	11–55	0.27 (0.08–0.48)			Rye 2000
T. sirtalis (4)	15	13–15		0.55 (0.46–0.64)	0.07	Bellemin et al. 1978
T. validus (1)	35	4–12			0.00	Lawson 1987

Viperidae, Crotalinae						
Agkistrodon piscivorus (5)	29	7–11	0.22 (0.16–0.42)		0.09 (0.07–0.10)	Merkle 1985
Bothriechis lateralis (1)	26	12		0.25 (0.15–0.44)	0.12	Werman 1992
Bothrops asper (6)	16	10–23	0.30 (0.15–0.77)		0.17 (0.06–0.25)	Sasa and Barrantes 1998
B. asper (1)	26	15		0.27 (0.11–0.46)	0.19	Werman 1992
Cerrophidion godmani (1)	26	14			0.00	Werman 1992
C. godmani (8)	15	13–22			0.00	Sasa 1997
Crotalus ruber (1)	29	15	0.16 (0.07–0.23)	0.16 (0.06–0.31)	0.17	Murphy et al. 1995
C. scutulatus (composite)	16	58		0.40 (0.28–0.49)	0.29	Wilkinson et al. 1991
C. willardi (1)	22	10		0.18 (0.16–0.20)	0.14	Barker 1992
Ovophis okinavensis (2)	26	12–13			0.15	Toda et al. 1999
O. tokarensis (1)	26	19		0.05	0.08	Toda et al. 1999
Protobothrops elegans (1)	26	12		0.25 (0.08–0.89)	0.38	Toda et al. 1999
P. flavoviridis (4)	26	11–21		0.30 (0.24–0.36)	0.20 (0.15–0.23)	Toda et al. 1999
Viperidae, Viperinae						
Vipera berus (4)	2	18–76	0.30 (0.09–0.41)		0.25 (0.07–0.33)	Madsen et al. 1995
Boidae						
Boa constrictor (2)	25	45–48	0.39 (0.36–0.42)	0.38 (0.35–0.41)	0.16	Rivera et al. 2005; Cardozo et al. 2007
B. Microsatellite DNA Loci						
Colubridae, Colubrinae						
Coronella austriaca (1)	16	16–23	0.51 (0.09–0.91)	0.55 (0.12–0.86)	0.07	Bond et al. 2005
Pantherophis obsoletus[a] (15)	6	6–43	0.56–0.75			Lougheed et al. 1999

TABLE 3.2—continued

Species	Loci	N	H_o	H_e	P	Reference
P. alleghaniensis x spiloides (hybrid zone)[a] (9)	6	6–43			0.00–0.13	Gibbs and Weatherhead 2001
P. obsoletus[a] (composite)	11	392–1227	0.65 (0.07–0.87)			Blouin-Demers and Gibbs 2003
Colubridae, Natricinae						
Natrix natrix (composite)	5	28–60	0.80 (0.62–0.96)			Hille et al. 2002
N. tessellata (2)	8	10, 19	0.45 (0.41–0.49)	0.60 (0.58–0.62)	0.25 (0.21–0.29)	Gautschi et al. 2000
N. tessellata (4)	8	10–22	0.37 (0.26–0.50)	0.51 (0.37–0.65)	0.27 (0.23–0.33)	Gautschi et al. 2002
N. tessellata (4)	6	11–28	0.28 (0.11–0.49)	0.39 (0.23–0.47)	0.30 (−0.02–0.53)	Guicking et al. 2004
Nerodia erythrogaster (7)	7	11–64	0.63 (0.53–0.70)	0.68 (0.57–0.76)	0.08 (−0.01–0.16)	J. Marshall, pers. comm.
N. fasciata (7)	4	15–20	0.58 (0.53–0.64)	0.54 (0.38–0.64)	−0.11 (−0.42–0.07)	Jansen 2001
N. sipedon (3)	8	50	0.71 (0.68–0.73)	0.72 (0.71–0.72)	0.02 (−0.01–0.05)	Prosser et al. 1999
N. sipedon (6)	6	9–35	0.68 (0.63–0.75)	0.79 (0.76–0.82)	0.13 (0.02–0.20)	J. Marshall, pers. comm.
Thamnophis elegans (20)	10	16–140	0.54 (0.47–0.61)	0.59 (0.52–0.64)	0.09 (−0.04–0.22)	Manier and Arnold 2005
T. gigas (9)	1	14–22		0.68 (0.58–0.76)		Paquin et al. 2006
T. radix (4)	4	9–12	0.57 (0.44–0.64)	0.60 (0.50–0.65)	0.03 (−0.28–0.31)	G. M. Burghardt, pers. comm.
T. radix (1)	6	72	0.75 (0.33–0.97)	0.70 (0.31–0.88)	−0.07	T. Wusterbarth, pers. comm.
T. sirtalis (1)	4		0.53 (0.27–0.82)			McCracken et al. 1999
T. sirtalis (10)	4	19–26	0.71 (0.61–0.83)	0.87 (0.84–0.91)	0.19 (0.08–0.29)	Bittner 2000

Species						Reference
T. sirtalis (13)	9	19–42	0.50 (0.45–0.55)	0.54 (0.50–0.58)	0.06 (−0.02–0.19)	Manier and Arnold 2005
T. sirtalis (5)	5	10–24	0.52 (0.44–0.56)	0.58 (0.49–0.62)	0.10 (0.07–0.15)	Garner 1998
T. sirtalis (1)	6	41	0.55 (0.49–0.61)	0.64 (0.47–0.78)	0.14	Garner et al. 2004
T. sirtalis (4)	7	12–41	0.62 (0.52–0.74)			T. W. J. Garner, pers. comm.
T. sirtalis (1)	6	15	0.67 (0.20–0.87)			T. W. J. Garner, pers. comm.
T. sirtalis (4)	5	21–24	0.47 (0.44–0.49)			T. W. J. Garner, wpers. comm.
Elapidae, Hydrophinae						
Aipysurus laevis (composite)	11	21–31	0.35 (0.04–0.86)			Lukoschek et al. 2005
Hoplocephalus bungaroides (composite)	7	10–14	0.39 (0.00–0.85)			Scott et al. 2001
H. bungaroides (composite)[b]	5	16	0.48 (0.00–0.75)			Burns and Houlden 1999
Notechis scutatus (composite)	7	62–70	0.53 (0.27–0.70)			Scott et al. 2001
Rhinocephalus nigrescens (1)	5	93	0.53 (0.24–0.82)	0.71 (0.46–0.91)	0.25	Stapley et al. 2005
Viperidae, Crotalinae						
Crotalus horridus (composite)	6	16–32	0.38 (0.10–0.69)			Villarreal et al. 1996
C. horridus (1)	6	69–85	0.53 (0.02–0.83)	0.47 (0.00–0.84)	0.08	Anderson 2006, pers. com.
C. horridus (composite)	6	54–82	0.41 (0.02–0.74)	0.35 (0.02–0.60)		Anderson 2006, pers. com.
C. horridus (14)	9	15–57	0.57 (0.49–0.63)	0.59 (0.55–0.66)		Clark et al. 2007
C. tigris (3)	6	14–62	0.61 (0.58–0.65)	0.69 (0.68–0.72)	0.12 (0.09–0.14)	Goldberg et al. 2003
C. viridis (composite)	5	182–212	0.44 (0.10–0.70)			Oyler-McCance et al. 2005
C. willardi (3)	9	18–54	0.68 (0.60–0.74)	0.69 (0.61–0.73)	0.01 (−0.01–0.03)	Holycross et al. 2002; Holycross and Douglas 2007

TABLE 3.2—continued

Species	Loci	N	H_o	H_e	P	Reference
Sistrurus catenatus (5)	6	20–81	0.51 (0.45–0.61)	0.63 (0.55–0.71)	0.19 (0.12–0.35)	Gibbs et al. 1997, 1998
S. catenatus (5)	3	10–32	0.55 (0.42–0.65)	0.69 (0.60–0.84)	0.21 (−0.01–0.35)	Andre 2003
S. catenatus (2)	6	21, 56	0.74 (0.70–0.76)	0.76 (0.74–0.80)	0.06 (0.05–0.06)	Holycross 2002
Viperidae, Viperinae						
Vipera berus (1)	6	53–62	0.63 (0.21–0.83)	0.67 (0.25–0.97)	0.06	Carlsson et al. 2003
V. berus (16)	7	5–63	0.50 (0.35–0.64)	0.52 (0.39–0.68)	0.04 (−0.04–0.26)	Ursenbacher et al. 2008
Boidae						
Epicrates subflavus (3)	9	17–36	0.48 (0.42–0.53)	0.60 (0.57–0.63)		Tzika et al. 2008a
C. RAPDs						
Colubridae, Colubrinae						
Pantherophis obsoletus[a] (10)	14	15–26		0.16 (0.09–0.24)		Prior et al. 1997
D. ISSRs						
Boidae						
Boa constrictor (2)	7	43, 72		0.05 (0.05–0.06)		Cardozo et al. 2007

Notes: Numbers in parentheses following species names refer to the number of populations surveyed. When data come from single populations, means (ranges) of H_o and H_e across loci are shown. When data come from a composite of several populations, only mean H_o (range across loci) is shown. When data come from more than one population, the mean of means (ranges of means) across populations are shown for H_o, H_e, and P or F_{IS}. Values calculated from information provided in the original references are shown in italics. So that results are comparable across marker types and studies, only polymorphic loci were included in generating estimates of H_e and H_o. Species names follow the Integrated Taxonomy Information System Catalogue of Life 2006 Annual Checklist (ITIS 2006). Family- and subfamily-level taxonomy follows Lawson et al. (2005).

F_{IS}, inbreeding coefficient (microsatellite DNA loci and RAPDs); H_e expected heterozygosity; H_o observed heterozygosity; ISSRs, inter-simple-sequence-repeats; Loci, the number of loci scored; N, number of individuals genotyped per population; P, proportion of loci that were polymorphic (allozymes); RAPDs, randomly amplified polymorphic DNA.

[a] Includes *Pantherophis obsoleta*, *P. alleghaniensis*, and *P. spiloides* of Burbrink et al. 2000 (but see Gibbs et al. 2006).

[b] Captive colonies at two zoos.

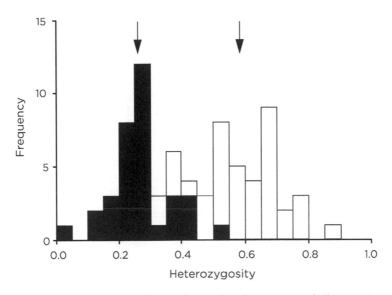

Fig. 3.1. Heterozygosity within snake populations based on surveys of allozymes (open histograms) and microsatellite DNA markers (filled histograms); arrows indicate median values (data in Table 3.2).

at three loci (Lukoschek et al. 2005). At eight other loci, H_o averaged 0.46, which is more in line with that observed in other studies.

Estimates of allozyme polymorphism were available from 50 studies and ranged from 0.00 to 0.49, with a median of 0.13 (Table 3.2, part A). Ten studies report estimates of $P = 0.00$, including *Nerodia cyclopion* (1 of 2 studies); *N. erythrogaster* (1 of 3 studies); *N. floridana* (two studies); *N. harteri* (2 studies); and *N. paucimaculata, Thamnophis validus,* and *Cerrophidion godmani* (2 studies; see Table 3.2 part A). Although the reasons for this lack of polymorphism are unclear, some of these species (e.g., *N. harteri, N. paucimaculata,* and *T. validus*) are characterized by relatively restricted geographic ranges. Unfortunately, information on other classes of molecular markers (e.g., microsatellite DNA) is unavailable to assess whether the low allozyme polymorphism reflects a general lack of genetic variation within these taxa.

Estimates of F_{IS} values were available from 25 microsatellite DNA-based studies and ranged from −0.11 to 0.30, with a median of 0.08 (Table 3.2, part B). All but one were greater than zero, indicating that outbreeding is not pervasive among snakes. Unusually high estimates of F_{IS} were observed in *Natrix tessellata* (Gautschi et al. 2000, 2002; Guicking et al. 2004), suggesting the occurrence of inbreeding. Again, this may reflect the history of (re)introduction of some of these populations.

Relatively large numbers of natricine and crotaline snakes are represented among taxa for which patterns of within-population genetic variation have been characterized (Table 3.2), allowing comparisons between these groups. An examination of the measures summarized in Table 3.2 suggest no clear differences; median allozyme heterozygosity equals 0.27 among Natricinae and 0.25 among Crotalinae, median microsatellite DNA heterozygosity equals 0.61 among Natricinae and 0.61 among Crotalinae, median proportion of polymorphic allozyme loci equals 0.11 among Natricinae and 0.15 among Crotalinae, and median F_{IS} equals 0.10 among Natricinae and 0.07 among Crotalinae. Remarkably, observed microsatellite DNA heterozygosity data are available for eight separate geographic regions or studies for the Common Gartersnake, *Thamnophis sirtalis*, and range from 0.47 among four Alberta, Canada, populations (T. W. J. Garner, pers. comm.) to 0.71 among 10 Lake Erie island and mainland populations (Bittner 2000).

Estimates of haplotype diversity and nucleotide diversity were available from 25 studies of mtDNA sequences and 3 studies of nuclear DNA sequences (Table 3.3 and Fig. 3.2). Haplotype diversity was highly variable within populations, ranging from 0.00 to 0.95 (median = 0.61). Nucleotide diversity was also highly variable, spanning three orders of magnitude (ranging from 0.0000 to 0.0300; median = 0.0023). Differences in haplotype and nucleotide diversity among genes (ND2, ND4, cytochrome *b*, ATPase 6 and 8, and D loop) or taxonomic groups (mostly Natricinae and Crotalinae) were not evident given the limited data available.

Haplotype and nucleotide diversity were equal to zero for cytochrome *b* and D-loop sequences in each of three severely reduced populations of *Vipera ursinii* (Újvári et al. 2005). This observation might suggest that low haplotype and nucleotide diversity can be used to identify populations in decline. Seven other studies, however, included one or more populations that had haplotype and nucleotide diversity equal to zero (Table 3.4) but that were not necessarily in decline. Small sample size may be one reason for such low values, and future studies based on larger samples may provide more objective criteria for identifying populations at risk. Haplotype and nucleotide diversity were unusually high in *Crotalus cerastes* and *C. mitchellii* (Douglas et al. 2006), perhaps partly due to the fact that sampling spanned large geographic regions (portions of several states in the western United States) and thus may overestimate within-population levels of variation. A subset of *Nerodia sipedon* and *Thamnophis sirtalis* populations for which ND2 sequences are available represent contact zones between genetic lineages that mostly fall east and west of Lake Michigan in the United States (Robinson 2005; Placyk et al. 2007). Nucleotide diversity is markedly higher in these contact-zone populations than in populations consisting entirely of just one lineage (for *N. sipedon*, π averages 0.0154 in 2 contact-zone populations compared to 0.0021 in 11 other populations; for *T. sirtalis*, π averages

TABLE 3.3
Patterns of haplotype and nucleotide diversity observed within snake populations based on mitochondrial DNA and nuclear sequence data

Species	N	Nh	h	π	Reference
A. Mitochondrial Gene Sequences					
ND2 (911–1101 bases)					
Nerodia sipedon (11)	3–7	1–4	0.73 (0.00–1.00)	0.0021 (0.0000–0.0044)	J. Robinson, pers. comm.
N. sipedon (2, contact zone[a])	3, 5	2, 3	0.70 (0.40, 1.00)	0.0154 (0.0110, 0.0198)	J. Robinson, pers. comm.
Storeria dekayi (2)	7, 8	3, 4	0.64 (0.52, 0.73)	0.0013 (0.0009, 0.0017)	J. Robinson, pers. comm.
Thamnophis radix (3)	4	1–3	0.28 (0.00–0.83)	0.0005 (0.0000–0.0015)	G. M. Burghardt, pers. comm.
T. sirtalis (2)	6, 8	2	0.38 (0.33, 0.43)	0.0004 (0.0003, 0.0004)	Janzen et al. 2002
T. sirtalis (12)	3–9	1–4	0.38 (0.00–0.67)	0.0006 (0.0000–0.0012)	J. Placyk, pers. comm.
T. sirtalis (5, contact zone[a])	9–12	3–5	0.67 (0.60–0.72)	0.0024 (0.0016–0.0034)	J. Placyk, pers. comm.
Crotalus viridis (1)	3	2	0.67	0.0083	Ashton and de Queiroz 2001
ND4 (486–876 bases)					
Thamnophis gigas (6)	20–51	1–4	0.41 (0.00–0.66)	0.0009 (0.0000–00026)	Paquin et al. 2006
T. sirtalis (2)	6, 8	3	0.61 (0.60, 0.61)	0.0015 (0.0012, 0.0017)	Janzen et al. 2002
Crotalus atrox (6 regional samples)	5–13	4–7		0.0026 (0.002–0.006)	Castoe et al. 2007b
Charina bottae (2)	3, 4	1, 4	0.50 (0.00, 1.00)	0.0026 (0.0000, 0.0051)	Rodríguez-Robles et al. 2001
Eunectes notaeus (4)	6–30	2–18	0.72 (0.20–0.89)	0.0089 (0.0008–0.0135)	Mendez et al. 2007
Cytochrome b (219–1239 bases)					
Thamnophis elegans (10)	3–4	1–2	0.18 (0.00–0.67)	0.0008 (0.0000–0.0044)	Bronikowski and Arnold 2001
T. sirtalis (2)	6, 8	2	0.48 (0.43, 0.53)	0.0012 (0.0009, 0.0015)	Janzen et al. 2002
Crotalus horridus (3 regional samples)	17–61	4–16	0.58 (0.33–0.84)	0.006 (0.002–0.009)	Clark et al. 2003
Vipera ursinii (3)	3–4	1	0.00	0.0000	Újvári et al. 2005
Epicrates subflavus (2)	41, 46	5, 7	0.61 (0.57, 0.64)	0.0024 (0.0014, 0.0034)	Tzika et al. 2008a
Eunectes notaeus (4)	3–42	3–15	0.85 (0.62–1.00)	0.0145 (0.0083–0.0207)	Mendez et al. 2007
ATPase 6 and 8 (676–865 bases)					
Crotalus cerastes (2 regional samples)	6, 13	5, 6	0.82 (0.77, 0.86)	0.019 (0.009, 0.029)	Douglas et al. 2006
C. mitchellii (3 regional samples)	2–84	2–27	0.95 (0.91–1.00)	0.030 (0.011–0.061)	Douglas et al. 2006
C. ruber (1 regional sample)	11	4	0.60	0.003	Douglas et al. 2006

TABLE 3.3—continued

Species	N	Nh	h	π	Reference
C. tigris (2 regional sample)	6, 28	1, 13	0.36 (0.00, 0.71)	0.002 (0.000, 0.003)	Douglas et al. 2006
C. willardi (4)	11–21	3–6	0.64 (0.44–0.78)	0.003 (0.0021–0.0035)	Holycross and Douglas 2007
D loop/control region (1239–1305 bases)					
Crotalus viridis (1)	3	2	0.67	0.0021	Ashton and de Queiroz 2001
Vipera ursinii (3)	3–4	1	0.00	0.0000	Újvári et al. 2005
mtDNA RFLPs					
Vipera berus (3)	19–47	2–8	0.25 (0.11–0.44)	0.0002 (0.0001–0.0003)	Carlsson and Tegelström 2002
B. Nuclear Gene Sequences					
Mc1r (945 bases)					
Thamnophis sirtalis (2)	2	2		0.0043 (0.0041, 0.0044)	Rosenblum et al. 2004
RAPDs					
Pantherophis spiloides (x alleghaniensis; hybrid zone)[b] (2)	10, 33			0.0028 (0.0033, 0.0022)	Gibbs et al. 1994
Sistrurus catenatus (2)	9, 9			0.0024 (0.0021, 0.0027)	Gibbs et al. 1994

Notes: Numbers in parentheses following species names refer to the number of populations sampled.

Values calculated from sequences obtained from GenBank based on information in the references cited are shown in italics. Only studies in which sequences were available for three or more individuals are included, except for the nuclear gene *Mc1r*, for which sample size = 2. π, nucleotide diversity (the average number of nucleotide differences per site between two sequences chosen at random); ATPase, adenosine triphosphotase; *h*, haplotype diversity (the probability that two haplotypes chosen at random within a population differ); mtDNA, mitochondrial DNA; N, number of individuals sequenced per population; Nh, number of haplotypes, RAPDs, randomly amplified polymorphic DNA; RFLPs, restriction fragment length polymorphisms.

[a] *Contact zone* refers to sites where haplotypes belonging to eastern and western clades both occur.

[b] See Gibbs et al. 2006.

0.0024 in 5 contact-zone populations compared with 0.0006 in 12 other populations; see Table 3.4). This observation emphasizes the need for a phylogeographic perspective in interpreting within-population patterns of nucleotide diversity (see Burbrink and Castoe, Chapter 2).

Estimation of Effective Population Size

The genetic characteristics of populations frequently correlate with effective population size (N_e). This is the size of an ideal population (1:1 sex ratio, random mating, constant size over time, and equal contribution of all adults to subsequent generations) having the same genetic characteristics as a real population of concern (Nunney and Elam 1994; Frankham 1995; Crandall et al. 1999; Nunney 2000). In cases in which the ways that real populations

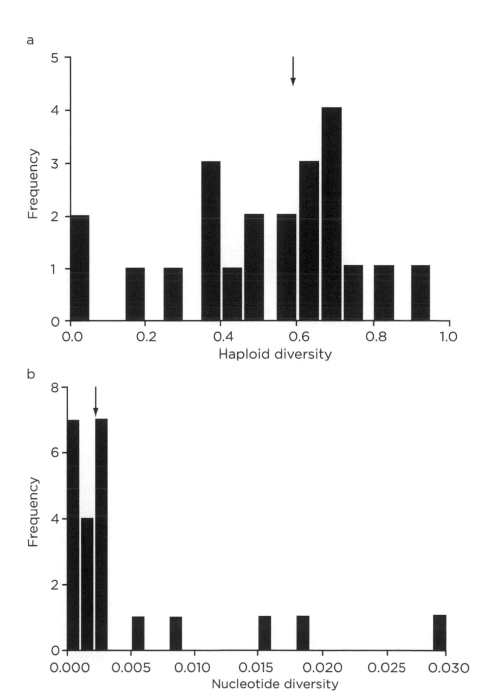

Fig. 3.2. Diversity within snake populations based on mtDNA sequence data. (a) Haplotype diversity (b) Nucleotide diversity. Arrows indicate median values (data in Table 3.3).

TABLE 3.4
Estimates of effective population size based on patterns of genetic variation in microsatellite loci

Method	Thamnophis sirtalis (Ohio and Ontario, 10 populations, 4 loci)	Thamnophis sirtalis (California, 13 populations, 11 loci)	Thamnophis elegans (California, 20 populations, 11 loci)	Thamnophis radix (Illinois, 1 population, 6 loci)
H_e (SSM)	76,741 (45,225–137,254)[a]	6,131 (2,708–4,444)[b]	4,606 (2,500–3,452)[b]	12,306[c]
H_e (IAM)	17,247 (12,744–23,816)[a]	3,575 (4,175–8,395)[b]	2,911 (3,750–5,836)[b]	5,733[c]
$4N_e\mu$	3,267 (2,285–4,675)[d]	325 (183–810)[e]	328 (150–765)[e]	
Linkage Disequilibrium	97 (35–275)[f]			25 (21–31)[c]
Mark-recapture	67–558[g]	148, 283[h]	204, 235[h]	65, 172[i]

Notes: Estimates based on H_e and $4N_e\mu$ assume mutation rate, $\mu = 0.0001$. For studies involving multiple populations, mean N_e (range) is shown. Estimates of adult population size based on mark-recapture techniques are shown for comparison. μ, mutation rate; H_e, expected heterozygosity; IAM, infinite alleles mutation model; N_e, effective population size; SSM, single-step mutation model.
  [a] Computed from data in Bittner 2000.
  [b] Computed from Manier and Arnold 2005.
  [c] Computed from data provided by T. Wusterbarth (pers. comm.).
  [d] Computed from Bittner and King 2003.
  [e] From Manier and Arnold 2005.
  [f] One site for which estimated $N_e = \infty$ (95% CI ranges from 99 to ∞) was excluded.
  [g] Nine populations (from Bittner and King 2003).
  [h] Two populations (from Manier and Arnold 2005).
  [i] Estimated by two methods (Stanford and King 2004).

deviate from an ideal population are known, the effective population size is typically smaller than the census population size (N) and the ratio N_e/N is less than 1. For example, among 56 comprehensive estimates (estimates that accounted for unequal sex ratio, variance in family size, and fluctuations in population size), N_e/N averaged just 0.11 (Frankham 1995). Unfortunately, estimates of N_e or N_e/N for reptiles generally, and for snakes in particular, are lacking. The only reptiles included in Frankham's (1995) review were two lizard species and only the variance in family size was incorporated into estimates of N_e/N for these taxa.

Recently, a comprehensive estimate of N_e/N was generated as part of recovery plan development for the Lake Erie Watersnake, *Nerodia sipedon insularum* (U.S. Fish and Wildlife Service 2003; King et al. 2006a). Using data on the proportions of reproducing males and females and annual survivorship for Lake Erie watersnakes and northern watersnakes (King 1986; Brown and Weatherhead 1999; Prosser et al. 2002), it was estimated that approximately 82% of adult females and 67% of adult males reproduced at least once in their lifetime, resulting in $N_e/N \approx 0.73$, due to the sex ratio. Observed litter sizes (King 1986; Prosser 1999) suggested that variation in offspring numbers among females should have little effect on N_e, but variation in the number of offspring among males (Prosser 1999; Prosser et al. 2002) resulted in $N_e/N \approx 0.7$. Finally, census data suggested perhaps a twofold change in population size in recent years, resulting in $N_e/N \approx 0.89$. Combining the information on sex ratio, variance in family size, and fluctuations in population size gives $N_e/N \approx 0.73 \times 0.70 \times 0.89 \approx 0.45$ (see U.S. Fish and Wildlife Service 2003 for details). A similar analysis of an isolated population of *Vipera berus* provides an estimated N_e of just 12.8 individuals (Madsen et al. 1995). The total number of adults in this population averaged 38.0 in 1980–1990 (Madsen et al. 1995), giving $N_e/N \approx 0.34$. These examples illustrate two things. First, effective population size can be markedly smaller than census size, and this has implications for population and conservation genetic characteristics of snake populations. Second, it is unlikely that the detailed demographic data needed to estimate N_e will become available for more than a handful of case studies; this makes other methods for estimating N_e appealing.

Emerging molecular genetic and analytical techniques provide methods for estimating N_e in the absence of detailed demographic data. Conceptually, the simplest of these techniques makes use of observed changes in allele frequency. If the population size is small, genetic drift should result in large changes in allele frequency over time, whereas if the population size is large, smaller changes in allele frequency are expected. Thus, it is possible to relate the magnitude of change in allele frequency to N_e (Nei and Tajima 1981; Waples 1989; Berthier et al. 2002; Tallmon et al. 2004a). An underlying assumption is that allele frequencies are unaffected by other evolutionary processes (selection, gene flow, and mutation), a reasonable assumption for microsatellite DNA loci in closed (isolated) populations over intervals of a few generations. A clever application of this technique was used in a population of *Crotalus willardi* by treating large individuals as one sample and smaller, younger individuals as a later sample, resulting in $N_e = 220$ (5% quantile = 103; 95% quantile = 293; Holycross and Douglas 2007). The corresponding census population size (based on mark-recapture techniques) was 300 (Holycross and Douglas 2007), giving $N_e/N = 0.73$.

Effective population size can also be estimated from a single sample using genotype or expected heterozygote frequencies. This can be done analytically (Nei 1982) or via maximum likelihood and coalescent theory, a retrospective approach that models the genealogy of alleles backward through time until they coalesce into a single ancestral allele (Nielsen 1997; Beerli 1998; Beerli and Felsenstein 1999, 2001; Hey and Nielsen 2004). Some methods allow flexibility in the choice of mutation model, but all require specification of the mutation rate, μ, in order to estimate N_e. Thus, accuracy of the estimates hinges both on the appropriateness of the mutation model (e.g., stepwise mutation, infinite alleles, or two-phase mutation) and accuracy of the estimated mutation rate. Some information on mutation rate is available (e.g., $\sim 10^{-5}$–10^{-6} for allozyme loci, 10^{-3}–10^{-4} for microsatellite DNA loci), but it is derived mostly from other vertebrate taxa (e.g., mammals) and, at best, provides only order-of-magnitude precision in estimates of N_e. This problem is partially avoided by estimating N_e in mutation-rate units—that is, by estimating $4N_e\mu$ (4 times the product of effective population size and mutation rate).

Finally, effective population size can be estimated from observed gametic disequilibrium (also referred to as linkage disequilibrium; Hartl and Clark 2006) within populations (Hill 1981; Peel et al. 2004). In an infinite, randomly mating population, gametic disequilibrium (D) and the correlation among alleles at different loci (r) equal zero, whereas in finite populations, D and r exceed zero by an amount proportional to the effective population size. This method assumes that the markers used are selectively neutral, mating is random, and gene flow and population substructure are negligible. Furthermore, estimates of N_e are downwardly biased by small sample size, particularly if N_e itself is small (England et al. 2006).

To illustrate the utility of these methods, estimates of the effective population size of three species of gartersnakes were computed or compiled from the literature and tabulated along with estimates of the census population size from mark-recapture work (see Table 3.4). Note that estimates based on expected heterozygosity appear unrealistically large, exceeding those based on $4N_e\mu$, D, and mark-recapture techniques by an order of magnitude or more. In contrast, estimates based on $4N_e\mu$ show reasonable agreement with mark-recapture estimates for California populations of *T. sirtalis* and *T. elegans* (Manier and Arnold 2005) but not for Ohio and Ontario populations of *T. sirtalis*. Estimates generated from observed gametic disequilibrium are somewhat lower than the mark-recapture estimates for Ohio and Ontario populations of *T. sirtalis* and for an Illinois population of *T. radix*. However, gene flow among populations (Bittner and King 2003) may have inflated D and r, resulting in lower estimates of N_e.

Population Bottlenecks and Population Trends

Population bottlenecks occur when population size is reduced dramatically for one or a few generations. Bottlenecks can have genetic consequences that last for tens or hundreds of generations but, unless observed directly, are unlikely to be incorporated into estimates of N_e based on demographic data. One solution is to use molecular genetic data to discern the occurrence and timing of population bottlenecks (Cornuet and Luikart 1996; Beaumont 1999; Garza and Williamson 2001; Williamson-Natesan 2005; but see Busch et al. 2007). Population bottlenecks result in reductions in both allelic diversity and heterozygosity; however, allelic diversity decreases more rapidly than heterozygosity. Thus, a population that is increasing from a past bottleneck has more heterozygosity than expected and a declining population has less heterozygosity than expected given the observed number of alleles (Cornuet and Luikart 1996; Beaumont 1999). Furthermore, for microsatellite loci, declining populations have fewer alleles given the range of allele sizes than constant or increasing populations (Garza and Williamson 2001). The time frame over which bottlenecks are detectable by these methods depends on population size; Cornuet and Luikart (1996) note that a bottleneck of $N_e = 50$ is detectable for approximately 25–250 generations.

Evidence for population bottlenecks has been reported in populations of *Crotalus willardi, Sistrurus catenatus,* and *T. sirtalis* (Bittner 2000; Holycross 2002; Bittner and King 2003; Holycross and Douglas 2007). Among three populations of *C. willardi,* 63-, 167-, and 1429-fold reductions in population size apparently occurred approximately 1688–9132 years ago. Unfortunately, the performance of methods for detecting population bottlenecks and population trends has not been well characterized and further verification is needed from cases in which there is independent information on population history (e.g., Holycross and Douglas 2007). In particular, the application of these methods to species with known histories of population declines (e.g., *Vipera aspis,* Jäggi et al. 2000; *Vipera berus,* Madsen et al. 1995; *Vipera ursinii,* Újvári et al. 2002), bottlenecks (*Natrix tessellata,* Gautschi et al. 2002), or increases (*Boiga irregularis,* Rodda et al. 1992) would be informative. Comparisons between declining and stable populations of the same species would also be instructive. For example, multiple snake populations at the University of Kansas Natural History Reservation have declined by one or two orders of magnitude as a result of successional processes, whereas populations at nearby experimental areas have remained stable (Fitch 2006).

Relatedness

If a pair of individuals are parent and offspring or full siblings, they should have one allele in common at any given locus (identical by descent). Thus, genetic similarity between individuals offers a clue to their relatedness. But allele sharing can also occur between unrelated individuals if a given allele is common in the population (identical by state). By combining information on allele sharing with estimates of allele frequency, relatedness between pairs of individuals can be generated (Blouin 2003). Such estimates are useful in several ways. If populations are small and fragmented, mean relatedness will be high even if mating occurs at random within subpopulations. Thus, high relatedness might indicate a situation in which inbreeding depression could be a concern. Estimates of relatedness can also provide information on population subdivision—if the population subdivision is strong, relatedness will be higher within subpopulations than among them. Finally, identification of highly related sets of individuals (e.g., littermates and parents and offspring) may be useful for investigations of mating systems and analysis of inheritance.

Distributions of relatedness between pairs of *Pantherophis [Elaphe] obsoletus* from 15 hibernacula were largely concordant with simulated distributions generated under the assumption that individuals were unrelated, suggesting that population sizes are large and mating occurs at random (Lougheed et al. 1999). This interpretation is further supported by F_{IS} values that do not differ significantly from zero (Gibbs and Weatherhead 2001). Low degrees of relatedness were also found in *T. sirtalis;* our research found that average relatedness within 10 Lake Erie populations ranged from 0.01 to 0.04, based on four microsatellite DNA loci. In contrast, a higher degree of relatedness was found in *Sistrurus catenatus;* our research found that average relatedness within five midwestern populations ranged from 0.03 to 0.17, based on three microsatellite DNA loci, with the highest of these values falling between those of first and second cousins (0.25 and 0.125, respectively).

Genetic Variation among Populations

Measures of Population Structure

Of equal interest to patterns of genetic variation within populations are patterns of population subdivision or population structure (the degree of genetic differentiation among populations). For neutral genetic markers, these patterns reflect the diversifying effects of genetic drift and the homogenizing effects of gene flow. When gene flow is rare, allele frequencies within populations drift independently of those in other populations and over time and large differences (including loss and fixation of alternative alleles) will arise. Such differences can arise rapidly if effective population size is small. In contrast, high rates of gene flow can prevent diversification by genetic drift, resulting in genetic uniformity among populations. For traits subject to natural selection, patterns of differentiation further reflect effects of environmental heterogeneity. Uniform selective regimes can prevent population differentiation even in the absence of gene flow, whereas varying selective regimes can promote differentiation even if gene flow is frequent.

A useful way to quantify population differentiation is provided by Wright's hierarchical F coefficients, introduced earlier (Wright 1931; Hedrick 2000). Of particular interest here is F_{ST}, which, in the case of a single locus with two alternative alleles, ranges from 0.0 when there is no population subdivision (i.e., identical allele frequencies among subpopulations) to 1.0 when subpopulations are fixed for alternative alleles. Calculation of F_{ST} can be extended to more than two alleles and more than one locus, but in such cases the maximum value is less than 1.0 and is inversely proportional to the amount of variation within subpopulations (H_S; Hedrick 1999, 2005b). Alternative formulations for estimating F_{ST} exist (e.g., θ of Weir and Cockerham 1984; R_{ST} of Slatkin 1995, a formulation specific to microsatellite loci) as do formulations designed for use with DNA sequence data (G_{ST}, γ_{ST}, N_{ST}, F_{ST}; Nei 1982; Lynch and Crease 1990; Hudson et al. 1992). In this review, F_{ST} is used in a generic sense to symbolize these related measures of population differentiation. F coefficients can also be extended to additional hierarchical levels. For example, in an allozyme-based analysis of watersnakes of the *Nerodia fasciata–Nerodia clarkii* complex, Lawson et al. (1991) used a five-level hierarchy, allocating genetic variability to local demes, regions, subspecies, and groups (freshwater and saltmarsh) within the total population.

Using Wright's island model, F_{ST} calculated from neutral molecular markers can be used to estimate $N_e m$, the effective number of immigrants per generation, from the relationship $F_{ST} \approx 1/(4N_e m + 1)$. Under this model, subpopulations are assumed to be in migration-drift equilibrium, equal and constant in size, and exchanging migrants symmetrically at rate m. Critics have noted that these assumptions are frequently violated (Whitlock and McCauley 1999), so caution is urged in interpreting estimates of $N_e m$ generated in this way. Note that for haploid markers (e.g., mtDNA sequences), $F_{ST} \approx 1/(2N_e m + 1)$.

As an alternative to F_{ST}-based methods, coalescent-based techniques provide estimates of gene flow under less restrictive assumptions. One approach is to jointly estimate population size (in mutation-rate units) and migration rate under the assumption that these remain constant over time but allowing for unequal population size and asymmetric gene flow (Beerli and Felsenstein 1999, 2001; Beerli 2006). Thus, in the case of two populations, four parameters are estimated: the size of each population and the rate of gene flow from each population to the other. Using this method, estimates of effective population sizes at 14 *Crotalus horridus* hibernacula in eastern North American averaged 156 (range = 37–317) and rates of gene flow averaged 1.0 (range = 0.3–3.1; Clark et al. 2007).

A somewhat more complex approach uses an isolation with migration (IM) model in which contemporary populations are assumed to have split from an ancestral population at some time in the past and population sizes have remained constant or are changing exponentially (Hey and Nielsen 2004; Hey et al. 2004). In the case of two populations of constant size, six parameters are estimated: the sizes of the two contemporary populations and the ancestral population, the rate of gene flow from each population to the other, and time since the populations split. With exponential population growth, an additional parameter is estimated that represents the proportion of the ancestral population from which one of the descendant populations was founded.

A third approach approximates recent rates of gene flow (over the last several generations) by inferring individual immigrant histories (immigrant, non-immigrant, and offspring of an immigrant and a non-immigrant; Wilson and Rannala 2003). This approach is complementary to other methods (which estimate long-term mean rates of gene flow) and makes fewer assumptions regarding population history, but it requires knowledge of multilocus genotypes (e.g., for microsatellite DNA loci). All these methods are computer intensive, using maximum likelihood, Bayesian inference, and Markov chain Monte Carlo (MCMC) simulation techniques that require some sophistication to implement and interpret. Given the ease with which large amounts of DNA sequence and multilocus genotype data can now be generated, however, these methods have the potential to provide remarkably detailed insights into contemporary and historical population processes.

Observed Patterns of Population Structure

F_{ST}-based patterns of among-population genetic variation have been documented for a variety of snake taxa using allozymes, RAPDs, microsatellite DNA, mtDNA sequences, and nonmolecular markers (Table 3.5 and Fig. 3.3). The examples presented here are limited to studies in which 10 or more individuals per population were included (three or more individuals for studies based on sequence data), unless they were a part of a larger study in which some samples did meet this criterion.

Estimates of F_{ST} were available for 19 allozyme-based studies (14 species, 2–13 populations per study, spanning distances of ~2–4000 km), 5 RAPD-based studies (3 species, 4–8 populations per study, spanning ~2–1500 km); 35 microsatellite DNA-based studies (11 species, 2–18 populations per study, spanning circa 1–1900 km), and 8 mtDNA-based studies (2–17 populations per study, spanning 2–1325 km) (see Table 3.5). The median F_{ST} was only slightly lower for microsatellite DNA (median = 0.05, range = 0.01–0.53) than for allozymes (median = 0.11, range = 0.00–0.46; see Figs. 3.3a, c) despite markedly greater heterozygosity observed for microsatellite DNA (see Fig. 3.1) (Hedrick 2005b). Furthermore, in the one study using both classes of markers to assess variation among the same populations, F_{ST} estimates did not differ between marker types (Bittner 2000; Bittner and King 2003). In contrast, F_{ST} was markedly higher for mtDNA sequences (median = 0.76, range = 0.42–0.97) than for other molecular markers (Table 3.5 and Fig. 3.3d). In three studies for which both mtDNA- and microsatellite DNA-based estimates are available, mtDNA-based estimates dramatically exceeded microsatellite DNA-based estimates (0.42 compared to 0.07 in *Thamnophis gigas*, 0.81 compared to 0.13 in *T. radix*, 0.52 compared to 0.10 in *Epicrates subflavus*; Table 3.5). Smaller effective population size for mtDNA-based markers and male-biased dispersal contribute to this pattern (Paquin et al. 2006), as does the more rapid coalescence of mtDNA markers as compared to nuclear markers (Zink and Barrowclough 2008).

Unusually high F_{ST} was observed in *Ovophis okinavensis* based on allozymes, equaling 0.46 between populations on two islands separated by just 40 km (see Table 3.3, part A). In contrast, F_{ST} for *Protobothrops flavoviridis* on these same islands equaled just 0.04 and averaged 0.23 across four islands spanning 310 km (Toda et al. 1999). Possibly, *O. okinavensis* represents a cryptic species pair. Regardless, an analysis of additional

TABLE 3.5
Estimates of FST and related measures of differentiation among snake populations for allozymes, RAPDs, microsatellite DNA, mitochondrial DNA, and nonmolecular markers

Species	Distance (km)	F_{ST}	Isolation by Distance?	Reference
A. Allozymes				
Colubridae, Colubrinae				
Elaphe dione (12)	*40–500*	*0.08*		Paik and Yang 1987
Colubridae, Natricinae				
Natrix natrix (13)	*~14–1200*	0.19		Hille 1997
Nerodia clarkii (4)	> 100	0.29		Lawson et al. 1991
N. fasciata (12)	10s–100s	0.21		Lawson et al. 1991
N. sipedon (7, 21)	1.3–73	0.03 (0.01–0.09)	Yes	King and Lawson 1995
Storeria dekayi (7, 21)	1.3–71	0.04 (0.01–0.11)	Yes[a]	King and Lawson 2001
Thamnophis proximus (4, 6)	*~100–600*	*0.03* *(0.02–0.05)*		Gartside et al. 1977
T. sirtalis (4)	120–270	*0.18*		Sattler and Guttman 1976
T. sirtalis (10, 45)	1–108	0.03 (0.01–0.09)	No	Lawson and King 1996; Bittner 2000; Bittner and King 2003
T. sirtalis (4, 6)	320–1220	0.14 (0.05–0.21)		Lawson and King 1996
T. sirtalis (5, 10)	14–692	0.07 (−0.04–0.12)	Yes[h]	Rye 2000
T. sirtalis (4, 6)	20–1014	0.02 (−0.03–0.04)	Yes[b]	Rye 2000
T. sirtalis (12, 49)	496–4476	0.32 (0.10–0.55)	Yes[b]	Rye 2000
Viperidae, Crotalinae				
Agkistrodon piscivorus (5)	*25–130*	*0.24*		Merkle 1985
Bothrops asper (6)	*16–140*	0.02		Sasa and Barrantes 1998
Cerrophidion godmani (8)	*~50*	*0.00*		Sasa 1997
Ovophis okinavensis (2)	*~40*	*0.46*		Toda et al. 1999
Protobothrops flavoviridis (4, 6)	*~40–310*	*0.23* *(0.04–0.50)*		Toda et al. 1999
Boidae				
Boa constrictor (2)	200	< 0.01		Rivera et al. 2005; Cardozo et al. 2007
B. RAPDs				
Colubridae, Colubrinae				
Pantherophis spiloides (x *alleghaniensis*; hybrid zone)[c] (4, 2)	1.2–1.6	0.02 (0.01–0.04)		Prior et al. 1997
P. spiloides x *alleghaniensis* (hybrid zone)[c] (5, 10)	15–50	0.04 (0.01–0.08)	Yes	Prior et al. 1997
P. obsoletus[c] (4, 6)	465–1500	0.16 (0.02–0.32)		Prior et al. 1997
Viperidae, Crotalinae				
Sistrurus catenatus (4, 6)	*100–600*	0.14 (0.09–0.26)		Lougheed et al. 2000

TABLE 3.5—continued

Species	Distance (km)	F_{ST}	Isolation by Distance?	Reference
Viperidae, Viperinae				
Vipera aspis (8)	~1–700	0.13		Jäggi et al. 2000
C. Microsatellite DNA				
Colubridae, Colubrinae				
Pantherophis spiloides (x *alleghaniensis*; hybrid zone)[c] (11, 11)	0.5–6	0.01 (0.00–0.04)		Lougheed et al. 1999
P. spiloides x *alleghaniensis* (hybrid zone)[c] (5, 10)	15–50	0.06 (0.00–0.13)		Lougheed et al. 1999
P. obsoletus[c] (4, 6)	465–1900	0.17 (0.08–0.23)		Lougheed et al. 1999
Colubridae, Natricinae				
Natrix tessellata (4, 6)	~10–450	0.53 (0.40–0.69)		Guicking et al. 2004
Nerodia erythrogaster (7, 21)	< 3 to 600	0.11 (0.01–0.23)		J. Marshall, pers. comm.
N. fasciata (7, 21)	20–320	0.19 (0.01–0.44)	Yes	Jansen 2001
N. sipedon (3, 3)	1.0–1.5	0.01 (0.01–0.01)		Prosser et al. 1999
N. sipedon (6)	< 3 to 600	0.05		J. Marshall, pers. comm.
Thamnophis elegans (20, 190)	1–50	0.02 (0.00–0.09)	Yes	Manier and Arnold 2005
T. gigas (14)	< 350	0.07		Paquin et al. 2006
T. radix (4, 3)	150 to > 1000	0.13 (0.08–0.22)		G. M. Burghardt, pers. comm.
T. sirtalis (5, 10)	14–150	0.04 (0.00–0.07)		Garner 1998
T. sirtalis (10, 45)	1–108	0.04 (0.01–0.08)	No	Bittner 2000; Bittner and King 2003
T. sirtalis (13, 78)	1–50	0.04 (0.00–0.10)	Yes	Manier and Arnold 2005
T. sirtalis (4, 6)	4–31	0.02 (0.00–0.05)	No	T. W. J. Garner, pers. comm.
T. sirtalis (4, 6)	45–219	0.05 (0.02–0.08)	Yes[a]	T. W. J. Garner, pers. comm.
T. sirtalis (9, 24)	> 1000	0.34 (0.12–0.50)		T. W. J. Garner, pers. comm.
T. sirtalis (7, 21)	16–233	0.03 (0.01–0.10)	No	Ridenhour 2004; Ridenhour et al. 2006
T. sirtalis (7, 21)	18–234	0.01 (−0.02–0.04)	No	Ridenhour 2004; Ridenhour et al. 2006
T. sirtalis (5, 10)	32–170	0.05 (0.01–0.08)	No	Ridenhour 2004; Ridenhour et al. 2006
T. sirtalis (18, 153)	16–780	0.04 (−0.05–0.16)	Yes	Ridenhour 2004; Ridenhour et al. 2006
Viperidae, Crotalinae				
Crotalus horridus (5, 10)	0.5–3.1	0.05 (0.00–0.12)	Yes[d]	Bushar et al. 1998
C. horridus (14,16)	1–11	0.02 (0.00–0.05)	Yes[e]	Clark et al. 2007
C. willardi (4, 6)	< 50	0.16 (< 0.01 to 0.21)		Holycross 2002; Holycross and Douglas 2007

TABLE 3.5—continued

Species	Distance (km)	F_{ST}	Isolation by Distance?	Reference
Sistrurus catenatus (2)	1.5	0.04		Gibbs et al. 1997
S. catenatus (2)	~5	0.03		Gibbs et al. 1997
S. catenatus (3)	< 35	0.18		Gibbs et al. 1997
S. catenatus (5, 10)	50–600	0.18 (0.09–0.26)		Gibbs et al. 1997
S. catenatus (4, 6)	100–600	0.20 (0.08–0.27)		Lougheed et al. 2000
S. catenatus (3, 3)	~5	0.01 (−0.01–0.01)		Holycross 2002
S. catenatus (2)	410	0.13		Holycross 2002
S. catenatus (3, 3)	~3	0.07 (0.02–0.15)		Andre 2003
S. catenatus (5, 7)	170–570	0.20 (0.10–0.32)		Andre 2003
Viperidae, Viperinae				
Vipera berus (16,120)	*< 1 to 600*	0.27 (0.03–0.50)	Yes	Ursenbacher et al. 2008
Boidae				
Epicrates subflavus (3)	< 200	0.10 (0.08–0.12)		Tzika et al. 2008a
D. mtDNA				
Colubridae, Natricinae				
Nerodia sipedon (13)	27–1325	*0.71*		J. Robinson, pers. comm.
Storeria dekayi (2)	596	*0.46*		J. Robinson, pers. comm.
Thamnophis elegans (10)	*72–1800*	*0.97*		Bronikowski and Arnold 2001
T. gigas (14)	< 350	0.42		Paquin et al. 2006
T. radix (3)	160–480	0.81		G. M. Burghardt, pers. comm.
T. sirtalis (2)	*1122*	*0.84*		Janzen et al. 2002
T. sirtalis (17)	*2–1086*	*0.92*		J. Placyk, pers. comm.
Epicrates subflavus (3)	< 200	0.52 (0.28–0.76)		Tzika et al. 2008a
E. Nonmolecular Markers				
Nerodia sipedon (7)—color pattern	1.3–73	0.49		Ray and King 2006
Thamnophis sirtalis (10, 45)—color pattern	1–108	0.15 (0.00–0.54)		Lawson and King 1996
T. sirtalis (6)—color pattern	1.3–20	0.28 (0.09–0.44)		M. K. Manier, pers. comm.
T. sirtalis (6)—scalation	1.3–20	0.25 (0.02–0.71)		M. K. Manier, pers. comm.

Notes: Numbers in parentheses following species names refer to the number of populations and number of pairwise F_{ST} estimates (if pairwise estimates are available). Distance refers to geographic distance between populations. Estimated distances (from published maps or locality information) and estimated F_{ST} values (from published genotype or allele frequencies) are shown in italics. Species names follow the Integrated Taxonomy Information System Catalogue of Life 2006 Annual Checklist (ITIS 2006). Family- and subfamily-level taxonomy follows Lawson et al. (2005). mtDNA, mitochondrial DNA; RAPDs, randomly amplified polymorphic DNA.

[a] Approaches significance ($P < .10$).

[b] Tests for isolation by distance included additional sampling locations with $N < 10$.

[c] Includes *Pantherophis obsoleta*, *P. alleghaniensis*, and *P. spiloides* of Burbrink et al. 2000 (but see Gibbs et al. 2006).

[d] Significant when two of ten population pairs are omitted.

[e] Significant when using a measure of geographic distance that accounts for basking habitat between hibernacula.

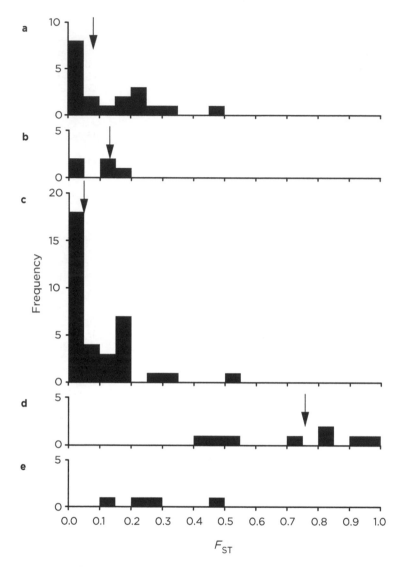

Fig. 3.3. Genetic differentiation among snake populations as measured by F_{ST} and related estimators. (a) Allozymes (b) RAPDs (c) Microsatellite DNA (d) mtDNA sequences (e) Nonmolecular markers. Arrows indicate median values (data in Table 3.5). RAPDs, randomly amplified polymorphic DNA.

molecular markers would be of interest. Unusually high F_{ST} was also observed in *Natrix tessellata* based on microsatellite DNA, equaling 0.54 across four populations separated by approximately 10–450 km. This snake is critically endangered in Germany (Guicking et al. 2004), and small population size and isolation may be responsible for elevated population subdivision. In contrast, unusually low F_{ST} was observed among populations of *Cerrophidion godmani* ($F_{ST} = 0.00$ among eight sites spanning ~50 km in Costa Rica; Sasa 1997) and *Boa constrictor* ($F_{ST} < 0.01$ between two sites separated by 200 km in Argentina; Rivera et al. 2005; Cardozo et al. 2007).

Relatively large numbers of Natricinae and Crotalinae snakes are represented among taxa for which patterns of among-population genetic variation have been characterized (Table 3.5). Median F_{ST} appears somewhat higher in Crotalinae than in Natricinae for both allozymes and microsatellite DNA (for allozymes, median $F_{ST} = 0.23$ among Crotalinae compared to 0.11 among Natricinae; for microsatellite DNA, median $F_{ST} = 0.10$ among Crotalinae compared to 0.05 among Natricinae). One interpretation of this pattern is that rates of gene flow are higher among Natricinae than Crotalinae. These groups are similar in mean F_{ST} (for allozymes, mean $F_{ST} = 0.19$ among Crotalinae compared to 0.13 among Natricinae; for microsatellite DNA, mean $F_{ST} = 0.11$ among Crotalinae compared to 0.10 among Natricinae), however, indicating that further investigation is warranted (e.g., of sympatric Natricinae and Crotalinae species).

The degree of population differentiation is expected to increase with increasing geographic isolation, a pattern referred to as isolation by distance. This pattern was found in 14 of 20 formal tests involving seven species (Table 3.5). Cases in which formal tests indicated that isolation by distance was lacking were restricted to *T. sirtalis* and involved transects spanning 234 km or less. Other analyses of this and related species, however, showed isolation by distance over transects spanning as little as 50 km. These results suggest that species differences in dispersal ability (e.g., King and Lawson 2001) and geographic features unique to a given transect (e.g., Manier and Arnold 2005) contribute to patterns of population subdivision.

Several analyses of snake population subdivision, all focused on *Thamnophis*, have made use of coalescent- and Bayesian-based methods. In Lake Erie, island and mainland populations of *T. sirtalis*, patterns of population subdivision were similar for allozyme and microsatellite DNA loci, but coalescent-based analyses showed greater population subdivision (lower rates of gene flow) than did F_{ST}-based analyses (Bittner 2000; Bittner and King 2003). Neither method showed a pattern of isolation by distance. In contrast, F_{ST}-based analyses of microsatellite DNA loci in sympatric *T. sirtalis* and *T. elegans* populations in northern California revealed clear patterns of isolation by distance (Manier and Arnold 2005). Furthermore, coalescent-based estimates of gene flow were highly asymmetrical, suggesting source-sink population dynamics and possibly a history of extinction and recolonization. Population genetic structures were correlated, suggesting that these two species were affected in similar ways by landscape features. Based on multiple linear regression, ecological correlates of inferred patterns of gene flow include population density (both species) and, for *T. elegans*, abundance of

T. sirtalis; distance; elevation; and habitat type (Manier and Arnold 2006). In addition, a 300-m escarpment was identified as a barrier to dispersal in both species. In contrast, the waters of Lake Erie represented no greater barrier to gene flow than did the intervening terrestrial habitat in *T. sirtalis* (Bittner 2000; Bittner and King 2003).

Estimates of recent gene flow were obtained for 18 Pacific Northwest *T. sirtalis* sampling locations along three transects (northern coast, southern coast, inland) representing clines in gartersnake resistance to newt (*Taricha granulosa*) tetrodotoxin (TTX; Ridenhour 2004, Ridenhour et al. 2006). Bayesian analyses indicated that two pairs of locations represented single panmictic populations (*m*, the proportion of recent immigrants constituting a given population, ≈ 0.50). After pooling these sites, estimated rates of immigration appeared bimodal with *m* averaging 0.04 (*n* = 5, range = 0.02–0.09) at low-immigration sites and 0.29 (*n* = 11, range = 0.25–0.31) at high-immigration sites. Migration rates were asymmetrical, indicating possible source-sink population dynamics as in northern California *T. elegans* and *T. sirtalis* (Manier and Arnold 2005).

Ideally, estimates of gene flow based on molecular genetic data are corroborated with direct observations (e.g., mark-recapture data). Such corroborative data are scarce, but interpopulation movements have been observed in *Thamnophis elegans* and *Nerodia sipedon* (Bronikowski and Arnold 1999; Bronikowski 2000; R. B. King, pers. obs.), species in which molecular methods indicate significant gene flow (see also Blouin-Demers and Weatherhead 2002b). Sex bias in rates of gene flow, which might result from sex-biased dispersal, can also be detected using multilocus genotypic data (Goudet et al. 2002; Prugnolle and de Meeus 2002; Proctor et al. 2004). For example, female *Rhinocephalus nigrescens* move shorter distances than do males, a pattern reflected in patterns of variation in five microsatellite DNA loci (Keogh et al. 2007). Similarly, sex-specific estimates of F_{ST} and Nm suggest that rates of gene flow were higher among male *Boa constrictor*, consistent with their active mate-searching behavior (Rivera et al. 2006).

Population Assignment

When allele frequencies differ among subpopulations, the probability of occurrence of a given genotype will also differ among subpopulations. For example, the probability that a heterozygous individual occurs in a randomly mating population with $p = q = 0.5$ is much higher ($2pq = 0.25$) than in a population with $p = 0.9$ and $q = 0.1$ ($2pq = 0.18$). When data are available for multiple loci, such probabilities can be used in likelihood tests of population assignment (Waser and Strobeck 1998; Berry et al. 2004; Paetkau et al. 2004; Manel et al. 2005). This information can be useful in assessing the degree of population subdivision. Assignment tests will correctly assign a large proportion of individuals to their source subpopulation when there is a high degree of subdivision. When subdivision is lacking, however, many individuals will be misassigned. This use is complementary to F_{ST}-based measures of population subdivision because it relies on less restrictive assumptions. This approach can also be used to assign individuals for which locality data are lacking (e.g., captive, rescued, or confiscated animals) to their likely source population.

Assignment tests applied to four *Sistrurus catenatus* populations separated by 50–600 km and three *Crotalus willardi* populations separated by approximately 50 km correctly assigned individuals to source populations 92–100% of the time (Lougheed et al. 2000, Holycross and Douglas 2007), suggesting low rates of gene flow and long periods of isolation. Among *Vipera berus* sampled across Fennoscandia, high proportions of correct assignments were observed on a coarse geographic scale (east vs. west of the Baltic Sea), but assignments were more ambiguous on finer geographic scales (Carlsson et al. 2004).

A phylogenetic analog to assignment methods is provided by an analysis of the origin of *Natrix maura* on the island of Mallorca (Guicking et al. 2006). Based on sequences of the mtDNA gene cytochrome *b* and ISSR band-sharing patterns, Mallorcan viperine snakes clearly group with European populations and not with Moroccan or Tunisian/Sardinian populations. Furthermore, viperine snakes appear to be relatively recent arrivals on Mallorca, perhaps representing an anthropogenic introduction by the Romans (Guicking et al. 2006).

Inferences Regarding Number of Subpopulations

In most studies of population structure, sampling locations are assumed to represent subpopulations and genetic data are used to make inferences about patterns of gene flow. An alternative is to make no assumption about the correspondence between sampling location and population structure and, instead, to use observed patterns of genetic variation to (1) test whether any subdivision exists and (2) infer the number of subpopulations (k) represented. The basic approach is to test for Hardy-Weinberg and gametic disequilibrium over the entire data set ($k = 1$). If null hypotheses of equilibrium are rejected, the process can be repeated assuming $k = 2$ or more subpopulations using an MCMC algorithm (Pritchard et al. 2000; Falush et al. 2003; Corander et al. 2004; Evanno et al. 2005; Chen et al. 2007).

Consistent with assignment test results described earlier, in adders (*Vipera berus*) sampled across Fennoscandia, support was strongest for the existence of two subpopulations occurring east and west of the Baltic Sea (Carlsson et al. 2004; further support comes from mtDNA sequence analysis; Carlsson and Tegelström 2002). Results were more ambiguous among *Crotalus horridus* sampled from six hibernacula in an area approximately 10 km in diameter in Missouri and characterized at six microsatellite loci (C. D. Anderson, pers. comm.). In this case, individuals could not be clustered into discrete subpopulations.

Time since Common Ancestry

Isolated populations become differentiated over time as a consequence of mutation and genetic drift. If population size remains constant, gene flow is lacking, and mutation-drift equilibrium has been reached, the degree of population differentiation can be used to infer time, t, since a single ancestral population became subdivided (Goldstein et al. 1995; Bahlo and Griffiths 2000). Using data on six microsatellite loci in five *Sistrurus catenatus* populations and assuming the mutation rate $\mu = 5.4 \times 10^{-4}$, Gibbs et al. (1997) estimated $t > 71,000$ years. They note that this number is unrealistically large (glacial retreat occurred from their study sites just 10,000 years ago) and interpret it as indicating that population isolation predates European settlement (Gibbs et al. 1997). In addition to requiring an estimate of mutation rate, this method assumes a stepwise mutation model that may fit some microsatellite DNA loci better than others (Zhivotovsky 2001).

Studies of Ecologically Significant Traits

Genetic Basis of Ecologically Significant Traits

In contrast to neutral molecular markers, ecologically significant traits affect fitness (survival and reproduction) and are thus likely targets of natural selection. A first step in the analysis of such traits is to establish their genetic basis. Among free-ranging animals, mode of inheritance may be inferred from conformity or nonconformity to Hardy-Weinberg expectations (e.g., sex-linked inheritance of fumarate hydratase variants in natricine snakes; King and Lawson 1996). Pedigree analysis based on inferred relatedness (e.g., from molecular genetic analysis) can also be used to establish mode of inheritance. More frequently, an analysis of litters or hatchlings obtained from wild-caught females is used. Such analyses have suggested major gene effects on *Nerodia sipedon* color pattern (King 1993a) and the occurrence of separate sets of genes influencing chemosensory responses to different prey types in *Thamnophis elegans*. Furthermore, an analysis of slug eating by *T. elegans* suggests inheritance via a major locus with dominance and incomplete penetrance (Ayers and Arnold 1983). In this example, frequency of the "slug-refuser" allele varies from 0.08 in a coastal population sympatric with slugs to 0.61 in an inland population where slugs are lacking (Ayers and Arnold 1983).

One shortcoming of the use of litter and hatchling data is that maternal and environmental effects may influence traits of interest, possibly leading to a misinterpretation of the mode of inheritance (Brodie and Garland 1993; King et al. 2001). This problem can be reduced through common garden experiments (rearing animals from different populations under uniform conditions). Common garden experiments were used to infer a genetic basis for ecotypic differences in growth rate in *Thamnophis elegans*

(Bronikowski 2000). In contrast, common garden experiments indicate that differences in growth rate among populations of Grass Snakes (*Natrix natrix*) represent phenotypic plasticity and are not genetically based (Madsen and Shine 1993b).

For species that can be bred in captivity, more detailed information on mode of inheritance can be obtained. Crosses between lineages of *Thamnophis elegans* differing in prey chemosensory behavior corroborates genetic interpretations based on analysis of litters born to wild-caught females (Arnold 1981). Similarly, crosses between lineages of ratsnakes suggest a genetic basis for coloration, pattern, and scalation (Sideleva et al. 2003). More extensive crosses have documented simple Mendelian inheritance of color pattern in gartersnakes (melanism is recessive to striped; Blanchard and Blanchard 1940; King 2003) and kingsnakes (ringed is recessive to striped; Zweifel 1981). Simple Mendelian inheritance of naturally occurring color pattern variants is also suggested by captive matings among *Nerodia sipedon* and *Elaphe longissima;* Cattaneo 1975; King 1993a). The more unusual color pattern variants favored by hobbyists also often show simple Mendelian inheritance (Bechtel 1995), but are of questionable ecological significance and, hence, are not included in this chapter.

Some of the most detailed information on the inheritance of ecologically significant traits in snakes comes from studies of venom proteins. In a number of cases, the genes encoding these proteins have been identified, allowing for an analysis of gene regulation and evolution in both a phylogenetic and an ecological context (e.g., Daltry et al. 1996; Chijiwa et al. 2000; Li et al. 2005; Sanz et al. 2006). Similarly, detailed analyses have been conducted on a gene controlling TTX-resistant sodium channels in the Common Gartersnake, *T. sirtalis* (Geffeney et al. 2005). Here, differences in TTX resistance can be attributed to differences in one or a few the amino acids constituting the *tsNaV1.4* sodium-channel protein.

Integrative Studies

A number of population genetic analyses of snakes have integrated information on inheritance, natural selection, and gene flow to provide unusually detailed demonstrations of microevolutionary processes. Four such examples are summarized next. Although all involve natricine snakes, these examples span a range of genetically determined traits and demonstrate the utility of combining analyses of neutral and ecologically significant traits. They also illustrate that adaptive differentiation among populations can arise rapidly and on fine geographic scales even in the presence of significant gene flow.

Among Lake Erie island and mainland populations of *T. sirtalis,* melanism occurs at unusually high frequencies (King 1988). Melanism in inherited in simple Mendelian fashion and is recessive to the typical striped color pattern (King 2003). Although melanism in many animals reflects allelic variation at the $Mc1r$ locus, this does not appear to be the case in $T. sirtalis$ (Rosenblum et al. 2004). Melanism provides a thermoregulatory advantage in cool lakeshore habitats, at least among adults (Gibson and Falls 1979; Bittner et al. 2002), but may carry with it a cost in terms of reduced crypsis to visual predators (Gibson and Falls 1988; but see Bittner 2003). Relatively little genetic differentiation in neutral genetic markers (allozymes and microsatellite DNA) is apparent among populations (Lawson and King 1996; Bittner and King 2003); mean $F_{ST} = 0.03$ for 12 allozyme loci in six island and four mainland populations (Lawson and King 1996), consistent with frequent gene flow. In contrast, $F_{ST} = 0.15$ for the color pattern locus in these populations (Table 3.5 and Fig. 3.3). This indicates that variation in color pattern allele frequency (ranging from 0.00 to 0.70) exceeds that attributable to the combined effects of genetic drift and gene flow, and it suggests that the strength and direction of natural selection on color pattern varies among populations, perhaps because of differences in microclimate and predator assemblages (Lawson and King 1996; Bittner 2000; Bittner and King 2003).

Color pattern is also highly variable in Lake Erie populations of watersnakes with island populations (*Nerodia sipedon insularum*) consisting of high frequencies of snakes with reduced patterning and mainland populations (*N. s. sipedon*) consisting only of regularly patterned snakes (King 1987). Variation within litters born to wild-caught females and the results of several captive matings suggest that color pattern is influenced by a major locus and that reduced patterning is recessive (King 1993a). Island and mainland habitats differ in the background against which snakes might be seen by visual predators (exposed rocky island shorelines vs. densely vegetated mainland marshes), and snakes differing in color pattern differ in degree of crypsis against these backgrounds (King 1992; 1993b). Among young snakes, greater crypsis is provided by a reduced pattern on island shorelines and a regular pattern in mainland marshes. In contrast, among adults, crypsis differs little among morphs. The predicted patterns of selection based on the match between snakes and backgrounds were confirmed by cross-sectional and longitudinal analyses (Camin et al. 1954; Camin and Ehrlich 1958; Ehrlich and Camin 1960; King 1993b)—the estimated survival of regularly patterned neonates is just 78–90% that of neonates with reduced patterns in island populations (King 1993b). As in gartersnakes, relatively little genetic differentiation in neutral genetic markers is apparent among watersnake populations (King and Lawson 1995). Mean $F_{ST} = 0.02$ for seven allozyme loci in five island mainland populations. F_{ST} for the putative color pattern locus in island populations is nearly identical (0.02), suggesting that differences in morph frequencies among islands are probably attributable to stochastic processes (Ray and King 2006). In contrast, when two mainland populations are included,

$F_{ST} = 0.07$ for allozyme loci and 0.49 for color pattern (Table 3.5 and Fig. 3.3a), reflecting the effects of selection on color pattern acting in opposite directions in island compared with mainland populations (Ray and King 2006).

Dramatic ecotypic differentiation involving color pattern, vertebral number, reproduction, growth, and survival occurs on a fine geographic scale among Wandering Gartersnake (*Thamnophis elegans*) populations in northern California (Bronikowski 2000; Bronikowski and Arnold 2001; Manier 2005; Manier et al. 2007). Common garden experiments and quantitative genetic analyses indicate that these differences have a genetic basis (Bronikowski and Arnold 1999; Arnold and Phillips 1999). Population differentiation appears to be the result of differing selective regimes associated with variations in food availability and predator assemblages (Bronikowski and Arnold 2001). Individuals from lakeshore populations, where prey and water are continuously available, exhibit fast growth, early maturation, high fecundity, and low adult survival. Those from meadow populations, where prey and water availability are more variable, exhibit slower growth, later maturation, lower fecundity, and higher adult survival (Bronikowski and Arnold 1999). Individuals from lakeshore populations also have higher vertebral numbers and are lighter in color than those from meadow populations (Manier et al. 2007). Differences have arisen between populations separated by short distances (a few kilometers) and experiencing frequent gene flow (Manier and Arnold 2005). The interpretation that differences among populations are the result of natural selection is supported by comparisons between F_{ST} for neutral molecular markers (nine microsatellite DNA loci) and Q_{ST}, a measure of quantitative trait differentiation, for 13 color pattern components and six aspects of scalation (Manier et al. 2007). Among six populations (four meadow and two lakeshore), F_{ST} averaged 0.04, whereas Q_{ST} averaged 0.25 for color pattern and 0.28 for scalation (Table 3.5 and Fig. 3.3). Furthermore, most population differentiation in color pattern and scalation could be attributed to differences between meadow and lakeshore sites as expected if selection favors different ecotypes in these habitats.

The importance of geographic variation in prey characteristics to the evolutionary diversification of snakes has been studied extensively in *Thamnophis sirtalis* populations in western North America, some of which co-occur with *Taricha granulosa* of varying toxicity (Brodie et al. 2002; Geffeney et al. 2002, 2005; Ridenhour 2004, Ridenhour et al. 2006). Resistance to newt TTX varies within and among gartersnake populations (Brodie et al. 2002), and this variation reflects allelic differences at the *tsNaV1.4* sodium-channel protein gene (Geffeney et al. 2005). Fine-scale geographic sampling, combined with phylogeographic analysis of mtDNA sequences, identifies two hotspots where TTX resistance has evolved independently, surrounded by clines of decreasing resistance (Brodie et al. 2002; Janzen et al. 2002;

Geffeney et al. 2005). An analysis of microsatellite DNA loci suggests little population subdivision at neutral marker loci (mean F_{ST} between pairs of populations = 0.04) and correspondingly high rates of gene flow among gartersnake populations (Ridenhour 2004, Ridenhour et al. 2006).

Population Genetics and Snake Conservation

Loss of Genetic Variation and Its Fitness Consequences in Declining and Fragmented Populations

Loss of genetic variation in small isolated populations has been especially well documented in snakes. Using RAPD markers, Jäggi et al. (2000) analyzed genetic variation within and among isolated Alpine populations of *Vipera aspis* in Switzerland and found higher levels of band-sharing (assessed using Jaccard distance) in small compared with large populations, suggesting a greater loss of genetic variation in small populations. Jaccard distance did not differ between connected and isolated subpopulations, however, nor did it correlate with distance to the next nearest population. Between-population levels of differentiation were not unusually high either (F_{ST} = 0.13; Table 3.5), perhaps because the isolation of subpopulations is recent, resulting from human-induced habitat loss and fragmentation (Jäggi et al. 2000).

The Meadow Viper, *Vipera ursinii*, persists in small numbers at isolated sites surrounded by farmland (Újvári et al. 2002). Using RFLP analysis of major histocompatability (Mhc) class I genes, levels of band sharing were compared among snakes from four Hungarian and two Ukrainian sites. Mhc band-sharing was 85–100% within Hungarian sites, compared to 57–63% within Ukrainian sites; also band number (reflecting allele diversity) was lower at Hungarian compared with Ukrainian sites.

An analysis of Swedish populations of *Vipera berus* differing in size and degree of isolation have been especially informative. Reduced allozyme heterozygosity, increased genetic similarity among individuals (as indicated by DNA fingerprint band-sharing), and an increased proportion of stillborn young are seen in an isolated population (mean H_o = 0.07, band-sharing D = 0.88, proportion of nonviable young = 0.32) compared to three intact populations (mean H_o = 0.31, band-sharing D = 0.75, proportion of nonviable young = 0.09; Madsen et al. 1995). Furthermore, among five Swedish adder populations, Mhc class I gene band-sharing (but not minisatellite DNA band-sharing) shows a steady decline along a gradient from the most isolated and smallest population (estimated population size = 15–40) to the most continuously distributed and largest population (estimated population size > 100; Madsen et al. 2000). In another study of Swedish adders, scale anomalies increased in frequency with increasing population isolation (Merilä et al. 1992).

A clear analysis of the impact of population bottlenecks on molecular genetic and morphological variation involves two native and two introduced populations of *Natrix tessellata* in Switzerland (Gautschi et al. 2002). Introduced populations were established at Lake Alpnach in 1944–1945 with 20–25 founders and at Lake Brienz in the late 1950s with approximately 60 founders obtained from Lake Alpnach (Gautschi et al. 2002). Thus, Lake Alpnach experienced one and Lake Brienz experienced two bottlenecks: one associated with the founding of the Lake Alpnach population and a second associated with the founding of the Lake Brienz population (Gautschi et al. 2002). Consistent with this history, introduced populations exhibited fewer alleles at eight microsatellite loci (2.4 per locus at Lake Brienz and 3.1 at Lake Alpnach compared to 4.9 and 5.4 for native populations at Lake Lugano and Lake Garda). Introduced populations also exhibited lower heterozygosity and more frequent scale abnormalities, suggesting that reduced genetic variability may have fitness consequences through its effects on developmental stability (Gautschi et al. 2002).

In an analysis of 16 island and mainland populations of tiger snakes (*Notechis scutatus* × *ater* complex), H_o at 23 allozyme loci was positively correlated with population size controlling for degree of isolation (Spearman rank partial correlation = 0.58, df = 13, P = 0.024) and negatively correlated with degree of isolation controlling for population size (Spearman rank partial correlation = –0.76, df = 13, P = 0.001; data from Schwaner 1990, Table 1). These results indicate that small isolated populations exhibit lower levels of genetic variation than do larger less isolated populations. In addition, frequency of scale and skeletal anomalies were negatively correlated with observed allozyme heterozygosity (Schwaner 1990).

Direct evidence of the fitness benefits of genetic diversity is provided by the observation that hatching success is positively correlated with levels of multiple paternity within clutches in *Vipera berus, Liasis fuscus,* and *Pantherophis obsoletus* (Madsen et al. 1992; Olsson and Madsen 2001; Madsen et al. 2005; Blouin-Demers et al. 2005). But such effects are not universal. In *Nerodia sipedon,* multiple paternity is uncorrelated with the proportion of infertile ovules or stillborn young (Prosser et al. 2002).

Together, these studies suggest that snake populations numbering on the order of a few tens of individuals may be at risk for loss of genetic diversity and reduced fitness. Examples involving the loss of variation in Mhc genes are of special interest because these genes play a central role in immune function (Madsen and Újvári 2006). Although a link between Mhc variation and disease resistance in snakes has not been established, wildlife disease ecology and genetics is an emerging subdiscipline and has become a standard component of some conservation programs (Travis et al. 2006; Whiteman et al. 2006). For example, parasite loads are higher and antibody levels are lower in more highly inbred Galapagos Hawk (*Buteo galapagoensis*) populations. Furthermore, hawk genetic diversity is positively correlated with island size, providing evidence of direct linkages among population size, genetic variation, and disease resistance (Whiteman et al. 2006). Comparable analyses of snake populations may reveal a greater generality of this phenomenon and identify important management concerns.

Genetic Restoration

One management strategy for countering losses of genetic diversity in small isolated populations is genetic rescue. Through genetic rescue, managers seek to increase fitness by introducing genes from other populations (Tallmon et al. 2004b). This is one component of a more general management strategy, termed genetic restoration, that seeks to address not just the detrimental effects of inbreeding but also patterns of neutral and adaptive variation (Hedrick 2005a). This more general strategy recognizes that efforts aimed at genetic rescue may result in the loss of adaptive alleles and genotypes (outbreeding depression) and neutral variation that may be adaptive in future environments.

One of the best-known cases of genetic rescue involves the isolated population of *Vipera berus* in Sweden discussed earlier (Madsen et al. 1995, 2000). In response to decreasing population size, loss of genetic variability, and an increasing proportion of nonviable young, 20 male adders from large and genetically variable populations were released into the isolated population in 1992, where they remained for up to four breeding seasons (Madsen et al. 1999). Evidence for increased genetic variation was clear. Band-sharing decreased from 0.96 among seven male adders sampled prior to the introduction to 0.84 among seven newly recruited males sampled in 1999 (calculated from Madsen et al. 1999, fig. 1). Furthermore, population size has shown a consistent pattern of recovery since then (Madsen et al. 1999, 2004).

Evolutionary Responses to Changing Environments

Recent analyses of microevolutionary processes demonstrate that evolution can occur rapidly and on fine geographic scales (reviewed by Reznick et al. 2004); this has implications for conservation and management (Crandall et al. 2000; Ashley et al. 2003; Stockwell et al. 2003; Stockwell and Ashley 2004). Snakes provide a number of examples of rapid evolution on fine geographic scales. Color pattern differences between Lake Erie island and mainland watersnake populations represent a dynamic balance between natural selection and gene flow over distances of 30 km or less and have arisen over a period of approximately 4000 years (King 1993a, 1993b; King and Lawson 1995; Ray and King 2006). Even more rapid evolution has been documented in an isolated Swedish population of *Vipera berus*. Individuals with an unusual color pattern ("blue" morphs) first appeared in this population in 1983 (Madsen and Shine 1992). This color morph was associated with higher growth rates and mating success in males, and in subsequent years, the frequencies of the blue morphs increased to 35% in males and 20% in females, with a concomitant 3-cm increase in snout vent length in both sexes. Neither the genetic basis of the blue morph nor its linkage to body size is known, but its rapid appearance and increase in frequency suggests a relatively simple (one-locus) inheritance (Madsen and Shine 1992).

Invasive species represent one kind of environmental change that can induce rapid evolutionary responses. The introduction and spread of *Rhinella (Bufo) marina* in Australia provides a case in point. This species is toxic to many vertebrate predators, including some snakes. Because snakes are gape-limited, individuals with small heads relative to their body size consume relatively smaller toads and thus receive lower doses of bufodienolide toxins (Phillips and Shine 2004). This has resulted in an apparent selection for increased body size and smaller relative head size in two toad-eating snakes (*Pseudechis porphyriacus* and *Dendrelaphis punctulatus*) that are sensitive to toad toxin but not in two other species that are either too small (*Hemiaspis signata*) to consume cane toads or are insensitive to toad toxin (*Tropidonophis mairii*; Phillips and Shine 2004). Furthermore, a comparison of *P. porphyriacus* populations that are sympatric and allopatric with toads suggests the evolution of an increased resistance to toad toxin (a standardized dose resulted in a 30% compared to an 18% reduction in swimming speed among snakes from allopatric vs. sympatric sites) and a decreased willingness to consume toads (100% compared to 50% among snakes from allopatric vs. sympatric sites; Phillips and Shine 2006). Given the short history of cane toads in Australia (since 1935), the rapidity of these evolutionary responses is remarkable. As the number and impacts of invasive species increase (see King et al. 2006b for another example involving snakes consuming invasive prey), such evolutionary responses may become commonplace. Anticipating these responses and identifying conditions under which evolutionary responses are unlikely and, hence, in which invasive species may, instead, result in population declines or extinction of natives represent important challenges to integrating evolutionary and conservation biology (Frankham and Kingsolver 2004).

Conservation Genetics of Captive and Exploited Populations

Human exploitation of populations in nature and propagation in captivity both can impose selection for traits not otherwise favored by natural or sexual selection. For example, the commercial harvest of fish and trophy hunting for large mammals have resulted in evolutionary changes in exploited populations (Coltman et al. 2003; Baskett et al. 2005). By extension, intense exploitation of some snake populations (pythons harvested for the skin trade and rattlesnakes harvested during roundups; Keogh et al. 2001; Shine et al. 1999a, 1999b; Fitzgerald and Painter 2000) might be expected to induce evolutionary changes in body size, behavior, or life history. Captive propagation can result in inadvertent selection that may affect the success of repatriation and augmentation efforts (e.g., Heath et al. 2003; Wisely et al. 2002; McPhee 2003; Frankham 2008). Thoughtful design of captive propagation programs (e.g., by minimizing the number of generations prior to release), however, can reduce such effects (see Kingsbury and Attum, Chapter 7).

Captive propagation can also lead to losses of genetic variability and result in inbreeding depression and loss of adaptability to future environmental change. These effects may impact captive populations directly as well as free-ranging populations reestablished or augmented with captive-bred individuals. Minimizing these effects involves acquiring a sufficient number of founders during establishment of a captive population, maximizing the rate of increase following establishment, implementing a breeding program that ensures that all individuals contribute equally to reproduction, and minimizing the number of generations in captivity (Allendorf and Luikart 2007; Leberg and Firmin 2008). Such considerations contribute to the design of Association of Zoos and Aquaria (AZA) Species Survival Plans (SSP), but to date only three snakes have been designated SSP species: the Aruba Island Rattlesnake (*Crotalus unicolor*), Eastern Massasauga (*Sistrurus c. catenatus*), and Louisiana Pinesnake (*Pituophis ruthveni*) (http://www.aza.org/).

Identification of Management Units

Molecular genetic data, combined with phylogeographic analyses (Burbrink and Castoe, Chapter 2), are sometimes used to identify genetic units in nature that might warrant separate conservation and management efforts. Questions sometimes arise, however, regarding the taxonomic status of these units and their utility as management units. For example, molecular and morphological data led Burbrink and coworkers (Burbrink et al. 2000; Burbrink 2001) to conclude that *Pantherophis obsoletus* probably consisted of three species, corresponding to three molecular clades. More recent analyses of populations from southern Ontario revealed an overlap between eastern and central clades with no detectable selective barriers to hybridization (Gibbs et al. 2006). Thus, although eastern and central clades differ in mtDNA, they do not appear to differ in ecologically significant traits. As a consequence, a management plan based solely on genetic lineages might jeopardize the occurrence of gene flow between units that serves as a significant source of genetic variation. Phylogeographic studies will continue to be important in identifying cryptic species and clarifying the historical processes responsible for observed patterns of genetic variation (see Burbrink and Castoe, Chapter 2). Management decisions also need to account for population-level processes contributing to local adaptation (Crandall et al. 2000; McKay and Latta 2002).

Management of Gene Flow among Fragmented Populations

The potential for loss of genetic diversity and reduction in fitness in small isolated populations can be counteracted by management practices that facilitate movements among populations via habitat corridors or translocation. A number of questions, however, arise in the implementation of such practices: How much gene flow is desirable? Which populations are appropriate sources of translocated animals? What risks (e.g., disease transmission and outbreeding depression) might be associated with such practices? Answering these questions is difficult (see Kingsbury and Attum, Chapter 7). The work of Wright (1931) resulted in the conventional wisdom that gene flow occurring at a rate of one individual per generation can serve to prevent the random loss and fixation of alleles via genetic drift. Recent analyses suggest that this is an oversimplification and that the actual rate depends on a variety of parameters, such as population size structure, age structure, and sex ratio (Mills and Allendorf 1996; Vucetich and Waite 2000, 2001; Wang 2004). In addition, a better understanding of patterns of gene flow in nature (e.g., the existence of natural corridors and barriers; Madsen and Shine 1998; Manier and Arnold 2005), the significance of anthropogenic barriers (e.g., roads; Bernardino and Dalrymple 1992; Rosen and Lowe 1994; Ashley and Robinson 1996; Andrews and Gibbons 2005; Roe et al. 2006), and the utility of specific kinds of corridors (Yanes et al. 1995; Roe et al. 2004) is needed.

Future Directions in Snake Population and Conservation Genetics

The empirical knowledge-base of snake population and conservation genetics is growing rapidly and, as the taxonomic and geographic breadth of studies increases, a much deeper understanding of the topics reviewed here will be gained. In addition, we can look forward to the application of new techniques to the study of snake population and conservation genetics. Among other things, a greater synthesis of quantitative genetic (Brodie and Garland 1993) and population genetic approaches could provide a better understanding of the evolutionary processes influencing ecologically significant traits. For example, quantitative trait locus analysis of well-studied traits such as vertebral number and defensive behavior (Brodie and Garland 1993; Burghardt and Schwartz 1999) would be of interest. Such analyses could provide insights into the number of genes involved and the mechanisms of gene expression and would aid in distinguishing among common ancestry, convergence, and parallel evolution as explanations for population- and species-level similarities.

Population genomics represents another avenue with great potential for increasing our understanding of population genetics (Luikart et al. 2003). This method allows researchers to distinguish evolutionary processes influencing the genome as a whole (e.g., genetic drift and gene flow) from processes that influence specific genome regions (e.g., natural selection and assortative mating). Furthermore, when combined with genome map information, population genomics can be used to identify candidate genes responsible for adaptive differences among populations. Although currently most feasible in model organisms for which large-scale sequence data exist, future applications to semi-model organisms are likely (Luikart et al. 2003). Among snakes, such an approach could be particularly informative in systems such as that provided by *Thamnophis elegans* in northern California, in which ecotypic differentiation in life history, color pattern, scalation, and diet exists on a fine geographic scale (Kephart and Arnold 1982; Bronikowski and Arnold 1999; Bronikowski 2000; Manier and Arnold 2005).

The interpretation of geographic patterns of variation in neutral and ecologically significant traits would benefit from the application of landscape genetics (Manel et al. 2003; Manni et al. 2004), geostatistical analysis (Thompson et al. 2005), and causal modeling (Cushman et al. 2006). Such techniques can aid in linking individual-level (vs. population-level) patterns of genetic variation with geographic and environmental features (Jenkins et al., Chapter 4) and provide a better understanding of the impacts of natural and anthropogenic features on patterns of gene flow and fragmentation. In one example of landscape genetic analysis, F_{ST} did not correlate with straight-line geographic distance among *Crotalus horridus* hibernacula but did correlate significantly with a measure of geographic distance that accounted for the amount of basking habitat between hibernacula (Clark et al. 2007).

The application of population genetic principles to snake conservation would benefit from comparative analyses of the genetic characteristics of continuously distributed and fragmented populations and of stable and declining populations. Furthermore, programs of genetic monitoring (Schwartz et al. 2007) might be initiated to provide ongoing assessments of the genetic health of populations of conservation concern. In addition, experimental analyses of the genetic consequences of potential management strategies are needed. Among other things, these might focus on the utility of habitat corridors and translocation in countering the genetic effects of habitat fragmentation and on the genetic consequences of captive breeding, augmentation, translocation, and reintroduction (see Kingsbury and Attum, Chapter 7). Finally, a broader understanding of population differentiation in ecologically significant traits is needed. Such information would complement analyses of neutral molecular markers and would aid in identifying biologically based management units (cf. Crandall et al. 2000; McKay and Latta 2002: Hedrick 2004), designing captive-breeding and translocation efforts, and assessing the potential for future adaptation to changing environmental conditions.

4

Modeling Snake Distribution and Habitat

CHRISTOPHER L. JENKINS, CHARLES R. PETERSON,
AND BRUCE A. KINGSBURY

Modeling the distribution of organisms is a diverse, active field (Morrison and Hall 2002; Scott et al. 2002a). In the last few decades, technological developments (e.g., geographic information systems [GISs], global positioning systems [GPSs], and remote sensing) and conceptual advances (e.g., application of scale, multivariate statistics, and geostatistics) have made modeling feasible. We use the term GIS to refer to a general technical approach (i.e., the combination of hardware and software) or to a specific system to store, process, analyze, and visualize spatial data. A GIS can be used to combine logical and statistical analyses of environmental and animal data to visualize spatial and habitat relationships. Such models can vary from fine to broad spatial scales (e.g., from an individual animal's activity area to the geographic range of a widespread species). The habitat relationships in the models may vary from simple to complex, for example, from the cover types used by a species to combinations of multiple environmental gradients (temperature, moisture, radiation, and so forth) and animal characteristics (developmental thermal tolerances and behavioral preferences). The overall modeling process results in algorithms, equations, and maps describing the distribution and habitat relationships of animal species that have a wide variety of conservation applications. The maps are also effective for communicating results to policymakers and land managers because they are often more easily understood than abstract formulae or qualitative text (Beauvais, pers. comm.). Thus, results from modeling help in the application of distribution and habitat information to actions benefiting conservation, but considerable care is required to select, assess, and apply the models appropriately.

Before we discuss in detail modeling snake distribution and habitat, let us first define what we mean by the terms *distribution* and *habitat* and highlight the difference between these terms and the term *range*. We follow the definitions used by Beauvais (pers. comm.).

> *Range*. The extent of the study area occupied by individuals, estimated as the sum of all map units with high likelihoods of occupation during a defined period of time with little or no consideration of the underlying environmental variation (rather coarse resolution).
>
> *Distribution*. The spatial arrangement of environments suitable for occupation by individuals, estimated as the subset of all environments in the study area that are regularly occupied during a defined period of time (finer than range maps).
>
> *Habitat*. Environments with the combination of resources and conditions that promote occupancy, survival, and reproduction by individuals during a given time period.

In this chapter we describe how models of snake distribution and habitat can be developed and applied to conserving snake populations. We focus on modeling that combines logical, statistical, and geostatistical analyses of environmental data and snake characteristics within a GIS to quantify and visualize the distribution and habitat of snakes (Fig. 4.1). This modeling is important to snake conservation because habitat loss, alteration, and fragmentation are among the most important causes of snake population declines (Gibbons et al. 2000). GIS models can be an important aid in visualizing changes in distribution, thus helping researchers to identify snake populations threatened with declines. Furthermore, changes in spatial distribution of snakes may prove useful in testing hypotheses concerning snake population declines, as changes in distribution have for some amphibian population declines (Davidson et al. 2002). Compared to the literature on birds and mammals, however, that on snake distribution and habitat modeling is quite limited. Consequently, this is an area in which important contributions to snake conservation can be readily made.

In this chapter we address seven questions that we consider important in developing and applying snake distribution and habitat models to conserving snake populations: (1) Why model snake distribution and habitat, (2) what are the appropriate scales for model development, (3) what snake and environmental data are useful for developing distribution and habitat models, (4) what are the main qualitative and quantitative approaches available for relating snake occurrence to spatial locations and habitat, (5) how is a GIS used to map distribution and habitat, (6) how can the appropriate models be selected and assessed, and (7) how can snake distribution and habitat models be appropriately used in snake conservation efforts? Representative examples and a case study are described to address these

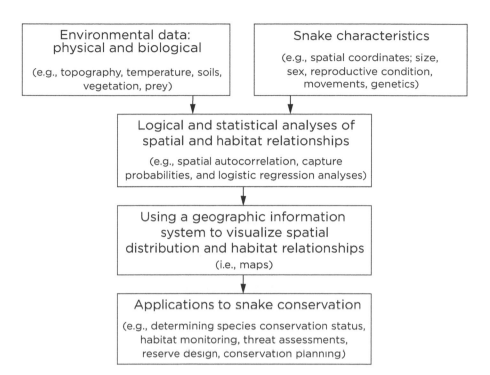

Fig. 4.1. Conceptual model for applying modeling to the conservation of snakes

questions. Throughout the chapter, we prioritize what we consider to be important areas for future research and some approaches likely to be useful in answering these questions.

Why Model Snake Distribution and Habitat?

A wide variety of reasons to model snake distribution and habitat exist, including providing a framework for summarizing and synthesizing information on the occurrence and effects of various factors on distribution; identifying gaps in the information needed for the modeling, thus helping to direct future research efforts; and generating and testing hypotheses concerning the effects of various factors on distribution.

Specific applications of snake distribution modeling include:

• Visualizing the spatial distributions of snakes and habitat. Knowing where species are likely to occur is a basic part of natural resource management (Rushton et al. 2004).

• Planning field inventories, surveys, and habitat studies (Dorcas and Willson, Chapter 1). For example, using a GIS to develop a stratified sampling scheme for reptiles on the Idaho Army National Guard training area allowed us to increase sampling efficiency as well as providing results that were used for species distribution models/maps (Peterson et al. 2002).

• Monitoring changes in distribution and habitat and inferring changes in snake populations (Beaupre and Douglas, Chapter 9). A broad picture of changes in snake distribution may be possible from modeling even if historical data are insufficient to empirically describe actual population changes. For example, over 30% of the native shrublands in southern Idaho have been lost in the last 120 years (Wisdom et al. 2000), but this has affected different snake species in different ways. Racers (*Coluber constrictor*) and Gophersnakes (*Pituophis catenifer*) appear to have lost less than 5% of their estimated suitable habitat, whereas Nightsnakes (*Hypsiglena torquata*) have lost about 33% and Groundsnakes (*Sonora semiannulata*) have lost over 50% (C. R. P. and C. L. J., pers. obs.; Svancara, pers. comm.).

• Analyzing the effects of landscape characteristics on gene flow and connectivity (King, Chapter 3). This application could aid in the identification of functional populations and the effects of natural and anthropomorphic landscape features on snake movements.

• Predicting the possible effects of various disturbances (e.g., fire, invasive species, and urbanization) on distribution. This is of obvious importance when researchers evaluate the consequences of planned or unplanned disturbances.

• Incorporating spatial relationships into population viability analyses (PVAs; Dorcas and Willson, Chapter 1; Shine and Bonnet, Chapter 6). For example, software programs such as RAMAS GIS integrate the results of distribution and habitat models to delineate local populations and to parameterize demographic and dispersal characteristics into spatially explicit PVAs (Kingston 1995).

• Identifying threats to snake populations. The Wildlife Conservation Society has developed an approach to conservation planning called the Landscape Species Approach that models and compares the distribution of species and threats to identify focal areas for conservation (Sanderson et al. 2002).

• Identifying species of conservation interest. This may include identifying species with limited ranges, with disjunct populations, on the edges of their range, or having suffered significant decreases in their distribution within the area of interest.

• Identifying and prioritizing areas for conservation efforts (e.g., hot spots, reserve selection, or the siting of development activities to minimize impacts on snake populations). For example, the Idaho GAP Analysis Project identified the Snake River Canyon and Plain in southwestern Idaho as a region with high snake species richness (Fig. 4.2), a relatively high number of species of conservation concern, and relatively little area in conservation reserves (Scott et al. 2002b).

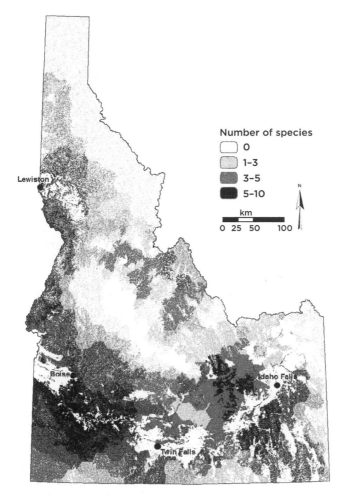

Fig. 4.2. Predicted snake species richness in Idaho from the GAP Analysis Project (logical modeling approach); darker areas represent areas with more snake species

Whatever the reason for modeling snake distribution and habitat, it is critical that the general goals and specific objectives of the project be clearly identified before model development. The purpose of the modeling heavily influences how the models are developed (Beauvais, pers. comm.). For example, broad-scale modeling to indicate regional snake distributions for coarse-scale assessments is more likely to rely on preexisting environmental and snake occurrence data and statistical approaches that require only positive snake occurrence data. In contrast, modeling snake distribution on a finer scale is more likely to rely on specially derived environmental data and a statistical approach that uses positive and negative snake occurrence data gathered in a dedicated field study (e.g., Lee and Peterson 2003).

What Are the Appropriate Scales for Model Development?

Before discussing the importance of scale in modeling snake distribution and habitat, we need to define *scale,* a term that has various definitions (Morrison and Hall 2002). For example, geographers define *scale* as associated with a map (e.g., the scale 1:24,000 indicates that 1 inch on the map equals 24,000 feet); thus, a larger scale means a smaller area is being represented. In contrast, ecologists typically associate larger scales with larger areas. One way to deal with these differences is to use the relative terms *fine* and *coarse scale* to refer to smaller and larger areas, respectively. Although it is typical to think about scale in terms of area, there are actually two distinct components of scale: grain and extent (Li and Reynolds 1995). The *grain* is the finest resolution of the data (e.g., a minimum grid or polygon size) and the *extent* is the scope of the data, typically the size of the study area. Grain and extent are often constrained by the availability of spatial information. In addition, it is important to consider the grain and extent from the perspective of the species or process of interest.

When we design a study to model snake distribution or habitat, an a priori scale should not be set. Rather, with the objectives of the study in mind, the appropriate scale should be determined to describe the relationship between environmental spatial heterogeneity and the snake species, population, or community of interest (Wiens 1989). Not setting an a priori scale is important because the spatial scale can have a significant impact on descriptions of landscape pattern (Trani 2002a) and, thus, how we assess snake distribution and habitat. Studies on a variety of other taxa have included multiple scales and have found that scale has an effect on researchers' detecting how organisms appear to respond to spatial heterogeneity (e.g., Thompson and McGarigal 2002). In some cases, the relationship between an organism and certain characteristics of the environment may change just by examining the relationship at a different scale. An example is the apparent effect of temperature on snake distribution. At broad scales, temperature (as indicated by spatial interpolations of the number of frost-free days) is important in predicting snake distributions (Scott et al. 2002a); at intermediate scales, temperature is not an important predictor of snake distribution (Jenkins 2007); and at fine scales, temperature is again an important predictor of distribution (Huey et al. 1989; Cobb 1994).

When modeling relates to habitat selection in snakes, it important to consider the hierarchical selections being made or being imposed on the animal. We have found Johnson's (1980) suggestions of a hierarchy of selection to be particularly useful in framing how to think about the interaction of scale and extent. The range of an organism within the context of the surrounding landscape is termed the first-order selection. For example, modeling the distribution of the Kirtland's Snake (*Clonophis kirtlandii*), endemic to the midwestern United States, would best be conducted by examining habitat availability across the Midwest rather than across all of North America. The second-order selection occurs when an animal establishes its home range within the local distribution of the species. A riverine snake in a particular watershed establishes a home range associated with the aquatic habitat within the watershed to satisfy its needs, such as hibernacula, feeding sites, and basking sites. Habitat construed as being available to the snake might be just the watershed or perhaps even the bottomlands along the river. Third-order selection is habitat preferences within the home range. Within that home range of the riverine snake, the snake may also use the habitat available disproportionately to address those same needs. The closed canopy forest between the basking areas and the aquatic foraging areas may see little use, but it may constitute a substantial proportion of the home range.

In this chapter, we focus primarily on spatial scale, but it is important to realize that the same issues apply when dealing with temporal scale. For example, if a biologist is interested in the influence of beaver ponds on watersnake (*Nerodia* spp.) distribution, the temporal scale related to wetland succession would be an important consideration, in addition to the spatial components of scale.

We close this discussion by emphasizing the importance of scale in modeling snake distribution. First, instead of setting an a priori research scale, snake ecologists should select a scale that is appropriate for the objectives of the study. For example, when modeling the distribution of Timber Rattlesnakes (*Crotalus horridus*), at coarse scales the presence of large patches of forest may be important to modeling their overall distribution; at intermediate scales certain combinations of geology, slope, and aspect may be important to model overwintering habitat; and at fine scales features such as particular rock formations or canopy gaps may be important for activities such as gestation and foraging. Second, within the scope of the study, snake ecologists should consider using a multiscale approach to model snake distribution and habitat. Like other animals, snakes relate to their environment at multiple spatial scales (Row and Blouin-Demers 2006) and, by taking a multiscale approach, researchers can often gain a better understanding of how animals relate to patterns in the environment (Urban et al. 1987). For example, Harvey and Weatherhead (2006) used a multiscale approach to examine habitat selection in Massasaugas (*Sistrurus catenatus*) and found that these snakes had apparent differential preference based on scale. These snakes selected areas close to retreat sites and shrubs at fine scales and used open habitats at coarse scales. Third, when possible, snake ecologists should conduct sensitivity analyses to determine the influence of scale on their study. These sensitivity analyses can be conducted by varying the grain and/or extent and examining changes in the relationships between snakes and their habitats.

Which Snake and Environmental Data Are Useful for Developing Distribution and Habitat Models?

The development of snake distribution and habitat models requires information about snakes and the mapped physical and biological characteristics of their environments. Here we describe the main types of data that can be used in modeling the distribution and habitat of snakes. We do not cover data management procedures, but simply note that the quality of the final model will depend on the quality of the data. Many of these data already exist and can be obtained from a variety of sources. Some types of data can be derived from existing sources (e.g., elevation, slope, and aspect can be derived from Digital Elevation Models, DEMs). Computer programs such as FRAGSTATS (McGarigal and Marks 1995) can calculate many landscape metrics (e.g., patch size and number) from cover-type maps. For some modeling, the investigator may need to acquire the data directly (e.g., remotely sensed imagery for creating a cover-type map, or a habitat study of the cover types in which snake species occur on a wildlife refuge). These data may be represented within a GIS in either raster or vector formats (Demers 2002). It is important to be aware of the spatial and temporal scales of each of the data layers used because they will affect the scales at which the model may be appropriately applied.

Snake Data for Modeling

The most commonly used snake data for modeling distribution are spatial coordinates. These data can be obtained from a variety of sources, including the literature, museum records, state or regional herpetological databases, field studies, and contributed observations. Other sources of information include Nature Serve (http://www.natureserve.org) and state Gap Analysis projects (e.g., http://www.gap.uidaho.edu/). Some sources provide only positive data (i.e., locations where snakes are known to be present), whereas other sources may provide negative snake data as well. Some statistical analyses require only presence data, whereas others require both presence and absence data.

If the snake locality data have not been digitized, the researcher will need to organize them into a tabular format and convert them into digital coordinates (e.g., decimal degrees or universal transverse Mercator coordinates). Increasingly, locations are being reported with coordinates measured with a GPS receiver. Since May 2000, when intentional degradation of GPS signals was discontinued by the U.S. Department of Defense, accuracies better than 20 m can usually be obtained with even relatively inexpensive GPS receivers. For older observations and museum specimen records, however, we typically use the written descriptions to find locations on digital topographic maps (digital raster graphics, DRGs) on a computer monitor. The use of a geographic feature index (e.g., U.S. Geological Survey Geographic Names Information System, USGS GNIS) greatly facilitates finding features on the map and reduces the possibility of confusing different features with the same names (e.g., 30 of 44 counties in Idaho have a feature named Spring Creek). We assign each record location an accuracy rating that represents the radius of a circle around the digitized point in which we are highly confident that the actual specimen location occurred. Detailed procedures for digitizing animal locations can be found on the HerpNet website (http://www.herpnet.org/).

Locality data can be used to generate point distribution maps for individual species. If reptile surveys have been previously conducted in the area, plotting the locations of all survey sites can help determine whether a lack of locality data for a particular species is due to species rarity or to insufficient survey effort. Including all locations where reptiles of similar detectability were observed can also help address this question.

Several other types of snake data also may be used for modeling. If the appropriate information exists, functional/behavioral classification (Weatherhead and Madsen, Chapter 5) of snake locations may be possible (e.g., overwintering, estivation, breeding, gestation, molting, and foraging areas) and can aid in modeling different types of habitat (Waldron et al. 2006; Jenkins 2007). Movement patterns and home-range estimates from mark-recapture and radiotelemetry studies may also be of value in distribution and habitat modeling (see Dorcas and Willson, Chapter 1). For example, we have used the distances that telemetered Western Rattlesnakes (*Crotalus oreganus*) migrate away from communal overwintering sites to develop finer-scale distribution maps on the Idaho National Laboratory in southeastern Idaho (C. R. P., pers. obs.). GIS landscape analyses of gene-flow estimates among snake locations (e.g., communal den sites) represent a type of data that may be useful for modeling distribution and habitat, but these techniques, to our knowledge, have not yet been applied to this purpose. Expert opinion is another source of information on snake locations, movements, and habitat use that is often used in developing logical models of snake distribution and habitat.

Environmental Data

A wide variety of environmental data is often used in modeling snake distribution and habitat. These data can be used to describe the environmental characteristics of grid cells or polygons across a study area and of snake locations. Some of the environmental data types used for modeling snake distribution and habitat include topography, hydrology, temperature, cover types (vegetation, water, talus, etc.), land use (urban, agriculture, nature reserve, etc.), land ownership, disturbances (fire, grazing, logging, etc.), and infrastructure (roads, buildings, power lines). In addition, distribution and habitat models for other species such as prey, predators, competitors, parasites, and pathogens may also be useful in modeling snake distribution.

When modeling snake distribution and habitat, it can be useful to combine multiple environmental data types to develop environmental type maps. Environmental type maps are combinations of environmental characteristics such as topographic classes and cover types. Environmental type maps can be used to stratify sampling and in predictive modeling.

Because snakes generally use their environments at a relatively fine spatial scale, identifying small habitat features (such as cave openings, vegetation patches, and ephemeral wetlands) is often important. Distinguishing such habitat features often requires data with both high spatial resolution (< 5 m) and high spectral resolution. Although such fine-scale data are becoming increasingly available, the lack of data appropriate for identifying and classifying reptile habitat is still a major limitation on modeling snake distribution. New types of remotely sensed data such as laser altimetry or light detection and ranging (lidar) (Lefsky et al. 2002) and hyperspectral imagery (Jia and Richards 1999) should prove useful for generating detailed habitat maps. Lidar can be used to generate fine (e.g., 20-cm vertical resolution) topographic maps that can be used to identify many habitat features (e.g., talus slopes and boulders) and measure vegetation height, cover, and canopy structure (Lefsky et al. 2002). Hyperspectral imagery provides hundreds of spectral bands, allowing researchers to distinguish fine-scale patterns in the environment.

Some final cautions include the need for consistency in the data used. Data derived for large areas, even when ground-truthed (checked on location) extensively, will have some degree of error. Maps constructed by the researcher for a study site will also have some degree of error. Most important here, such spatial representations must be constructed with equivalent effort across the study site. Second, it is also important to realize that the availability of data will often constrain our ability to use the appropriate scale for the given objectives. Finally, if assigning the environmental characteristics to snake locations from a data layer in a GIS (e.g., cover type), be certain to consider possible changes in the environmental feature from the time when the snake location and the environmental features were quantified.

What Are the Main Qualitative and Quantitative Approaches Available for Relating Snake Occurrence to Spatial Locations and Habitat?

A variety of logical, statistical, and geostatistical approaches to modeling distribution exist (Table 4.1). Logical models are based on the presence or absence of snakes in an environmental feature. Most statistical approaches relate the probability of snakes occurring in a location with a particular

TABLE 4.1
Modeling types that can be used for modeling snake distribution and habitat, with examples of their applications

Model Types	Examples and Sources	Environmental Data Used Data	Snake Data Used
Rule-based	Idaho GAP analysis (Scott et al. 2002b)	Cover type map	Museum records, literature, survey data, expert opinion
Trapping/observation probability	Modeling probability of occurrence for Racers at the Orchard Training Area, Idaho (Peterson et al. 2002)	Environmental type map (based on topography and cover)	Probability of capture for environmental types
Mahalanobis distances	Modeling Timber Rattlesnake hibernacula in Arkansas (Browning et al. 2005)	Multivariate spatially explicit data set (e.g., elevation, slope)	Presence
Logistic regression	Modeling rattlesnake hibernacula on the Idaho National Laboratory (Cooper-Doering 2005)	Multivariate spatially explicit data set (e.g., slope, geology)	Presence/absence
CART	Modeling snake occurrence in Michigan Standora and Kingsbury 2002)	Soils, geology, presettlement and current habitats, and anthropogenic features	Presence/absence
Maximum entropy	None known		
Compositional analysis	Modeling habitat selection by Massasaugas in Indiana (Marshall et al. 2006)	Cover-type map	Radio telemetry locations
Euclidean distance analysis	Habitat use by Eastern Massasaugas (Bieser 2008; DeGregorio 2008)	Cover-type map	Radio telemetry locations
Geostatistics	Modeling Gartersnake occurrence at Craters of the Moon National Monument, Idaho (Lee and Peterson 2003)	None	Presence/absence

Notes: CART, classification and regression trees.

environmental value or combination of values. Geostatistical approaches are based on the spatial autocorrelation of snake occurrences rather than on correlations with environmental conditions.

Logical (Rule-Based or Boolean) Approaches

Given some understanding of a snake's ecology, a set of rules may be derived from combining the data available for environmental variables with species presence data, analyzing the variables, and combining only those with high explanatory power to derive a predictive map. This Boolean approach uses simple additive or multiplicative strategies and the commands AND, OR, and NOT. The output of each rule test is a true or false binary map in which the true areas (value of 1) met the criteria of the request and false areas (value of 0) did not contain all the requested variables. The values of habitat variables chosen as good indicators of species presence can be known a priori or derived from survey findings or radiotelemetry. Each layer (e.g., wetlands) will have several categories (e.g., shrub-scrub, emergent) that can be analyzed to determine which aspect of each layer is preferred. Knowing the category/value of each layer that is suitable makes it possible to create a Boolean string across many layers requesting a selection of areas within the study area that meets all good habitat criteria. Boolean models are relatively intuitive to create and understand, and they have straightforward interpretations. Consequently, they have particular value as an exploratory tool. They also have the benefits that they can be created using presence data only and may use qualitative or quantitative data. Although they tend to make commission errors (i.e., false positives), they are excellent for excluding the unsuitable areas of a landscape. They are also good in cases in which a commission error is desired to protect rare species whose presence is difficult to detect.

Statistical Approaches

Statistical models are often important in predicting snake distribution and habitat because they provide an understanding of how snake distribution and habitat relate to environmental characteristics. These relationships can be used subsequently to produce spatially explicit predictions of snake distribution if the environmental data exist in a GIS.

Trapping/Observation Probability
Calculating trap or observation probabilities for sampling locations is a straightforward technique for assigning probabilities of snake occurrence to sampling locations when the sampling design was based on environmental characteristics, such as environmental types (see data section). These probabilities can then be combined with an environmental type map of the study area to generate a map of the probability of snake occurrence. The number of probability categories is determined by the number or replicates per environmental type. This is a particularly useful technique when the number of sampling sites with positive data is too small to be used with other statistical techniques. For example, even though we captured only two Long-Nosed Snakes (*Rhinocheilus lecontei*) at 36 trapping sites on the Idaho National Guard Orchard Training Area, we still were able to generate a predicted distribution map for this species that identified important habitat areas (Peterson et al. 2002).

Mahalanobis Distances

Mahalanobis distance is a technique that uses the characteristics at locations where snakes are present to predict their distribution in areas where it is not known whether they are present (Mahalanobis 1936). Mahalanobis distances are based on distances between any vector describing habitat variables at a given pixel and the mean vector for sites the species used. Thus, it provides the user with a measure of how similar the combination of independent variables (e.g., slope, aspect, or vegetation) at a given location is to the mean value of all used locations. An advantage of using this approach for modeling is that it uses only positive data. This is beneficial because often negative data do not exist and it can be difficult to characterize an area as not used. In addition, the approach allows for the inclusion and interaction of multiple independent variables.

One variation of this modeling approach, partitioned Mahalanobis distances, has been used to model the distribution of snake hibernacula. Partitioned Mahalanobis distances allow the user to select a set of informative principal components that relate to the species's requirements. Principal components that represented slope, aspect, and elevation were used to model areas that had high probabilities of occurrence for Timber Rattlesnake (*Crotalus horridus*) hibernacula in northwest Arkansas (Browning et al. 2005).

Logistic Regression

Logistic regression is a form of regression that uses the presence/absence data of snakes and a series of independent variables to predict snake distribution (Hosmer and Lemeshow 1989). Results from logistic regression analyses provide probabilities of snake occurrence based on environmental characteristics. One advantage of logistic regression is that it can model the probability of getting a given state for the binary dependent variable (e.g., present) given the values of the environmental conditions at a particular location. A number of studies have used this technique to model snake habitat selection (e.g., Cross and Peterson 2001) and distribution (Peterson et al. 2002).

One recent advance in logistic regression modeling is the recognition that it is important to incorporate the probability of detecting an animal into the models (MacKenzie et al. 2006; Dorcas and Willson, Chapter 1). If detectability is not incorporated into the modeling, the results will probably underestimate the occurrence of snakes because the probability of detecting a snake is rarely 1.0. In addition, detectability can vary depending on a variety of factors (e.g., observers, weather conditions, and habitat conditions). Ways to estimate detectability include sampling each location multiple times in locations where snakes are known to occur. Future studies modeling snake distribution should incorporate detectability into the models. Software (e.g., PRESENCE) exists that allows scientists to model species occurrence using logistic regression and to incorporate detectability into these models.

Classification and Regression Trees

Classification and regression trees (CART) models (Breiman et al. 1984) can be used to predict the distribution of snakes using a series of IF THEN statements. CART models discriminate between sites where snakes are known to occur and where they are assumed to not occur based on independent variables. This process repeats itself multiple times, splitting the data set into smaller subsets based on the values of independent variables. For example, when using vegetation characteristics to predict rattlesnake distributions a CART analysis might predict that rattlesnakes generally occur in areas with shrub cover greater than 25%. In areas with shrub cover equal to or less than 25%, rattlesnakes may occur more frequently in areas with native grass understories than in areas with exotic grasses. CART models have a number of advantages, including the simplicity of the results and that they are nonparametric. These characteristics make CART models especially valuable for exploratory data analysis or data mining where there may be no a priori prediction of how vegetation characteristics, for example, relate to snake distribution. Our case study later in the chapter used CART to model the distribution of the Massasauga habitat on the lower peninsula of Michigan (Standora 2002). CART analysis was also used to determine that variables such as urban cover, isolation, and elevation were important for predicting snake distribution in Illinois (Cagle, pers. comm.).

Maximum Entropy

Maximum entropy models can be used to determine a probability distribution for snakes or habitats using prior information. The use of prior information in maximum entropy is similar to Bayesian techniques. This prior information is used to make predictions about incomplete information (i.e., the sampled distribution). This modeling technique has a variety of advantages, including that it generates a probability map, is robust to low sample size, and does not require absence data. To our knowledge, this technique has not been used to model snake distributions and habitats.

Compositional Analysis

Compositional analysis is a technique presented by Aitchison (1986) and Aebischer et al. (1993) to model preference for discrete habitat patch, or macrohabitat, types. (Historically, researchers often used variations of chi-squared analysis to analyze macrohabitat data. Unfortunately, such analyses are inappropriate because of the heteroscedasticity of the macrohabitat data, the nature of the repeated measures inherent in radiotelemetry data, and the variation in what is available to each study animal. Compositional analysis overcomes these challenges and is a technique commonly used to model macrohabitat preference in snakes (e.g., Barlow 1999; Coppola 1999; Hyslop 2001; Harvey and Weatherhead 2006; Marshall et al. 2006; Moore and Gillingham 2006).

Compositional analysis is based on comparisons of the log ratios of the available resources to used resource (Aitchison 1986) and was proposed for habitat use analysis by Aebischer et al. (1993). The approach renders the study animal as the sampling unit, somewhat addressing the repeated measures constraint. Compositional analysis addresses the interplay of habitat avoidance and preference (e.g., the use of A resulting from avoidance of B, or use of A because of the preference for A relative to all other habitats) by using the log-odds ratio of habitat proportions (Aebischer and Robertson 1992), making each habitat-use comparison within an individual independent.

When using compositional analysis to model habitat selection by snakes, we recommend examining preference at different scales, equivalent to second and third orders of habitat selection (Johnson 1980). Such an approach will potentially reveal different levels of habitat selection. Investigators must also not be capricious in identifying the habitat of the snake. Unless there are clear boundaries to a site, such as the shoreline for an aquatic snake, boundaries will necessarily be arbitrarily defined according to researcher-imposed constraints. We have relied on defining the study site as the rectangle that includes all observations, but many alternatives exist. The key is to be consistent and to consider definitions that maximize comparability with other research.

The use of compositional analysis has its limitations. For some species, individuals may have idiosyncratic solutions to microhabitat needs, dramatically weakening the discriminating power of the approach (Sage 2005). Habitats that are relatively rare can have disproportionately large impacts on the results. Harvey and Weatherhead (2006) removed habitats constituting less than 3% of the study site to address this problem. Positional error is also a concern with respect to the accuracy of macrohabitat boundaries and snake locations near edges. We recommend reverting back to the habitat assigned in the field while tracking rather than relying later on the convenience of GPS coordinates and the computer.

Euclidean Distance Analysis

Euclidean distance analysis (EDA) is an emerging technique that provides an alternative to compositional analysis. Historically EDA was used to analyze distance to specific locations and linear features, but Conner and Plowman (2001) pointed out that the approach has utility for analyzing proximity to macrohabitats as well. The approach they present examines the distances between animal positions and the available habitats as opposed to randomly selected locations and their distances to the available habitats. The technique does not require scaling all habitats to a common habitat because the distance data do not have a unit-sum constraint and unavailable habitat types do not place challenges on the analysis. It does have the constraint that it does not differentiate between linked habitats; an attraction to one habitat may cause the researcher to infer an attraction to neighboring ones as well. Analytical outcomes are also influenced by the local juxtaposition of habitat patches, such that most or all habitats may appear "preferred" if animals select areas with high patch heterogeneity as opposed to more homogenous areas.

Geostatistical Approaches

Geostatistics is a field of various modeling techniques that use patterns of spatial autocorrelation to predict spatial distribution (Isaaks and Srivastava 1989). Geostatistics differs from many of the other modeling techniques in that it does not require habitat information to predict distribution. In addition, it is based on the spatial dependence of the data, whereas most traditional statistical techniques assume independence of the data.

One example of a geostatistical modeling technique is kriging, which uses patterns of spatial autocorrelation in occurrence data to predict values in areas that were not sampled. Kriging produces a smooth interpolated map of predicted values. These maps are produced using the relationship between the differences in values at pairs of sample points and separation distances. A figure representing these relationships is called a semi-variogram (Lantuéjoul 2002).

One important characteristic and advantage of geostatistical models is that they allow the user to take advantage of spatial patterns to predict the distribution of snakes and their habitats. Geostatistical models can incorporate presence/absence data or continuous data such as abundance values. In addition, co-kriging can be used to incorporate other information that may be valuable in predicting the distribution of the dependant variable. For example, if vegetation is an important factor influencing the distribution of a snake species, co-kriging can be used to incorporate both spatial autocorrelation patterns in the snake distribution data and the relationship between vegetation and the occurrence of snakes.

Geostatistical models are underused in studies of snake ecology; they have been used so far only to model the distribution of snakes at Craters of the Moon National Monument (Lee and Peterson 2003). For a number of snake species, geostatistical models were more accurate at predicting snake distribution than habitat-based approaches (Lee and Peterson 2003). Future studies on snake distributions and habitat should consider spatial autocorrelation patterns in their data and the use of geostatistical modeling.

How Is a Geographic Information System Used to Map Distribution and Habitat?

One of the most important and useful results from any modeling project with snakes is the development of output maps (e.g., see Fig. 4.2). These maps are useful because most snake conservation issues have a spatial component. In addition, wildlife and land managers and conservationists often have experience working with maps and have experience applying the information presented in them to management and conservation issues. It is important to remember, however, that any output map is only as good as the model used to create it.

Maps can be created in a GIS using the results from logical, statistical, or geostatistical models. Specifically, a GIS can be used to create maps from logical models by combining species cover-type matrices (i.e., matrices that describe the cover types that a species occurs in) and a cover-type map. The cover types that species occur in are then displayed as having the species present, and others are displayed as having the species absent. A GIS can be used to create maps of snake distribution from statistical models by combining the statistical equation from the model with environmental cover types. For example, with logistic regression a number of environmental layers can be overlaid and the combinations of values at each pixel (i.e., location) can be exported into a spreadsheet program. Using statistical analysis software, the model can then be created. The coefficients for each environmental variable in the model can then be entered into a GIS extension (Spatial Analyst) and the GIS will create a map displaying the probabilities of snake occurrence across the study area. Finally, a GIS can be used to create maps of distribution using geostatistical models. Using snake location data in a GIS extension (Geostatistical Analyst) allows the user to develop a spatial autocorrelation model and then use the model to predict snake distribution.

How Can the Appropriate Models Be Selected and Assessed?

Model selection and assessment are important aspects of any study modeling snake distribution. The outputs from models are only as good as the information going in, and thus models that are not appropriately evaluated have the potential to have negative effects on snake conservation. For example, the GAP models produced for Ringneck Snakes (*Diadophis punctatus*) in Idaho indicate that the species has a very patchy distribution. But Ringneck Snakes are fossorial and can be difficult to detect during surveys in Idaho; as a result, this model may have high omission errors (i.e., false negatives). Thus, if the predicted Ringneck Snake distribution map was used in a conservation effort, it would likely not do an adequate job of conserving the species. In addition, the use of inaccurate models is one of the primary reasons modeling has often been criticized.

When creating models of snake distribution, it is important to use model selection criteria for selecting the best model. Differences in what is meant by *best* are often the core differences among approaches to model selection.

There are a number of approaches to selecting the best model, but in this chapter we focus on two: (1) maximum likelihood–based approaches that have been widely used for many years and (2) the information theoretical approach that has become much more widely used in recent years. The first approach (e.g., stepwise selection) involves maximum likelihood techniques such as a series of F tests in which each independent variable is evaluated separately for significance to determine whether it should be included in the model (Sokal and Rohlf 1995). There are a number of criticisms of techniques such as stepwise regression, including the influence of sample size on p values, issues of multiple comparisons when series of F tests are run on the same data set, and the difficulties of interpreting p values given that each is dependent on the previous test. Information theoretical approaches such as Akaike information criteria (AIC) are based on log likelihoods and balance the complexity of the model with the model fit (Burnham and Anderson 2002). There has been a recent emphasis on the importance of their use in herpetology (Mazerolle 2006). Similarly, the use of information theoretical approaches to model selection is becoming widespread in studies of snake ecology (Matthews et al. 2002; Webb et al. 2002b; Thompson and Burhans 2004; Webb et al. 2003; Gregory and Issac 2004; Phillips and Shine 2004; Luiselli 2006; Keogh et al. 2007). We suggest that snake ecologists continue to apply information theoretical approaches when modeling snake distribution and habitat.

Another important aspect of modeling snake distributions is assessing the accuracy of the model. Without accurate models, we cannot produce useful output maps. It is also important to understand that it is not always the overall accuracy of a model that is most important. In some cases, the most useful model may have increased commission (false positive) or omission error. For example, if a conservation plan is being developed for a rare species such as the Eastern Indigo Snake (*Drymarchon couperi*), a model that increases commission error may be a better tool because it will predict a broader area and have a higher probability of achieving conservation goals. Regardless of whether a model is developed to maximize overall accuracy or to introduce bias toward omission or commission error, it is important to state what the accuracy is, how it was estimated, and the logic behind decisions such as introducing a bias toward commission error.

One of the most common approaches to assessing model accuracy is by comparing a predicted snake distribution to an actual data set. It is important that the data used to develop the model (i.e., training data) are independent from the data used to test the model (i.e., testing data; Henery 1994). In most cases, however, one data set is collected and it is then partitioned. Algorithms have been developed to determine the optimal proportion of the data set that should be partitioned into training and testing data sets (see Schaafsma and van Vark 1979). In addition, techniques such as bootstrapping, jackknife procedures, k-fold cross validation, leave-one-out cross

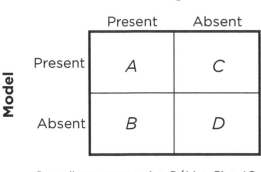

Testing data

		Present	Absent
Model	Present	A	C
	Absent	B	D

Overall accuracy = $A + D/(A + B) + (C + D)$

Producer's accuracy = $A/(A + C)$

Omission error = 100% – Producer's accuracy

User's accuracy = $A/(A + B)$

Commission error = 100% – User's accuracy

Fig. 4.3. Diagram of an error matrix with equations for calculating accuracy and error rates.

validation, and Monte Carlo permutation tests can be used to partition the training and testing data.

The comparison of training and testing data is typically made in an error or confusion matrix. An error matrix plots the presence and absence of a species based on the model against its presence and absence based on testing data (Fig. 4.3). Thus, the matrix displays errors as false positives (Fig. 4.3, box C) and as false negatives (Fig. 4.3, box B). Using error matrices, producer's accuracy (the probability that a true positive is correctly classified) and user's accuracy (the probability a true negative is correctly classified) can be calculated using the equations in Figure 4.3. In addition, omission and commission error rates can be calculated from the matrix to estimate the rate of false positive and false negative errors (Fig. 4.3). (For a more detailed description of error matrices and other error metrics, see Congalton and Green 1999.)

Another important issue when assessing the accuracy of a model is that the probability of occurrence (POC) threshold has a large impact on omission and commission errors. For example, we will have higher commission and higher omission errors when using a POC threshold of .5 (i.e., characterizing a pixel as having a given species present if the probability of occurrence is .5 or higher) than a threshold of .8. One way to determine the optimal threshold is to calculate the area under the curve (AUC) from receiver operating characteristic curves (ROC plots) (Green and Swets 1966; Zweig and Campbell 1993). An ROC plot shows the relationship between the true-positive probability of detection (i.e., sensitivity) and true-negative probability of detection (i.e., specificity) across a range of POC thresholds. The POC threshold with the greatest AUC is the threshold that maximizes the fraction of correctly predicted positives and minimizes the fraction of incorrectly predicted positives.

How Can Snake Distribution and Habitat Models Be Used Appropriately in Snake Conservation Efforts?

One of the most important aspects of any project that models snake distribution and/or habitat is its application to conservation. Snake ecologists should strive to make their modeling results available and, when possible, should work with organizations to ensure that modeling efforts are used to conserve snake populations. In this section, we present a case study that illustrates some of the modeling approaches presented in this chapter and discuss how the results from the studies have been applied to snake conservation.

Modeling Massasauga Distribution and Habitat in the Lower Peninsula of Michigan

The Massasauga is a candidate species for federal listing by the U.S. Fish and Wildlife Service as threatened. The lower peninsula of Michigan has been identified as the area with most of the remaining Massasauga populations in the United States. However, the status of the populations and the extent of their habitat are unclear, the area of interest is large and diverse, and the Massasauga is cryptic and secretive. B. A. K. employed modeling to help predict where to look for the snake and to prioritize areas to protect. The initial, simplistic efforts largely failed, leading to the approaches detailed in this case study, derived from work conducted principally by Michelle Standora (2002; Standora and Kingsbury 2002).

Objectives of the Model
There were a variety of objectives for the modeling efforts. Specifically, models were developed to (1) determine the landscape-level features in Michigan that are indicative of suitable habitat; (2) determine the areas in which Massasaugas should and should not be a management consideration; (3) determine which specific areas are particularly good for the species, both in terms of quality and extent; (4) provide guidance regarding subsequent surveying efforts aimed at delineating remaining populations; and (5) learn about the strengths and weaknesses of the different modeling approaches for use in herpetofaunal conservation efforts.

Issues of Spatial Scale

The extent of the areas examined varied, depending on the modeling approach used. In all cases, however, the areas covered were large, ranging from clusters of counties to the entire lower peninsula of Michigan. The landscape data used was also relatively coarse, often based on 30 × 30-m grid cells. Conversion of vector data into raster data often also resulted in grids that had cells that were 100 × 100 m. From the standpoint of the snake, we were relatively content with the scale of the data we had available. Studies on Massasaugas (reviewed in Johnson et al. 2000) indicated that Massasaugas routinely make moves on the order of this scale. Furthermore, observations of Massasaugas over the years at most locations were incidental, such as a roadkill, and probably not actually within the ideal habitat. Consequently, fine-scale data may be misleading, for example, characterizing a road as selected habitat.

Data Availability

The environmental data used in this model was obtained from the state of Michigan, which maintains an extensive library of GIS layers in a common projection for soils and geology, presettlement and current habitats, and a variety of anthropogenic features. In terms of the snake locality data, Michigan Natural Features Inventory (MNFI) has been monitoring reliable localities of Massasaugas for some time. Consequently, an extensive data set of localities was available dating from as early as 1858 but mostly from the last few decades. Given concerns about declines of Massasaugas in Michigan, field survey and outreach efforts were expanded in the mid-1990s (Legge 1996). The MNFI locality database was supplemented with records from the University of Michigan Museum of Zoology and with surveys we conducted. Replicate observations from the same sites were omitted, so ultimately we arrived at 376 point occurrences.

Modeling

Approach.   We used three modeling techniques, each of which had different strengths and weaknesses: Boolean, classification tree, and partitioned Mahalanobis distance. Boolean models were advantageous because they require only presence data, are simple to construct, and are simple to interpret. We did have a relatively good understanding of which landscape features should influence suitability for Massasaugas (i.e., wetlands: shrub-scrub, emergent, and so forth).

The classification tree method provides a hierarchical classification system that can lead to multiple solutions. Reasonably, many organisms may occur under differing combinations of landscape features, so a model that can identify those different sets of features is particularly valuable when considering extensive, variable landscapes. This approach also has the advantage of producing a clear set of intuitive rules that have been statistically derived.

It does require absence data, which we generated by selecting random locations at least a kilometer away from any known localities.

The Mahalanobis distance technique is another approach that relies on presence data only. It forms of a comparison of the distance in a multivariate space between the suitability of a site versus the ideal habitat. We used the Mahalanobis D^2 method (Duncan and Dunn 2001), which allows the identification of the variables that statistically contribute to the model, so the model can be restricted to only those variables with predictive power. We tested the accuracy of our predictions with subsets of our observations, and this was accomplished by retaining one-third of our locality points as a test set rather than using all the observations in model development.

Output.   The Boolean model proved to be an excellent tool for a preliminary examination of habitat distribution. It identified clear associations with certain landscape variables that encouraged further investigation, identified regions of conservation interest, and helped shape our approach with the classification tree. The model for the entire lower peninsula identified 19% of the state as containing six or all seven of the suitable habitat features and 72% of the test locations fell within those areas (Fig. 4.4). Splitting the lower peninsula into north and south portions led to an increase in area included (32% north and 26% south), but improved prediction rates (77 and 77.5% correct, respectively). The recognition of the distinct differences between the upper and lower peninsulas influenced how we approached subsequent modeling efforts.

The classification tree outputs allowed the derivation of intuitive rule sets to predict the presence of Massasauga habitat. These had value in and of themselves because they were easily interpreted (e.g., "IF wetland type variety is greater than 2.2 within 1 km, AND quaternary geology is NOT of types 12–15, 5, or 6, AND the number of wetlands is greater than 0.15/km², THEN snakes present"). These rule sets were also easily converted into a graphic depicting whether these rules had been satisfied, providing much more specific delineations of quality habitat than the Boolean models (Fig. 4.5). Less than 1% of the state met all the presence rules, but 27% met at least one. The predictive power of the model was about the same as the Boolean models (77%), however, because of increased omissive errors.

The Mahalanobis outputs were the most specific of the three model outputs. Visualization was achieved by mapping the model output as a grid of cells with the dimensionless output values. To produce an output that had intuitive utility, we mapped target areas with values distributed such that the landscape was divided into decilees (tenths) and the best tenth was distinguished from the next best tenth, and so on.

Model Limitations
Each approach had its benefits and limitations. The limitations of the Boolean model were that the output was quite coarse and subject to commission

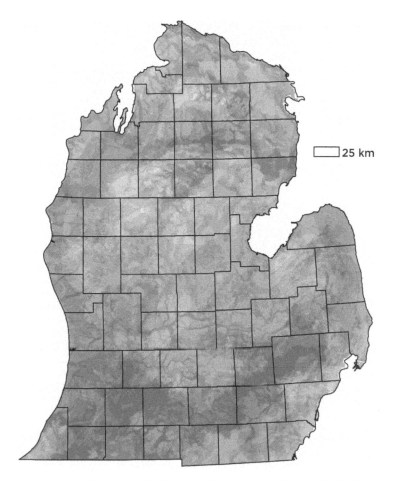

25 km

Fig. 4.4. A representation of the Boolean model output depicting suitable habitats for the Eastern Massasauga in the lower peninsula of Michigan. Amount of shading depicts the addition of up to seven binary habitat layers, with the most suitable habitats shown as the darker shading and the less suitable habitats shown as lighter shading. (Adapted from Standora and Kingsbury 2002)

error. It should, therefore, not be used for teasing out small patches of suitable habitat from a matrix of unsuitable habitat, or visa versa.

The classification tree required both presence and absence data. Although we can be sure of presence when we see a snake, we cannot be certain of absence when we do not see one. So we had to create the absence data to run the model. Although this solution was not ideal, we felt it did not discount the results, only weakened their omissive strength. The Mahalanobis approach created results that were not as intuitive as the other models because the output values are the distance values taken from a multivariate space. Although mapping the grid values provided a good means for viewing the

Fig. 4.5. A representation of classification tree results for southern Michigan. Shading indicates areas where at least one presence node rule has been met. Up to three presence node rules can be met in an area (darkest shade), indicating the highest-quality habitat. The insert provides a closer view of the same model for a single county. (Adapted from Standora and Kingsbury 2002)

results, there was no way to decide where good habitat ended and bad began, only a gradient of better or worse. We arbitrarily divided the landscape into deciles, as mentioned previously. Thus, our compromise was that we could at least provide a mechanism to prioritize conservation efforts.

The Mahalanobis approach was also computer intensive and very regionally sensitive. Consequently, we evaluated only portions of the lower peninsula at any one time, focusing on areas with the greatest number of snake locations. In applications in which such constraints lead to the inclusion of only a limited number of snake locations, modeling will lose power.

Conservation Applications and Implications

Ultimately, the modeling had several conservation uses and largely met or exceeded our expectations. The Boolean models identified regions of Michigan in which to focus conservation efforts. More accurately, these models showed us where not to be concerned about Massasaugas. They also led to a list of features associated with Massasauga habitat and motivated us to consider the north and south portions of the lower peninsula separately. Although they missed some of the localities, the classification tree and Mahalanobis models allowed the delineation of much smaller areas on which to focus conservation efforts and in which to survey for new snake populations. Ultimately, model outputs did lead to locating new populations. Model outputs were also used to propose areas within many of the public properties of Michigan that should be managed with Massasaugas in mind. Such delineations are an important component of the Candidate Conservation Agreement with Assurances (CCAA) that is under development for the state of Michigan.

The models have the potential to contribute even more toward local decision making, although a combination of political and communication circumstances preclude this at times. A lesson from our experiences applying these models to conservation is that the use of models is only as good as the ability of all parties to convey the information to the decision makers on the ground and their willingness to consider it.

Areas for Future Research

Overall, we feel that the spatial representation of snake distributions and habitats is an important area of future research that can have a significant influence on snake conservation efforts. It is important to stress that the types of modeling approaches presented in this chapter are underused in studies of snake ecology and that more studies should use these types of approaches. We have described a number of aspects of modeling that would benefit from additional research. First, there is a need to have environmental data represented in a spatial context. Many studies examine how snake distribution relates to environmental characteristics, but few acquire or create the data needed to look at these relationships in a spatial context. Second, there is particular lack of fine-scale environmental data that could be very useful for understanding many aspects of snake distirbution and habitat. Future studies should take advantage of high spatial and spectral resolution data to examine how snakes relate to fine-scale patterns in the environment. Third, there is little information on spatial autocorrelation patterns and snake distribution, and little use of geostatistics as a tool for predicting snake distribution. Future studies with snakes should also use multiscale approaches when feasible.

We think that the field of snake spatial ecology, particularly modeling distribution and habitat, is wide open. With the recent attention on the conservation of snakes (e.g., Gibbons et al. 2000), ecologists should strive to include spatial aspects into their studies. This is particularly valuable given that land and wildlife management agencies are using these types of models with a variety of other taxa. In addition, we think that there are a number of ways that studies modeling snake distribution and habitat could have broader impacts. For example, using spatial models to conserve snakes can protect a broader range of species because many snake species have wide-ranging movements (e.g., Cobb 1994). Furthermore, snakes are tightly linked to their prey and, by conserving snake habitat, we are probably protecting a variety of species that are the prey and predators of snakes (Jenkins 2007).

Acknowledgments

We thank Steve Mullin and Rich Seigel for all their inputs and efforts during all stages of preparing this chapter. We especially acknowledge the ingenuity and energy of Michelle Standora, the graduate student who did the work highlighted in the case study. We also thank Jeremy Shive for discussions that helped in writing this chapter and Leona Svancara for developing the figure of reptile species richness in Idaho. Finally, we thank Gary Beauvais for extensive discussions and for providing an unpublished report (from the Wyoming Natural Diversity Database at the University of Wyoming) that contains a great deal of important information used in writing this chapter.

5

Linking Behavioral Ecology to Conservation Objectives

PATRICK J. WEATHERHEAD AND THOMAS MADSEN

In the introduction to one of the first books linking behavioral ecology and conservation, Caro (1998) pointed out why, on the one hand, these two fields appear to have little in common while, on the other hand, making the case that behavioral ecology can contribute meaningfully to conservation. The focus of behavioral ecology is on individuals and on how their behavior and morphology affect their survival and reproductive success. The focus of conservation biology is on populations and species and on the ecological factors that affect their abundance and persistence. The link between these two disciplines arises from how the actions of individuals affect populations. In principle, these links should be abundant. Examples that clearly demonstrate population consequences arising from individual behavior are few, however, and scarcer yet are cases in which understanding a behavior leads to practical conservation solutions (Caro 1998). Our first goal in this chapter is to review studies of snakes that link behavioral ecology to conservation. Our second and broader goal is to identify areas of study in which the potential for such links seems high and to discuss the kinds of data and research approaches that seem most likely to produce practical conservation measures.

Behavioral ecology is a broad discipline; we address only parts of it here. Some areas of behavioral ecology are either largely irrelevant to snakes or poorly studied because snakes are usually not tractable subjects for such investigations (e.g., parental care and parent-offspring conflict). Snake reproductive biology, by contrast, is becoming increasingly well studied (Shine and Bonnet 2000; Shine 2003) and is certainly relevant to conservation (e.g., Madsen et al. 1999, 2004). Reproductive biology (covered in Shine and Bonnet, Chapter 6) is not considered here. Our focus is on two general areas of snake behavioral ecology that have great potential relevance to conservation: thermal ecology and predator-prey interactions. For the first, we concentrate primarily on the conservation implications of the relationship between thermal ecology and habitat selection, with particular attention to how that relationship will be affected by global warming. In the predator-prey section, we consider topics such as the effects of introduced species and direct human impacts on snakes. One of the human impacts we consider in detail is the effect of roads on snakes. Although this is not a conventional predator-prey relationship, it does involve snake mortality, and predator-prey theory can be helpful in understanding how snakes respond to roads and vehicles (e.g., Shine et al. 2004; Andrews and Gibbons 2005). Although the prevailing theme throughout this chapter is that snakes are increasingly in need of conservation efforts, we do consider several instances in which snakes are the problem and actions directed against snakes might be required to conserve other taxa (e.g., Wiles et al. 2003; Weatherhead and Blouin-Demers 2004b).

Thermal Ecology and Habitat Selection

The central importance of body temperature to reptiles accounts for herpetologists' enduring interest in thermal ecology (e.g., Heath 1964; Avery 1982; Huey 1982; Tracy and Christian 1986; Lillywhite 1987; Huey and Kingsolver 1989; Peterson et al. 1993). Temperature affects everything from specific functions such as ecdysis and embryonic development to more general functions such as digestion, recovery from injury, growth, and locomotion (Huey 1982; Peterson et al. 1993). Thus, the maintenance of appropriate body temperatures affects both the fitness of individual snakes and the viability of their populations. Given that thermoregulation is probably the most important factor affecting habitat selection in snakes (Reinert 1993; although less important in tropical species, Shine and Madsen 1996), any environmental perturbation that alters snakes' abilities to maintain preferred body temperatures will be a conservation concern.

Habitat loss and fragmentation are the greatest general threats to conserving biodiversity (Wilson 1992; Meffe and Carroll 1997; Wilcove et al. 1998) and there is every reason to expect these to be the major threats for snakes as well (Shine 1991; Gibbons et al. 2000). As habitat is lost, thermal ecology could be relevant in evaluating the value of the habitat that is left, particularly if only some of it can be saved. Furthermore, understanding snake thermal ecology could help us evaluate the effects of fragmenting the remaining habitat. The essence of behavioral thermoregulation is that snakes can move between habitats or microhabitats to find appropriate temperatures. The one exception is snake eggs. Although female snakes use temperature as a cue in choosing where to lay their eggs (Blouin-Demers et al.

2004), once the eggs are laid they must develop in the conditions that prevail at that site. Therefore, we discuss thermal aspects of nesting separately from other aspects of habitat selection.

Thermal Quality of Nest Sites

Incubation temperature can affect hatching success (Burger and Zappalorti 1988a; Ji and Du 2001a, 2001b; Lin et al. 2005); incubation period (Shine et al. 1996; Ji and Du 2001a, 2001b; Blouin-Demers et al. 2004); developmental stage at hatching (Lin et al. 2005); hatchling morphology, including the occurrence of abnormalities (Shine et al. 1996; Ji and Du 2001a; Blouin-Demers et al. 2004); hatchling behavior and locomotor performance (Burger 1998a, 1998b; Webb et al. 2001; Blouin-Demers et al. 2004); and sex differences in these traits (Burger and Zappalorti 1988; Webb et al. 2001). All these effects seem likely to influence the subsequent performance of hatchlings. For example, neonatal size can affect the survival of snakes early in life (Jayne and Bennett 1990; Bronikowski 2000; Kissner and Weatherhead 2005), and conditions experienced by snakes when young can affect their performance later in life (Madsen and Shine 2000a). The manifold effects of incubation temperature in combination with the snakes' inability to modify the thermal quality of their nests (with the exception of shivering thermogenesis documented in some pythonid snakes; Hutchison et al. 1966; Van Mierop and Barnard 1978; Shine et al. 1996) make it likely that the thermal properties of nests are disproportionately important to the conservation of oviparous snakes.

It should be possible to assess the thermal quality of nesting habitat by determining how it affects hatching success and hatchling quality. As an example, female Water Pythons (*Liasis fuscus*) use two types of nests that differ markedly in their thermal quality (Madsen and Shine 1999b). Hollows in paperbark tree roots are cool and require females to remain with their eggs for 2 months to assist incubation, whereas varanid lizard burrows are warm and female pythons abandon their eggs shortly after laying them there. Because nest attendance generally precludes feeding, using cool nests results in females being emaciated when they leave the nest, which reduces their chances of survival. Furthermore, these females take 2 years to build up sufficient reserves to breed again, unlike the single year that is normal for females that use warm nests. Warm and cool nests also affect hatchling phenotypes differently (Shine et al. 1996), which is likely to affect subsequent performance of hatchlings. Thus, the thermal quality of nests has far-reaching effects on demography. Quantifying how these demographic effects influence population size and, hence, viability remains a challenge for future studies.

Once the quality of a snake population's nesting habitat is determined, several conservation scenarios are possible. If the best nesting habitat is abundant and occurs within the snakes' general habitat, conserving nesting habitat becomes synonymous with conserving general habitat. If the best nesting habitat occurs within the general habitat but is scarce or patchy, we could enhance nesting habitat (see Shine and Bonnet, Chapter 6) while protecting general habitat. Finally, if nesting habitat is specialized and separate from general habitat, that special habitat needs to be conserved in addition to maintaining the connections with the general habitat.

Ratsnakes (*Pantherophis* [*Elaphe*]) in Ontario provide an example with elements of the last two scenarios (Blouin-Demers and Weatherhead 2000; Blouin-Demers et al. 2004). The natural nesting habitat for the snakes is hollow trees. Because virtually all forests in Ontario have been clear-cut several times since European settlement, the large dead or dying trees that provide this nesting habitat are scarce, even though the second-growth forests currently present appear suitable for the rest of the snakes' needs. Ratsnakes have adjusted to the scarcity of hollow trees by nesting in human-made habitats such as leaf piles and compost piles that provide the warm moist environments their eggs require. As a consequence of this change in behavior, female ratsnakes have greater contact with people and often have to cross roads to access these human-made nesting sites, both of which are likely to increase mortality of female snakes. By creating suitable leaf piles within the snakes' habitat and away from roads and human activity, it should be possible to compensate for the loss of natural nesting habitat while reducing the negative consequences of nesting in human-made habitats. Ultimately, the goal should be to allow some forests to return to their natural state or, at least, to manage them so that some trees are allowed to grow old and die. Artificial nests could at least serve as an interim measure to increase the availability of nest sites for ratsnakes.

Grass Snakes (*Natrix natrix*) provide a similar example. In southern Sweden, most Grass Snakes rely on manure piles for nests, presumably because their natural nest sites no longer exist (T. M., pers. obs.). A single manure pile regularly produces more than 600 eggs. A change in Swedish law that requires manure piles to be surrounded by a 1-m-tall concrete wall has made those sites unavailable to Grass Snakes, with dire consequences for the population. Modifying these structures to allow access by snakes seems to be the immediate conservation solution.

Thermal Quality of Other Aspects of Habitat

Just as with eggs, embryonic development in viviparous snakes is affected by temperature during gestation (Weatherhead et al. 1998, 1999; Blouin-Demers et al. 2000; Arnold and Peterson 2002; Lourdais et al. 2004; O'Donnell and Arnold 2005). Consistent with these effects, female snakes often modify their body temperature during gestation (Peterson et al. 1993; Graves and Duvall 1993; Charland 1995; Dorcas and Peterson 1998; Brown and Weatherhead 2000; but see Isaac and Gregory 2004) and change their use of microhabitats (Reinert and Zappalorti 1988; Reinert 1993; Harvey and Weatherhead 2006). Therefore, just as it should be possible to conserve or improve nesting habitat for oviparous snakes, it should also be possible to do this for the gestation habitat of viviparous species.

It is possible that, for some snake species, the thermal properties of other aspects of their habitat could be limiting and amenable to conservation actions. The endangered Broad-headed Snake (*Hoplocephalus bungaroides*) is such an example. This nocturnal snake uses exposed sandstone rocks as retreat sites during the day, as does the gecko that is its principal prey (Webb and Shine 2000). The removal of these rocks for landscaping has degraded the snakes' habitat and contributed to their decline, but artificial paving stones appear to provide a suitable replacement for natural rocks (Webb and Shine 2000). A second problem with the rocks that snakes use as retreat sites is that shading by tree canopies makes many otherwise suitable rocks thermally suboptimal (Pringle et al. 2003), a problem exacerbated by fire suppression (Webb et al. 2005a). Thinning the canopy was sufficient to restore the appropriate thermal properties of the rocks and increase their use by Broad-headed Snakes (Webb et al. 2005a).

Thermal Ecology, Habitat Fragmentation, and Edge Effects

A common outcome of habitat loss is that the remaining habitat is not only reduced in area but also fragmented into patches, separated by a new type of habitat that might be totally unsuitable for species that used the original habitat (Ujvari et al. 2002). Some snake species might benefit from the increased availability of the new habitat (e.g., Urbina-Cardona et al. 2006), but those that depend on the original habitat will not. Snakes in the remaining habitat can be further negatively affected depending on their response to the size of the remaining habitat patches (i.e., area effects; Luiselli and Capizzi 1997; Hager 1998) and their response to the interface between the original and new habitats (i.e., edge effects). It is the effect of thermal ecology on the latter that we consider here.

When an area of forest is cleared, the remaining forest is affected by the adjacent cleared area. The greater penetration of light and wind modifies the microclimate of the forest edge, which in turn affects plant communities. These effects can extend more than 50 m into the forest (Murcia 1995; Harper et al. 2005). For species that avoid the modified habitat in the forest edge, the suitable area of a given forest fragment is reduced, presumably increasing their risk of local extinction (Lehtinen et al. 2003). For some species, however, the forest edge can be beneficial because the range of microclimates available for thermoregulation is much greater than in either the forest or the open habitat (Blouin-Demers and Weatherhead 2001b). Ratsnakes preferentially use forest edges (Durner and Gates 1993; Blouin-Demers and Weatherhead 2001b; Carfagno and Weatherhead 2006).

This preference appears unrelated to prey abundance (Blouin-Demers and Weatherhead 2001b; Carfagno et al. 2006), but it is consistent with the snakes' preferring edges because of their superior thermal quality (Blouin-Demers and Weatherhead 2001b, 2002a). When ratsnakes were fed experimentally in the field, individuals fed while in edges remained there, whereas those fed in forest moved to edge habitat (Blouin-Demers and Weatherhead 2001a). Because snakes require higher body temperatures to digest a meal efficiently, the preference for edges by snakes that had just eaten was consistent with edges being preferred for thermal reasons.

Ratsnakes are listed as threatened in Canada, where habitat loss has greatly restricted their distribution (Prior and Weatherhead 1998). Conservation efforts for the snakes are likely to be most successful if some edges are retained in forests, given the snakes' preference for them (Weatherhead and Charland 1985; Blouin-Demers and Weatherhead 2001b).

Ratsnakes are also a conservation concern at the southern edge of their distribution, but in this case, the snakes are the problem. Video cameras placed at nests of Black-Capped Vireos (*Vireo atricapillus*) and Golden-Cheeked Warblers (*Dendroica chrysoparia*), two endangered bird species, revealed that Texas Ratsnakes (*P. obsoletus*) are the principal nest predator for both species (Stake and Cimprich 2003; Stake et al. 2004). Efforts to conserve these birds could involve modifying the habitat to make it less attractive to ratsnakes. If Texas Ratsnakes have the same affinity for edges as their northern cousins, reducing habitat fragmentation (and thus the amount of edge) could be effective, unless the birds also prefer edges (Weatherhead and Blouin-Demers 2004b). Other snake species also prefer forest edges (Henderson and Winstel 1995; Carfagno and Weatherhead 2006), so managing the amount of edge could prove to be of general importance in snake conservation.

Climate Change

Climate generally and temperature in particular have long been recognized as important determinants of the distribution, abundance, and activity of animals (Merriam 1894; Andrewartha and Birch 1954; Gaston 2003). Thus, the warming of global climates is causing growing concern about how climate change affects natural populations (Hughes 2000; McCarty 2001). Predicting how a population will respond to warmer climates requires knowledge of how the important demographic features of the population vary in response to current climatic variation. Responses to global warming include short-term effects on populations (e.g., changes in abundance) and long-term effects that result in shifts in species' distributions (Currie 2001).

Research on the thermal ecology of reptiles indicates some of the important ways in which snakes are likely to be affected by global warming. For example, snakes display considerable plasticity in life-history traits. In response to increased food intake, snakes can grow faster, mature at larger sizes, and be more fecund (Parker and Plummer 1987; Ford and Seigel 1994; Beaupre 1996; Lindell 1997; Luiselli et al. 1997). Warmer climates are likely to affect food intake by altering digestion times, the length of time snakes can forage per day, and the number of foraging days per season. Global warming seems likely to have similar direct and indirect effects on most aspects of snake behavior and ecology.

Most of what we know about how climate change has already affected animal populations comes from long-term studies conducted for reasons other than to document climate change effects (Hughes 2000; McCarty 2001; Walther et al. 2002; Parmesan and Yohe 2003; Root et al. 2003; Krajik 2004). For example, birds are migrating earlier (Oglesby and Smith 1995; Bradley et al. 1999) and nesting earlier (e.g., MacInnes et al. 1990; Winkle and Hudde 1997; Dunn and Winkler 1999), and butterflies are appearing earlier in the spring (Sparks and Yates 1997) and shifting their ranges northward (Parmesan et al. 1999). Thus far, only a handful of studies has been published documenting climate change effects on ectothermic vertebrates (e.g., Beebee 1995; Pounds et al. 1999; Blaustein et al. 2001; Gibbs and Breisch 2001), and none of these is on snakes. Weatherhead et al. (2002) did document a long-term decline of *Pantherophis spiloides* that appeared to be climate driven, but they were unable to identify the specific climate features that were responsible. Clearly, snake biologists need to be more aggressive in using their long-term data sets to identify population trends and determine if these trends are associated with changes in climate.

The first step in predicting how populations will be affected by climate change is to determine how they respond to contemporary climate variation. In general, we expect that if conditions become warmer, both the costs and benefits of thermoregulation should decrease for temperate-zone snakes. This should allow snakes either to thermoregulate more effectively or to spend less time thermoregulating, either of which should be beneficial. Available evidence is indirect but consistent with this expectation. European Adders (*Vipera berus*) grow faster in years with warmer, sunnier active seasons (Forsman 1993; Lindell 1997). Aspic Viper (*Vipera aspis*) population dynamics appear to be driven by juvenile survival, which in turn varies strongly with winter weather (Altwegg et al. 2005). Variation in temperature during the active season affected growth and, thus, age at maturity of watersnakes (*Nerodia;* Brown and Weatherhead 2000). Substantial overwinter mortality of Red-sided Gartersnakes (*Thamnophis sirtalis parietalis*) was attributed both to flooding and to freezing associated with light snow cover (Shine and Mason 2004). The fact that snakes control their body temperatures behaviorally should make it easier for them to adjust to warming climates. At some point, however, conditions could become too warm for snakes to adjust behaviorally. Local extirpation and range shifts will be the likely outcome (e.g., Currie 2001; Kling et al. 2003).

To predict how much a species's range might shift we need to know how it is currently limited by climate (e.g., Humphries et al. 2002). An effective approach to understanding how an ectotherm is affected by climate is to study the species's thermal ecology at the limit of its range (Peterson et al. 1993). The geographical limit might occur where the animals' thermal tolerances are occasionally exceeded, where resources limit reproduction, or where thermal conditions are often inadequate for development (Peterson et al. 1993). The distinction among these three hypotheses is artificial in some respects, however, because all aspects of the thermal ecology of animals are interconnected. For example, for an oviparous temperate-zone snake, temperature should affect the time it takes a female to find and assimilate enough food to produce eggs and, thus, when eggs are laid. Temperature will also affect how long it takes the eggs to develop. Therefore, temperature could affect hatching time through its effects on both laying time and embryo development. In turn, when the eggs hatch will determine how long hatchlings have to find a hibernation site adequate for avoiding lethal winter temperatures. This, again, underscores the need for comprehensive studies of thermal ecology, something currently lacking for snakes at a species's latitudinal limit (Gaston 2003).

Predictions that snakes' ranges will shift in response to global warming assume that the snakes will be able to disperse from their current ranges to their future ranges. If current habitats shift in a gradual fashion, it is reasonable to expect snakes to shift with them. But this scenario may be unrealistic for many species. For example, montane species with narrow elevational ranges may be able to follow their habitat upward but only to the extent that the mountains they occupy are high enough to accommodate the habitat shift (Greene 1994). More broadly, human development (e.g., roads, agriculture, and urbanization) will present barriers to dispersal that will restrict range shifts. Even where barriers do not exist, snakes need to disperse quickly enough to keep pace with habitat changes. At present, we know little about snake dispersal, including such basic questions as how far snakes disperse and whether dispersal distances differ by sex, as occurs in other taxa (e.g., Greenwood 1980). Measurements of which segments of the population move at which times of the year can be obtained from mortality patterns (e.g., Bonnet et al. 1999b), and some measure of juvenile dispersal can be obtained from mark-recapture studies (e.g., Webb and Shine 1997a). These approaches should be pursued; however, molecular genetic analyses of populations may provide a more reliable, albeit indirect, approach to assessing dispersal (Gibbs and Weatherhead 2001; King, Chapter 3). Genetic evidence from several species suggests dispersal is quite limited (Villarreal et al. 1996; Gibbs et al. 1997; Lougheed et al. 1999; Prosser et al. 1999), although more species need to be studied before general patterns become apparent (Gibbs and Weatherhead 2001).

Predator-Prey Interactions

Predation is a major ecological force influencing population dynamics, species distributions, and community structure (Kerfoot and Sih 1987). The most obvious direct effect of predation is the killing of prey, but a variety of indirect effects of predators on the behavior and life histories of prey and on dynamics at other trophic levels have been recognized (Kerfoot and Sih 1987). Thus, understanding predator-prey interactions is critical for management and conservation of species, and it relies in part on understanding the ecological and evolutionary processes at work.

As with all predators, many aspects of snake biology are affected by the abundance of their prey. Low prey availability may result in reduced growth rate and low reproductive output and, hence, low population densities (Fitzgerald and Shine 2004). Conversely, areas with high prey densities have been demonstrated to harbor very large snake populations (Madsen and Osterkamp 1982; Bonnet et al. 2002b; Madsen et al. 2006). Compared to mammalian predators, the lower metabolic rate and, hence, reduced energy requirements of snakes may make populations less sensitive to temporal changes in prey numbers (Madsen and Shine 1999a). Nevertheless, large temporal variation in prey abundance can still have a dramatic impact on snake population demography. In *Vipera berus,* a massive reduction in prey density resulted in substantially higher adult mortality (Madsen and Stille 1988; Forsman and Lindell 1997).

The declines of amphibian populations reported worldwide (Houlihan et al. 2000) will almost certainly have important negative effects on predators specialized for feeding on this group of vertebrates, such as many natricine snakes. In the Sierra Nevada, United States, amphibian declines are well documented (Knapp and Matthews 2000), and the decline of amphibians has indeed resulted in a concomitant decline of Terrestrial Gartersnakes (*Thamnophis elegans;* Matthews et al. 2002). Furthermore, food availability early in life has a disproportionate effect on later growth and maximum body size in some snakes (Madsen and Shine 2000a). Because body size influences many aspects of a snake's interaction with the environment—including food habits, vulnerability to predation, and reproductive output—such "silver spoon effects" (Grafen 1988) seem likely to have an impact on snake population demography.

In contrast to many other predators, snakes often show an ontogenetic shift in diet (e.g., Brito 2004; Quick et al. 2005; Webb et al. 2005b). Due to morphological constraints, juveniles often feed on small prey such as lizards or juvenile frogs, whereas adults often feed on large prey such as mammals (Mackessy et al. 2003). The habitats used by such disparate prey are often very different. Hence, to maintain viable snake populations it is not sufficient to protect the habitat needs of only one of the major types of prey.

The habitat needs of all the important prey species (for both juveniles and adults) must be addressed.

Habitat Destruction and Prey Numbers

One of the greatest threats to conserving biodiversity is anthropogenic change and destruction of habitats. For most snake taxa, such alterations will almost certainly have a negative impact on prey abundance, reducing the probability of long-term survival of this group of predators (Ujvari et al. 2000). For some species, however, human alteration of habitats may in fact have positive effects. In Sweden, human-made landscapes that are mosaics of agricultural and natural habitat support large populations of small rodents and also large populations of European Adders, one of their main predators (T. M., pers. obs.).

Similar positive effects of human-made habitat alterations have been recorded in Africa. In the Kakamega District in western Kenya, small fragments of secondary rainforest habitats exist as patches in an area used for intensive agriculture. A survey was conducted to compare the abundance of common rainforest snake taxa in the forest fragments with numbers encountered in the Kakamega Forest National Park. Approximately ten times more snakes were encountered per night in the forest fragments than in the national park (T. M., pers. obs). Similar results were also obtained in a survey of snakes in Nigeria. In the swamp forest in the Niger Delta and in the dry forest in the Cross River State, approximately five times more snakes were encountered in secondary habitats compared to primary forest habitats (L. Luiselli, pers. comm.). These results suggest that several snake taxa were more abundant in the secondary forest fragments than in primary forests.

One of the reasons for the higher snake densities in forest fragments could be that prey densities are higher. The agricultural landscape in the Kakamega District supported large numbers of rodents that were often observed during the nocturnal surveys in the forest fragments, whereas no rodents were observed in the park. Another factor explaining the higher abundance of snakes in the fragments could be the virtual absence of snake predators, such as mongoose and birds of prey, which are killed by the farmers because these predators also prey on their chickens. Unfortunately, since 1984 when the survey was conducted in the Kakamega District, these forest fragments have been cleared at an alarming rate and converted to agriculture, and hence, the future survival of the snake populations is doubtful.

Climate Change and Prey Numbers

We have discussed the conservation implications of the direct effects of global warming on snakes; climate change is also likely to affect snake populations through its effects on prey populations. In the tropical areas of Australia, annual variation in rainfall has been demonstrated to have a dramatic impact on prey population numbers and, hence, also on the dynamics of snake populations (Madsen and Shine 2000b; Madsen et al. 2006). Therefore, assuming that global climate change includes altered rainfall patterns, the demography of prey populations and their predators will also change. Furthermore, global climate change has also been suggested to result in increased stochasticity in weather patterns, such as increases in flooding and prolonged droughts (Hughes 2003). For Massasaugas (*Sistrurus catenatus*), flooding not only resulted in habitat destruction but also severely reduced prey availability (Seigel et al. 1998). On the other extreme, a severe drought in South Carolina, United States, in the mid-1980s resulted in long-term changes in aquatic snake abundance and species composition, most likely due to a dramatic reduction in suitable aquatic prey (Seigel et al. 1995). Subsequent monitoring found that snake species varied in their response to the end of the drought, with some recovering quickly and others showing longer-term effects (Willson et al. 2006). Behavioral attributes beneficial to surviving the drought included the ability to estivate during inclement conditions (e.g., *Seminatrix pygaea*) and the natural propensity to migrate in response to varying prey availability (e.g., *Agkistrodon piscivorus;* Willson et al. 2006).

Predicting how climate change will affect prey population demography and, in turn, how this will affect predators such as snakes will require improved climate models and a much better understanding of the ecology of snake predator-prey relationships than is currently available. For example, amphibian populations were quite resilient to a 2.5-year drought, even in an isolated wetland (Gibbons et al. 2006), which would have been beneficial to snakes that preyed on those amphibians. Because the quality of the terrestrial habitat surrounding the wetland almost certainly contributed to the resiliency of the amphibians (Gibbons et al. 2006), proper habitat management is likely to be effective in mitigating the climate effects on snakes and their prey. Ultimately, however, the only general conservation recommendation that is likely to emerge from an improved understanding of how climate change affects snakes is that we need to change the factors that produce climate change. Concerns about snakes are unlikely to influence the ongoing debate on that issue.

Invasive Prey and Predators

The impact of invasive species on native species, communities, and ecosystems has been widely recognized for decades (Elton 1958; Diamond 1989; Lodge 1993), and invasive species are presently regarded as a significant component of environmental change (Vitousek et al. 1996). Invasive species often have rapid and far-reaching negative impacts on populations, ecological communities, and biodiversity (Sakai et al. 2001). For example, since its

1935 introduction into tropical and subtropical Australia, the highly toxic Cane Toad (*Rhinella [Bufo] marina*) has expanded its range at an alarming rate (Freeland 1985). These amphibians can reach astounding densities in suitable habitats (> 2000 individuals/ha; Freeland 1986). To snakes that prey on amphibians, Cane Toads represent an abundant but potentially deadly prey. As many as 49 snake species are at risk from toads (Phillips et al. 2003). Recent research has demonstrated, however, that Red-Bellied Blacksnakes (*Pseudechis porphyriacus*) living in toad-exposed areas show increased resistance to Cane Toad toxin and a decreased preference for toads as prey (Phillips and Shine 2006). Laboratory experiments suggest that these changes are not a result of learning or of acquired resistance but most likely reflect the effects of rapid selection on snake behavior and physiology (Phillips and Shine 2006). It remains to be determined whether other snake species might have a similar ability to respond to the Cane Toad invasion and what aspects of a species's ecology determines how it is affected by Cane Toads.

Another example of an invasive species causing the decline or extinctions of snake fauna is the release of the Indian Mongoose (*Herpestes javanicus*) into some of the Lesser Antillean Islands. This notorious snake predator is thought to have caused the extinction of snakes such as *Alsophis* spp. in some of the islands (Henderson 2004). Presumably, when relatively small populations of snakes are confronted with an effective predator against which they have no defense, extinction is a much more likely outcome than the evolution of novel defensive behaviors.

Among invasive species, ants may be the most widespread threat to snake populations, although at present almost everything we know about this threat is anecdotal. Three species of fire ants (*Solenopsis invicta, S. geminata,* and *Wasmannia auropunctata*) are among the most widespread, abundant, and damaging invasive ants that present multiple potential threats to snakes (Holway et al. 2002). Although native to South and Central America, collectively these three species have been introduced to Africa, Australia, New Zealand, North America, and numerous islands in the Caribbean and the Indian and Pacific Ocean (Holway et al. 2002). Invasive fire ants have the potential to harm snakes indirectly through negative effects on their prey but also directly by predation facilitated by their potent stings. Fire ants have been implicated in the declines in lizard populations in New Caledonia (Jourdan et al. 2001) and Southern Hog-nosed Snakes (*Heterodon simus*) in the United States (Tuberville et al. 2000). Observations of direct effects of fire ants on reptiles include predation on the eggs and young of Rough Greensnakes (*Opheodrys aestivus;* Conners 1998b), American Alligators (*Alligator mississippiensis;* Allen et al. 1997; Reagan et al. 2000), and a number of turtle species (e.g., Moulis 1997; Conners 1998a; Allen et al. 2001; Buhlmann and Coffman 2001).

At present we lack information on the extent to which fire ants destroy eggs and hatchling snakes, but the potential impact is enormous given the high densities these ants achieve where they are invasive (Holway et al. 2002). Adult vertebrates appear to be able to escape fire ants unless confined (Holway et al. 2002), so adult snakes probably face little predation risk from fire ants. Nonetheless, if fire ants force snakes to move more often than they would do otherwise, the resulting increased energy expenditure and exposure to other predators could still be costly. There is an obvious need for systematic study of how invasive ants affect snakes.

There are several cases in which invasive species appear to have had positive effects on snake populations. The threatened Lake Erie Watersnake (*Nerodia sipedon insularum*) has benefited from feeding on the introduced Round Goby (*Neogobius melanostomus*) by growing faster, achieving larger body size, and increasing production of offspring (King et al. 2006b). Similarly, the introduction of Marsh Frogs (*Rana ridibunda*) in the southeastern United Kingdom may have resulted in an increased abundance of Grass Snakes (*Natrix natrix*) in this area (Gregory and Isaac 2004). The introduction of the House Mouse (*Mus domesticus*) in Australia has resulted in extremely high mouse densities in some of the agricultural areas of the southern part of the continent, up to 2500/ha during plague years (Boonstra and Redhead 1994). A principal predator of mice in these areas, the Eastern Brownsnake (*Pseudonaja textilis*), appears to have benefited from the high prey abundance by reaching very high population densities (Shine 1989; Whitaker and Shine 2003).

Snakes too can be invasive species and can have devastating effects on native fauna. On the island of Mallorca, the introduced Viperine Watersnake (*Natrix maura*) has been implicated in the decline of the endemic Mallorcan Midwife Toad (*Alytes muletensis;* Moore et al. 2004). The rapid expansion of a *Boa constrictor* population on Aruba Island since its introduction in 1999 has resulted in concerns about its impact on the local fauna (Quick et al. 2005).

One of the most devastating and best-documented cases of an invasive species affecting an ecosystem resulted from the accidental introduction of the Brown Treesnake (*Boiga irregularis*) to the island of Guam (Wiles et al. 2003). The Brown Treesnake is a nocturnal, arboreal, mildly venomous colubrid that can reach a total length of up to 2.3 m and can weigh as much as 2 kg (Rodda et al. 1999b). The species's native range included Sulawesi through New Guinea and the humid northeastern rim of Australia to the Santa Cruz Islands (Rodda et al. 1999b). During the 1950s, Brown Treesnakes were inadvertently transported from New Guinea to Guam (Savidge 1987). By the mid-1980s, snake densities were estimated at 50–100/ha (Rodda et al. 1999c).

Brown Treesnakes are dietary generalists and have been observed to eat chicken bones, cooked spare ribs, lizards, birds, rodents, domestic fowl hatchlings, puppies, piglets, rabbits (in hutches), and pet birds (in cages; Rodda et al. 1999a). It seems that their catholic diet and flexible foraging modes (constriction and envenomation) account for the breadth of the harm caused by Brown Treesnakes (Greene 1989). During the last half of the twentieth century, predation by Brown Treesnakes has devastated the avifauna of Guam, causing the extirpation or serious reduction of most of the island's 25 resident bird species (Wiles et al. 2003). The Guam population of Marianas Fruit Bat (*Pteropus marianus*), already impacted by hunting, has been further decimated by the snakes (Wiles et al. 1995), and many of the native species of lizards have also been negatively affected by snake predation (Rodda and Fritts 1992a). In addition to the ecological damage, snakes move along powerlines in search of prey, resulting in frequent power blackouts and extensive damage to power transmission equipment, causing millions of dollars in economic losses (Fritts et al. 1987).

Numerous techniques have been tried to reduce the impact of Brown Treesnakes, including the creation of barriers to restrict snake movements (Campell 1999), trapping (Engeman and Linnell 2004), and chemical control using acetaminophen-treated mouse baits (Johnston et al. 2002). Although Brown Treesnakes are highly susceptible to this toxin, nontarget species such as the endangered Marianas Crow (*Corvus kubaryi*) are also put at risk (Johnston et al. 2002). A promising and much more targeted approach is to use species-specific sex pheromones to capture the snakes (Mason 1999), as has been used successfully to attract and destroy insect pests (Ridgeway et al. 1990). Sex pheromones induce responses in males such as trailing and courtship (Greene and Mason 1998) that could be used to attract them into traps during the mating season. Alternatively, permeating the environment with pheromones could confuse males, thereby reducing the number of males that locate females (Mason 1994), although active mate searching by females could undermine the effectiveness of this approach. The management outcome from implementing these ideas is unclear at this time (R. Mason, pers. comm.).

Although relatively few snakes are invasive species, the accidental introduction or intentional release of pet snakes has the potential to result in new cases. For example, several snake species have become established in Florida by this route (Dalrymple 1994; Meshaka et al. 2004). Features that seem likely to make a snake species a successful invader include being both a habitat and prey generalist. Species attaining larger sizes may also be more likely to succeed because large size facilitates diet breadth (Arnold 1993). To survive in a novel locale, the climate must be similar to that in the snake's native range (tropical species will not survive in temperate areas, and desert species are unlikely to survive in humid areas). All these factors clearly facilitated the establishment of Brown Treesnakes in Guam (Greene 1989), and this case illustrates how profound an effect an invasive snake can have. These attributes can also be used together with importation data to assess the risk posed by species imported as pets that could escape or be released (Reed 2005). Knowing the traits that facilitate a successful invasion could help us avoid such occurrences in the future by restricting importation of species with the greatest risk of becoming invasive. Also, knowledge of risky attributes can help determine the seriousness with which we confront incipient introductions that are detected. For example, a number of treesnake species (*Boiga*) share the attributes that have allowed the Brown Treesnake to have such a devastating impact on the fauna of Guam, so the introduction of any of those species to other Pacific islands should be treated extremely seriously (Greene 1989).

Humans as Predators

A news report released by World Wildlife Fund—Australia in 2003 estimated that 89 million reptiles die each year as a result of broad-scale clearing of vegetation in Queensland. Although most of the reptiles killed in Queensland are probably lizards, the number of snakes killed must still be very high. Extrapolated globally, an enormous number of snakes must be lost to habitat destruction. By comparison, the collection of snakes for human consumption and for the skin and pet trade must pose a lower threat, but it can nonetheless be substantial. In some parts of the world, snakes are consumed in large numbers; for example, approximately 1 million snakes are harvested in the northeastern part of China each year (Zhou and Jiang 2005). The total volume of snakes traded each year in China is estimated to be as high as 9×10^6 kg (Wan and Fan 1998, as cited in Zhou and Jiang 2005). The huge consumption of snakes for food and use in traditional medicine has resulted in grave concerns for some species involved in this trade (Zhou and Jiang 2004, 2005). Largely to support the Chinese food market, an estimated 8500 snakes per day of five species of homalopsine watersnakes are harvested from Tonle Sap Lake and Tonle Sap River in Cambodia (Stuart et al. 2000). Even more staggering, more than 500,000 Reticulated Pythons (*Python reticulatus*) are harvested each year from Southeast Asia (Groombridge and Luxmoore 1991; Jenkins and Broad 1994). Although the number of snakes being removed might appear to be unsustainable, Shine et al. (1999) studied the harvesting of Reticulated Pythons in northern Sumatra and concluded that it was unlikely to extirpate these snakes from their Indonesian range, although they pointed out the need for careful monitoring. Unlike large temperate-zone snakes, the rapid growth, early maturation, and relatively high reproductive output (Shine et al. 1999; Madsen and Shine 2000a) of tropical species might allow them to survive higher harvesting levels. Clearly this hypothesis requires much more rigorous testing before it should be used to justify extensive harvesting of tropical snakes.

The most infamous example of human hunting of temperate-zone snakes is the rattlesnake roundup, versions of which are held in eight U.S. states (Fitzgerald and Painter 2000). Although up to 18,000 snakes have been reported to be killed in a single weekend event (Weir 1990), the biological

1994), longer-term responses to humans can be detected. In a Canadian park, gravid female *Sistrurus c. catenatus* in areas with higher human activity were less visible to observers and the movement of all snakes decreased (Parent and Weatherhead 2000). No effects of human activity were found on the condition, growth, or litter size of the snakes, but the possibility that such effects exist could not be ruled out (Parent and Weatherhead 2000). These effects were tested, however, in ratsnakes that carried transmitters for a year (Weatherhead and Blouin-Demers 2004a). These snakes gained less mass and produced relatively lighter clutches than snakes without transmitters. One possible proximate mechanism for this transmitter effect was that human observers spent more time near the snakes with transmitters because they were tracking them, causing those snakes to modify their behavior in ways that reduced their food intake (Weatherhead and Blouin-Demers 2004a).

The way in which snakes respond to people appears to vary due to multiple factors. Environmental variables such as temperature and habitat affect responses (Prior and Weatherhead 1994; Shine et al. 2000). What matters more from a conservation perspective, however, are the factors that can potentially be managed. For example, when basking on a canal bank adjacent to a pedestrian path, Northern Watersnakes (*Nerodia sipedon*) and *Thamnophis sirtalis* were disturbed in response to both the greater number of people and the closer pedestrian proximity (Burger 2001). By determining threshold distances and the number of people that disturbed the snakes, Burger (2001) was able to make explicit recommendations about where paths should be located relative to snake basking areas to reduce the impact that people had on the snakes.

Individual and Population Effects of Roads on Snakes

During the last century, many snake habitats have become intersected by roads (Mader 1984), and these structures may impact up to 20% of the total land area of some densely populated countries (Reijnen et al. 1995). For this reason, many snake populations are likely to come in contact with roads and therefore have the potential to be affected by roads. Our including a review of snake-road interactions under the broader heading of predator-prey interactions is more appropriate than it might first appear. First, as with predators, snake interactions with roads result in large numbers of snakes being killed (Dodd et al. 1989; Bernardino and Dalrymple 1992; Rosen and Lowe 1994; Ashley and Robinson 1996; Smith and Dodd 2003; Andrews and Gibbons 2005; Roe et al. 2006; Row et al. 2007). Much of that mortality may be accidental, but some is almost certainly intentional (Langley et al. 1989; Ashley et al. 2007). Second, an understanding of predator-prey ecology can help us predict and interpret how snakes respond to roads (e.g., Shine et al. 2004a; Andrews and Gibbons 2005).

Constructing a road within any snake habitat will almost certainly have negative consequences for local snake populations. It is possible that some minor benefits could result, such as creating edge habitat that is preferred by snakes (Blouin-Demers and Weatherhead 2001b), but even in these cases, the net effect is still likely to be negative. If snakes cross roads, road mortality will certainly occur. Alternatively, if the road is a barrier to snake movement, dispersal will be reduced, resulting in population fragmentation and reduced gene flow (Shine et al. 2004a; Andrews and Gibbons 2005). Despite the well-documented snake mortality on roads, biologists have only recently begun to study how snakes behave when they encounter roads—yet knowledge of this behavior is essential for understanding the impact of roads on snake populations. Studies of how different snake species (and ages and sexes) respond to roads and vehicles (e.g., Shine et al. 2004a; Andrews and Gibbons 2005; Row et al. 2007) can help us predict which species will be at greatest risk from roads. Using mortality data in population models (e.g., Row et al. 2007) can help predict the magnitude of this risk. The general assumption is that snakes are likely to avoid roads to reduce their exposure to predators, and the trade-off between the perceived risk and the benefit from crossing a road will shape crossing behavior. Species at less risk of predation (larger, venomous snakes) may be more likely to cross roads than species with higher predation risks. The benefits of this behavior (e.g., finding mates) might cause males (and females; Blouin-Demers and Weatherhead 2002b) to cross roads more frequently during the mating season.

Consistent with the prediction that snakes perceive roads as dangerous habitat, Red-sided Gartersnakes (*Thamnophis sirtalis parietalis*) that encountered roads during dispersal often changed direction to move parallel to the road (Shine et al. 2004a). Subjects that crossed roads did so at right angles, thereby minimizing the distance traveled on the road. Furthermore, male *T. s. parietalis* were unable to follow female scent trails across a gravel road, suggesting the additional cost of lost mating opportunities (Shine et al. 2004a). Similar responses were recorded following the experimental release of nine species of snakes along the margin of a paved road (Andrews and Gibbons 2005). Not only did snakes cross at right angles to the road (consistent with Shine et al. 2004a), but smaller snake species were less likely to cross the road. Venomous snakes crossed the road more slowly than non-venomous species, potentially reflecting both the lower risk of predation and intrinsically slower locomotor ability of stout-bodied, venomous species (Andrews and Gibbons 2005). Furthermore, most Eastern Racers (*Coluber constrictor*), *Pantherophis alleghaniensis,* and Timber Rattlesnakes (*Crotalus horridus*) froze in response to a vehicle approaching or passing them on the road. Andrews and Gibbons (2005) proposed that this freezing response could explain the belief that snakes bask on roads because freezing involved snakes flattening themselves against the road surface, as has been described for basking behavior (Sullivan 1981). The implications for how roads affect snakes are different if snakes actively move on to roads to bask versus snakes assuming a basking-like posture when confronted while crossing a road. It is important for researchers to differentiate between these behaviors and to assess the circumstances under which each occurs.

The experimental approaches used in studies of how snakes respond to roads (Shine et al. 2004a; Andrews and Gibbons 2005) documented the immediate responses of snakes encountering a road, but they did not assess the longer-term effects. For example, if a snake stops at a road margin and does not cross in a trial lasting several minutes, does that mean that the snake never crosses the road? Telemetry allows us to assess how snakes respond to roads in a longer time frame. Richardson et al. (2006) found that Prairie Kingsnakes (*Lampropeltis c. calligaster*) almost never crossed roads, although one individual used both sides of a road by moving back and forth under a bridge. In addition, females were often found near roads and their home ranges abutted roads, consistent with roads acting as barriers to movement. In contrast, data from an 8-year telemetry study of ratsnakes produced no evidence that roads affected movement, either overall or when data were analyzed separately by gender or season (Row et al. 2007). Subjects crossed roads just as frequently as random walks generated using matching starting points, distances moved, and movement frequencies. These results contrast with those of Andrews and Gibbons (2005), suggesting either that ratsnakes behaved differently in their experiments than they do naturally or that short-term reluctance to cross roads does not lead to less frequent road crossing.

In addition to increasing our understanding of how snakes respond to roads, we need to quantify the impact of roads on snake populations. For example, using a previously derived estimate of population size and the observed rate of road mortality for their radio-tracked ratsnakes, Row et al. (2007) used population viability analysis (see Dorcas and Willson, Chapter 1) to model the effect of the estimated mortality rate on the population. Even though the estimated risk of mortality was only 0.026 deaths per crossing and individual snakes were estimated to cross the road less than once per year on average, the resulting mortality was still sufficient to increase the probability of extinction for their study population from 7 to 99% over 500 years. In their population of 400 snakes, road mortality of just three adult females per year increased the extinction probability to over 90% over 500 years (Row et al. 2007). In long-lived species with delayed sexual maturity, such as ratsnakes, populations are expected to be sensitive to adult mortality, particularly adult female mortality (Brooks et al. 1991; Congdon et al. 1994). In turtles, nesting behavior results in substantially greater mortality of adult females on roads (reviewed in Steen et al. 2006). The differences in mobility and nesting substrate between turtles and oviparous snakes makes it unlikely that snakes face similar adult female-biased mortality.

The conservation actions available to lessen the negative effects of roads on snakes are limited to erecting barriers to prevent snakes from crossing roads and creating underpasses designed to allow snakes to move under roads. These topics are discussed by Shoemaker et al. (Chapter 8). From a behavioral perspective, however, additional research is required to explore how snakes respond to barriers and underpasses so that these structures can be designed to maximize their effectiveness.

Future Research

Linking behavioral ecology to conservation requires identifying how individual attributes (behavior or morphology) affect population performance and how that effect can be used to manage the population. We found only a few clear cases in which all these links have been identified. Part of the problem lies in the historic rarity of snakes as subjects of behavioral ecology research, a pattern that is fortunately changing. Another part of the problem is that linking individual behavior to population performance is challenging, regardless of the taxonomic group being studied. Nevertheless, there are areas where it seems likely that behavioral ecology can inform conservation efforts. For example, the importance of thermal ecology to snakes, coupled with the link between habitat selection and thermoregulation, suggest that an improved understanding of behavioral thermoregulation will be important in identifying critical habitats or microhabitats, which can then be managed. Thermal ecology is also important for understanding how snakes will be affected by climate change, particularly global warming. Whether that leads to conservation recommendations other than the obvious—reversing global warming—remains to be seen.

We now discuss areas where we think research in snake behavioral ecology will contribute to snake conservation. We begin with a general argument in favor of long-term behavioral ecology studies of snakes. Because behavioral ecologists are interested in individuals, their research is likely to produce measures of individual rates of survival and reproduction that can be critical to models aimed at analyzing population viability. In some instances, behavioral studies can also identify behaviors that contribute directly to population viability (LeGalliard et al. 2005; Gerber 2006). By studying the same populations through time, behavioral ecologists are likely to accumulate data that allow us to identify how populations are changing through time and to test hypotheses about the factors that are responsible. Even though the data may not have been collected with these types of analyses in mind, researchers need to recognize the value of using long-term data for conservation purposes. Favoring long-term studies in ecology is akin to extolling the virtues of motherhood. Therefore, it needs to be stressed that long-term research for conservation purposes must be seen as a means to an end (i.e., published analyses of trends) and not simply as something intrinsically worthwhile (see Dorcas and Willson, Chapter 1; Seigel and Mullin, Chapter 11).

Among the specific aspects of snake behavioral ecology that warrant more attention are the thermal ecology and habitat selection of reproductive females. We need to know more about the features of nests or gestation sites that enhance reproductive success and offspring quality so the best sites can be protected or so the features of those sites can be mimicked in the design of artificial nest and gestation sites. To enhance use of artificial sites we need to know how female snakes locate nests or gestation sites. For example, female Common Keelbacks (*Tropidonophis mairii*) preferentially lay their eggs where there are empty eggshells (Brown and Shine 2005b) and daughters return to nest in the same location as their mothers (Brown and Shine 2007). If other snakes behave similarly, placing eggshells in artificial nests could promote their use, as would moving clutches from threatened nests into safe artificial nests.

General studies of thermal ecology and habitat use can also provide information useful to snake conservation. The preferential use of particular sites for basking, shedding, or retreat could identify opportunities for habitat modification. Furthermore, if our goal is to understand the thermal factors that limit snake ranges and thus predict how snake ranges might shift in response to global warming, we need to know more about how temperature variation affects rates of reproduction, hatching success, and recruitment. If environments appropriate for particular snakes are shifting, we need to know much more about snake dispersal to know whether snakes can shift with their habitats. In the past, snakes successfully colonized areas that had been glaciated, but the speed of habitat shifts associated with global warming and the effects of barriers to dispersal (e.g., roads and urbanization) have created new conditions. Because we rely on radiotelemetry to study snake movement and because juvenile snakes are the probable dispersers, the size limitation on the transmitters remains a major technological problem, even for species in which the adults are large enough for telemetry.

A better understanding of the interactions of snakes with their prey species would be valuable in several ways. Often we know little about what snakes eat, and yet diet information can be important in interpreting population fluctuations, responses to invasive species, and predicting how changes in prey communities (e.g., associated with climate change) might affect snake populations. Understanding the hunting behavior of snakes could also help identify ways to protect endangered prey species to which snakes pose a threat (Weatherhead and Blouin-Demers 2004b).

Research on snakes as invasive species is necessarily very targeted, but control techniques such as using pheromones to capture snakes could have general utility. Where snakes are at risk from invasive species, conservation measures are likely to be specific to a particular circumstance (e.g., snakes and Cane Toads in Australia). One exception that could prove to be more widely relevant is the effect of invasive fire ants on snakes, only because fire ants have been introduced in so many places. At present we do not even know whether fire ants have a significant impact on snakes, only that the potential exists. We need to know much more about the extent to which snake eggs or neonates are preyed on by fire ants. Similarly, we need to know how juvenile and adult snakes avoid fire ants and the cost of such avoidance behavior.

There are many research needs and opportunities in the area of understanding how humans affect snakes. In the case of hunted species, what selection pressures does hunting place on populations, and what are the behavioral and life-history consequences? We are beginning to know more about how snakes respond to roads, but we need to expand the number of species examined. We also need to know how snakes respond to barriers to keep them off roads and to conduits intended to allow snakes to pass under roads, so both kinds of devices can be made more effective. Finally, research is needed to assess how snakes are affected by human activity. As more snake populations become restricted to reserves and parks it is essential that we determine whether recreational activities as apparently benign as hiking or wildlife observation actually cause snakes to modify their behavior, with detrimental long-term consequences.

6

Reproductive Biology, Population Viability, and Options for Field Management

RICHARD SHINE AND XAVIER BONNET

Because reproductive individuals are more at risk,
they are often the key for the population viability.

SHINE AND BONNET, 2000

No individual organism is immortal, so the viability of any biological population ultimately depends on reproductive success. Over a broad spatial scale, the balance between the rates of production of new individuals (reproduction) and the rates of loss (mortality) determines the number of animals within any population. Thus, the challenge for conservation biologists is to understand both sides of this equation. Frequently, attention is focused largely on the determinants of mortality and, thus, on processes such as predation, depletion of resources, disease, competition from invasive species, and anthropogenic sources of mortality. All are major threats to population viability in a world subject to intense and widespread human disturbance. It is important to recognize, however, that processes affecting the recruitment of new individuals may be equally sensitive to human activities and, hence, that any comprehensive understanding of threats to population viability needs to incorporate a clear understanding of the reproductive biology of the taxon in question. In this chapter, we review (1) research and ideas about the link between reproductive biology and conservation in snakes and (2) actual reproduction-focused management manipulations designed to achieve conservation objectives in snakes.

We begin by defining two main terms. Under the phrase *reproductive biology,* we include the life-history parameters tightly linked to reproductive output: offspring number and offspring size; rates of growth and thus ages at maturation; and more obvious traits such as reproductive mode, reproductive frequency, and effort. We also include aspects of the mating system (e.g., the presence or absence of male-male combat or extensive mate-searching) and other reproduction-specific behaviors (e.g., nesting migrations, aggregations, and courtship and mating behaviors). We define *conservation* in a similarly broad fashion, focusing on the development of rational, information-based approaches to conserving biodiversity. Some of those approaches might focus on the preservation of declining populations, whereas others might instead aim to reduce the numbers of an invasive species of snake or estimate the sustainable levels of harvest for commercial utilization of wild populations. Although the aims are very different, the underlying philosophy is the same—that an understanding of the reproductive biology of the study organism can provide a robust basis for developing effective management tactics. This field is in its infancy with snakes, but the opportunities are exciting and the initial results encouraging.

Our primary focus in this chapter is on understanding the reproductive biology of animals in the field. This is an essential prerequisite to providing clear advice to wildlife managers, to manipulating (generally enhance) reproduction *in situ*, and to setting up educational programs. Alternative approaches whose aim is to establish links between reproductive biology and snake conservation, such as endocrinological investigations, captive breeding, and re-introduction, are also important (Kingsbury and Attum, Chapter 7). We believe, however, that field-based programs based on reproductive phenology under natural conditions are essential for two major reasons. First, conservation programs will be more feasible logistically if they do not have to overcome the challenges associated with the long-term housing of captive snakes (but, see Chiszar et al. 1993) or await the outcomes of complex academic research. We cannot afford the luxury of allocating the limited resources for snake conservation (Clark and May 2002) to processes unlikely to protect snake populations in any short time frame. Second, educational programs conducted in the field can be more effective than are those performed with captive animals. Field experiences (even short ones) are very effective in educating people about the necessity to conserve both habitats and species. The framework we have adopted for this purpose is oriented toward the ways in which information on reproductive biology might be used for conservation-relevant tasks: to predict endangerment; to clarify and ameliorate the processes causing endangerment; and to identify the critical resources, times, and places at which reproductive activities create points of vulnerability for the study population.

Given that we know so little about most snake species, especially threatened taxa, how can we use information on reproductive biology to inform management decisions? Funds for conservation are extremely limited, especially for unpopular organisms such as snakes (Clark and May 2002; Seigel and Mullin, Chapter 11). Snake biologists have a role to play in modifying that situation, but realistically, we cannot afford to invest heavily in the conservation of all potentially threatened taxa. Unfortunately, the problem is compounded by our lack of detailed information about actual population status for most species of snakes, even in parts of the world with a long history of detailed ecological research on these animals (Fitch 1999; Shine and Bonnet 2000). For many regions on the planet, especially in tropical and subtropical areas, we still have little real understanding of the taxonomy and phylogenetic relationships of the snake fauna, let alone their detailed ecology (Zhao and Adler 1993; Greer 1997; Branch 1998). Thus, wildlife management authorities are faced with the challenge of allocating resources to conservation based on very incomplete information about which taxa are most under threat.

Here, we attempt to provide useful practical advice for nonspecialist wildlife managers. Because this field is in its infancy, we are forced to rely heavily on anecdotes and the results of unpublished experiments. We believe that successful initiatives, even if limited in their impact, should be reported to encourage further testing in the field; in the long term, some of these ideas might well prove broadly applicable for snake conservation.

Background of Snake Reproductive Biology

Because the current chapter is aimed toward providing practical information, we limit this discussion to some major features of snake biology and ecology. In the vast majority of snake species, most individuals are extremely secretive and remain largely invisible (at least to humans) except during reproduction or during periods of emergence from hibernation (which are also associated with mating in many cases). More than in most other vertebrates, reproduction generates behavioral shifts that can place the snake in danger. Although nonreproductive individuals typically retain their cryptic habits throughout the year, reproduction frequently generates vulnerability, for example, intensive displacements for mate-searching in males and migration to oviposition (egg-laying) sites in females with long periods of exposed basking during follicular growth (vitellogenesis) and pregnancy. This situation has at least two major consequences: (1) we must understand the process of reproduction if we want to minimize mortality, notably via the improvement of disturbed habitats, and (2) we can exploit the conspicuous behaviors of snakes during reproduction as a monitoring tool (e.g., to determine population status), again calling for an understanding of reproductive biology to correctly interpret the field data.

In addition, the support (and it is hoped, participation) of local people is critical to the success of most conservation initiatives. To set up efficient educational plans, it is important to involve the targeted population (e.g., schoolchildren) in practical activities; improving habitats is one such action that should be based on a thorough knowledge of snake ecology. Similarly, to educate people about snake ecology rather than focusing on the sensational aspects typically emphasized in the popular media, opportunities for the general public to observe free-ranging animals can be very effective. For these purposes, reproduction is probably the easiest aspect of snake natural history that can be observed and explained. Because this is also one of the most risky periods for the snakes, it is also a time when conservation actions are likely to be overtly effective and, hence, rewarding for the people involved.

The diversity of snake reproductive strategies combined with the flexibility of these animals generates an immense range of behaviors (Shine 2003). Each population is unique and there is no standardized information or management recommendation that is likely to apply to all species. Nevertheless, several broad characteristics of snake reproductive biology can be identified. Neonates (newly hatched or born offspring) are independent at birth and do not receive any form of food provisioning from their parents. In males, reproduction is apparently limited to the fertilization of the female's ova. The female reproductive task is far more demanding because females invest extensive resources to produce mature follicles. Therefore, in field populations, three main categories directly associated with reproduction can be distinguished: adult males, adult females, and neonates. Each category requires specific conservation actions.

1. Males are at risk during the mating season. There is a strong competition for access to females. In temperate and cold areas, early in the active season, males tend to bask frequently and are vulnerable to predation. Later, whatever the climate, males often undertake extensive and frequent displacements to locate reproductive females. This is probably the most perilous stage for adult males. Unfortunately, the fittest individuals (large males in good body condition) may be most at risk because they travel over longer distances (Madsen et al. 1993). Limiting male mortality associated with mate-searching activities is a priority.

2. Females develop large, and sometimes numerous, follicles. Vitellogenesis is a long process (requiring weeks or even months; Aldridge 1979), during which the metabolic rate of the female must be elevated. In temperate or cold countries, vitellogenic females bask in the sun. In warmer areas, reproductive females may attempt to maintain high body temperature during the night by selecting hot substrates. In many tropical regions, this behavior results in gravid females lying on paved roads at night and, hence, being subject to high levels of mortality due to vehicles. Providing alternative (and low-risk) sites that provide suitable thermoregulatory opportunities can help to protect reproductive females during vitellogenesis. After vitellogenesis is completed, an important distinction arises between oviparous and viviparous species. In the latter group, precise thermal requirements for embryonic development favor careful thermoregulatory behavior in females. To conserve viviparous snakes,

the availability of appropriate thermoregulatory sites remains a key factor during pregnancy. The risks to oviparous females are different. They do not need to thermoregulate as carefully for long periods (because they lay rather than retain the eggs), but they must find oviposition sites that provide safety for their clutch, as well as suitable thermal and hydric conditions throughout incubation. Females sometimes travel over long distances to such sites and may face major predation risks in the process (Seigel et al. 1987). Thus, the availability of appropriate laying site close to the usual activity range is a priority for oviparous species.

3. Whatever the reproductive mode, neonates tend to disperse rapidly and are often vulnerable to predation early in life. Thermoregulatory sites for gravid viviparous females, and oviposition sites for oviparous females, thus must be adjacent to (and preferably, surrounded by) microhabitats favorable for the neonates.

Based on these broad characteristics, we can examine conservation priorities and field management tactics to suggest useful directions.

Determining Conservation Priorities

Even at a crude scale, collecting ecological information on the reproductive biology of each snake species is essential. Analyzing and interpreting that kind of information relies on academic research programs. Can we afford to wait for the outcomes of these (often long) research processes to set up conservation actions (Seigel and Mullin, Chapter 11)? There is a strong argument from hands-on managers that, in practice, the critical issue generally boils down to the simple preservation of habitat. If the habitat is retained, the snake (or other) population will generally persist also; whereas if the habitat is lost, so is the population of interest. We have sympathy for this approach and suspect that it contains much empirical validity. Thus, we strongly advocate the vigorous protection of representative habitats as the essential core of any broad conservation initiative (see Jenkins et al., Chapter 4; Weatherhead and Madsen, Chapter 5).

Ultimately, however, it is not enough for at least three reasons. First, even when habitat is retained and humans are excluded, anthropogenically influenced processes may threaten population viability. For example, invasive species may kill and consume the threatened animals, or invasive vegetation may significantly modify ambient thermal environments or prey availability. Even in the absence of exotic species, vegetation changes associated with ecological succession (following the modification of agricultural practices or fire regimes) can eliminate suitable habitat via the removal of direct human intervention (Flannery 1994; Fitch 1999, 2006; Shoemaker et al. Chapter 8).

Second, snake populations can persist even in highly modified habitats, and conservation in such systems requires a far more detailed understanding of the nature of the interaction between people and snakes than is currently available. Minor changes to agricultural practices (e.g., maintenance of shrubby borders around fields) can be critical for successful reproduction (e.g., Madsen 1984; Shine and Fitzgerald 1996; Shoemaker et al. Chapter 8).

Finally, public education to change attitudes toward snakes has the potential to dramatically reduce incidental mortality due to roadkills or malicious killing (Bonnet et al. 1999b; Andrews and Gibbons 2005; Burghardt et al., Chapter 10). In the absence of detailed information on the reproductive ecology of the snakes, there is little chance to set up educational programs. Several snake species become visible to humans only during reproduction, especially when the animals exhibit conspicuous behaviors such as courtship, mating, or male-male combat and when they lay their eggs in sites created by human construction (e.g., walls and terraces). Education about the ecological context of the behaviors involved can play a major role in persuading people to tolerate the snakes by helping residents to understand why snakes sometimes visit their gardens. The visibility of reproductive snakes also provides an opportunity to develop school programs, including the construction of appropriate oviposition sites (X. B. is currently developing such a program with several primary schools, which is generating great enthusiasm from both the students and the teachers). Thus, conserving snakes is not simply about conserving habitats.

Captive-Raised Snakes

One link that most members of the general public would see as important is the idea that we can "save" an endangered population by bringing animals into captivity. These snakes can then be bred so that their offspring can replenish the wild population, presumably after management efforts in the field have removed or minimized the otherwise fatally threatening process (Kingsbury and Attum, Chapter 7). Although this approach has been advocated widely (especially with high-profile mammalian and avian taxa, Sarrazin and Legendre 2000; but also with reptiles, Pedrono and Sarovy 2000; Tuberville et al. 2005), there are few examples of success (e.g., with *Epicrates subflavus;* http://www.durrellwildlife.org; Kingsbury and Attum, Chapter 7). The reality is that there are often no places into which the captive-bred animals can be released. We have no objection to the practice; it can play an important role in public education (Mittermeier et al. 1992; Kingsbury and Attum, Chapter 7), and there may be circumstances in which the maintenance of animals in captivity can contribute to population viability. For example, it is possible to capture reproductive females during late vitellogenesis or pregnancy and release them just after laying or parturition. In this way, the animals can be protected from predators during the phase of the reproductive cycle when they are most vulnerable and we can maintain them in optimal conditions (of temperature regimes and food and water supply) during the phase when offspring development is most sensitive to ambient conditions.

Many authorities believe that the mortality rates of neonate snakes are often high and, if so, head-starting offspring may be a useful conservation tool (King and Stanford 2006). We have conducted a preliminary test of this idea in Australian Tigersnakes (*Notechis scutatus;* Aubret et al. 2004) by capturing adult females during late vitellogenesis, maintaining them in captivity until they gave birth, and then raising the neonates for 2 years before releasing them into the field. Fed on mice during this 2-year period, the young snakes increased in mass from approximately 10 g to more than 100 g, therefore escaping a perilous life-history stage before being released into the wild population. Three of these seven animals were recaptured 2 years later in good condition (mean body mass = 283 g), demonstrating that they adapted well to their novel environment despite a long period of captivity. King and Stanford (2006) used a similar procedure with the Plains Gartersnake (*Thamnophis radix*) and found no differences between captive raised offspring and those from the wild. These results indicate that captive breeding may allow successful reintroduction of individuals in the field, at least for some species.

Use and Limits of Current Data Sets

Species-Focused Field Studies
Next we consider the situation of a threatened population that has already been the subject of detailed ecological and demographic studies. With extensive data sets of this kind, it is possible to develop mathematical models to investigate the numerical trajectory of the study population (Is it stable, growing, or declining?), the reasons for any such trend (e.g., the relative roles of recruitment and mortality rates), and the probable consequences of changes to any of the main parameters within the model (e.g., Would population viability be enhanced more by a 10% increment in juvenile survival rate than by a 10% increase in mean litter size?—see Webb et al. 2002b). Empirical studies suggest that population viability analyses (PVAs) of this type have been used successfully in predicting subsequent trends in population numbers for turtles, lizards, and other vertebrates (Heppell 1998; Berglind 2000; Brook et al. 2000, 2002; Mitro 2003), although less commonly for snakes (but see Dorcas and Willson, Chapter 1). Given the long-term mark-recapture data sets becoming available on several snake species worldwide (e.g., Timber Rattlesnakes, *Crotalus horridus,* Brown 1993; Martin 2002; European vipers, *Vipera aspis, V. berus,* Madsen and Shine 1994; Bonnet et al. 1999b; Australian Water Pythons, *Liasis fuscus,* and Broad-headed Snakes, *Hoplocephalus bungaroides,* Madsen and Shine

1999b; Webb et al. 2003), it would be interesting to see PVA applied to those data. These might reveal generalities about the demographic processes structuring snake populations.

Unfortunately, this approach has rarely been applied to snakes because the necessary detailed field-based data sets generally have not been available or because snake biologists have not availed themselves of these methods (but see Dorcas and Willson, Chapter 1). In addition, some long-term studies have revealed remarkably high levels of temporal as well as spatial variation in major life-history parameters such as reproductive frequency, litter sizes, costs of reproduction, and survival rates of adults and juveniles. Such annual variation can generate massive fluctuations in population size, structure, and recruitment (Madsen and Shine 2000b). For example, within a single population of *Vipera aspis* the abundance of adult females fluctuated greatly among years, whereas the numbers of adult males remained relatively stable (Bonnet et al. 2001; Lourdais et al. 2002). Similarly, important reproductive parameters, such as relative litter mass, exhibit substantial temporal and spatial variation in this species (Bonnet et al. 2003). Most dramatically, comparisons between *V. aspis* populations at the northern limit of the distribution range with those in the south reveal semelparity in the former (most females produce only a single litter during their lifetime) but iteroparity in the latter (Zuffi et al. 1999; Bonnet et al. 1999c, 2002a). Thus, short-term data sets may provide only a weak basis for any inference about long-term population viability.

One factor contributing to this problem is a rarely noticed artifact arising from the ways that researchers select study populations—typically, they choose to study organisms at a place where those animals are common and easily found. That decision is motivated by the ease of observing free-ranging animals and of collecting abundant (hence, publishable) data. In recent decades, most of the ecological studies carried out on snakes have been based on carefully selected populations. In any system with marked interannual variations in abundance, this means that any study is more likely to record a population decline than a population increase (because the study probably began when the population was at its peak). Thus, we need long-term data sets in order to overcome such methodological biases. How long is long enough? Unfortunately, the answer to that question depends on the life history of the study organism as well as the time scale of fluctuations in the ambient conditions.

Adding to this difficulty is that even if we have detailed information on a single study population over a long period, it may be difficult to extrapolate the results widely on a spatial scale. Studies on geographically widespread snake species have revealed an unexpectedly high diversity in mating systems and reproductive biology. For example, populations of Australian Carpet Pythons (*Morelia spilota*) differ in whether or not rival males engage in combat bouts during the mating season and, thus, in the direction of sexual size dimorphism (Shine and Fitzgerald 1995). Populations of an Asian colubrid, *Amphiesma pryeri*, differ in whether females reproduce by laying eggs or giving birth to live young (Ota et al. 1991). Many other traits, for example, reproductive frequency, display a high level of phenotypic plasticity so that local variation in prey abundance or climatic conditions can generate massive local variation in life-history traits (Seigel and Fitch 1985; Madsen and Shine 1999b, 2000b). Thus, even if we have detailed information about one population, this may provide only a weak basis for inferring ecological traits of other populations of the same species.

Climate and Population Dynamics

Broad biological principles and general patterns in snake life histories provide a basis for inference even when detailed information is unavailable or unreliable. Temperature determines the rate of almost all biological processes, including demographic traits such as growth rates, ages at attainment of sexual maturity, energy acquisition rates, and reproductive frequencies (Adolph and Porter 1993; Weatherhead and Madsen, Chapter 5). Thus, we can confidently predict that snake populations in relatively cold climates will tend to have slower life histories (and, hence, lower rates of population recovery after any crash) than genetically similar populations inhabiting warmer climates that allow longer activity seasons each year. For example, female Aspic Vipers (*Vipera aspis*) reproduce on an approximately 1-year cycle in southern Europe, but are biennial in their northern distribution in France (Naulleau 1984; Zuffi et al. 1999) and triennial in the Alp mountains (Saint Girons and Kramer 1963). Thus, the sustainable level of harvesting from wild populations will likely be higher, as a general rule, for tropical taxa than for temperate-zone species (Shine et al. 1995, 1999b). Other factors exacerbate this vulnerability of cool-climate populations, notably the higher incidence of viviparous rather than oviparous reproduction (and, hence, a reduced clutch frequency per female). Gravid females also tend to select more exposed positions (thus, increasing their vulnerability to predators) because of the difficulty of maintaining optimally high incubation temperatures for their developing embryos.

Another potential solution to the problem of limited data and dubious extrapolation from very detailed single-population studies is to work with more superficial data sets at a broader taxonomic scale. We can use available information to look for any consistent ecological differences between threatened and nonthreatened taxa. Such an analysis has two potential benefits in terms of conservation planning. First, traits correlated with endangerment can suggest hypotheses about the causal reasons for the observed population declines—for example, correlations between endangerment and slow life histories (i.e., late maturation, infrequent reproduction, and small litters) suggest a causal role for low recruitment rates in population decline (Webb et al. 2002b). Second, such an analysis may identify species that, although not currently recognized as endangered, share many of the same life-history attributes as threatened taxa. Given the parlous state of our knowledge about snake populations, many of these suspect taxa also may be in danger and, hence, warrant conservation-oriented research to test this prediction. To our knowledge, the only such analysis for snakes was that by Reed and Shine (2002) on Australian snakes, enabled by a large quantitative database on these animals arising from examination of museum specimens (Shine 1994). These authors detected nonrandom associations between endangerment and a variety of traits, including ambush predation (linked to slow life histories because of low and inflexible feeding rates; Webb et al. 2003) and mating systems. Snake taxa that do not display male-male combat tend to have females that attain larger body sizes than conspecific males, and such large animals may be more vulnerable to anthropogenic disturbances (including direct predation by humans). Hence, there may be strong and nonintuitive links between population vulnerability and ecological traits such as foraging biology and mating systems. Reed and Shine's (2002) analysis also identified several nonthreatened taxa that share such traits and, hence, warrant more detailed investigation.

Adaptive Responses to Environmental Changes

There is at least one other potential application of reproductive data to broad species-level issues about conservation priorities. Conservation biologists generally have focused on immediate threats to population viability and have ignored the possibility that evolutionary changes to their study populations may play a significant role in helping the species cope with anthropogenic assaults. There is increasing evidence, however, that snakes can evolve rapidly enough that possible adaptive responses are relevant to predicting the long-term impact of threatening processes. For example, frog-eating Australian snakes have experienced significant population declines due to the spread of feral Cane Toads (*Rhinella [Bufo] marina*) because the toxins in these anurans are deadly to the snakes (Phillips et al. 2003). The snakes have rapidly evolved, however, in ways that reduce the impact of toads. Over the 70-year history of toad colonization in Australia, at least two species of snakes have evolved relatively smaller heads, thus decreasing their ability to ingest a toad large enough to make a fatal meal (Phillips and Shine 2004). Similarly, Redbellied Blacksnakes (*Pseudechis porphyriacus*) in toad-infested areas display a shift in feeding habits (they refuse to eat toads) and an increase in physiological tolerance to the toads' toxins (Phillips and Shine 2006). Microevolutionary theory tells us that the rate at which such changes can occur is constrained by factors such as the level of variation in relevant traits and demographic features such as fecundity (hence, opportunity for selective mortality). Studies on such reproductive traits can provide a basis for predictions about the longer-term rather than immediate impacts of invasive species and, hence, inform management priorities.

Identifying Ecological Requirements for Population Viability

Food Resources

The reproductive biology of a population affects the kinds of resources that it will require to remain viable. Imagine two snake species with identical adult body sizes, mating systems, and total energy allocation to reproduction, but differing in the packaging of that output. One species has large litters of very small offspring, whereas the other has small litters of large offspring. Comparisons among snake species, and between snakes on the one hand and lizards on the other, suggest that a wide range in body sizes from hatching to adulthood (as is inevitable with a small relative offspring size) is a strong predictor of ontogenetic niche shifts (Shine and Wall 2007). Many snake species produce small offspring that feed on ectothermic vertebrates (frogs and lizards) during juvenile life and switch to mammals as they grow larger (Mushinsky et al. 1982; Mushinsky 1987). In some species that attain very large maximum body sizes, mammals are the preferred prey items throughout life, but the types of mammal eaten changes from rodents to much larger animals. For example, Reticulated Pythons (*Python reticulatus*) in Sumatra take rodents almost exclusively until the snake attains about 3 m in length, at which point it shifts to larger prey such as pangolins, monkeys, and wild pigs (Shine et al. 1998a). In contrast, snake species that produce relatively large offspring (and thus, fewer of them) often specialize on one or a few types of prey throughout the snake's lifetime because these prey are ingestible even by newborn individuals. Aspic Vipers feed mainly on voles (*Microtus* spp.; Naulleau 1984) throughout their life, with neonatal vipers taking juvenile voles, gradually increasing the size of the voles consumed as the snake grows larger. *Notechis scutatus* on Carnac Island exhibit a more complex ontogenetic transition—the neonates appear to feed only on newborn mice; during the juvenile phase, they shift to adult mice and lizards and, finally, as adults, are able to swallow seagull chicks (Bonnet et al. 1999a). Important intra- and interspecific flexibility may occur in these broad patterns, but there is little doubt that ontogenetic shifts in diet are inevitable for many species.

A snake species with an obligate ontogenetic shift in prey types depends on the persistence of multiple prey types within its habitat, whereas a snake species that lacks such a shift may depend on only one prey taxon. A population of *Natrix natrix* may require both amphibians and rodents, whereas a population of Aesculapian Snakes (*Elaphe longissima*) can persist with one of these resources (e.g., voles only). Accordingly, successful management of the former species will require careful management of multiple prey taxa, often (as in the case of many broadly sympatric anurans and mammals) requiring quite different microhabitats and resources. In turn, conservation zones may need to incorporate an array of habitat types to ensure continued viability for all size classes of predators within the snake population.

An even more direct conservation implication of relative size at birth comes from the case of a toxic invasive species such as the Cane Toad. Within Australia, snakes that feed on frogs during juvenile life are potentially vulnerable to the toad invasion, even though the adults of some of these taxa specialize in mammalian prey. For example, Roughscaled Pythons (*Morelia carinata*) are mammal-eating snakes restricted to a small area of tropical Australia that is about to be invaded by Cane Toads. Recent captive breeding of this threatened taxon suggests that juveniles feed primarily or exclusively on frogs, thereby raising concern about the probable impact of the imminent arrival of toxic toads (R. S., pers. obs.).

Identifying Critical Habitats to Protect

In many species, reproduction induces a shift in habitat use. Thus, information on the spatial location of individuals over the course of the reproductive cycle can help to identify critical habitat elements that should be incorporated in management planning. Such reproduction-related habitat requirements may occur at any stage of the reproductive cycle. For example, courtship and mating in an assemblage of nocturnal saxicolous elapid snakes in southeastern Australia occur mostly during daylight hours in the cooler months of the year while the snakes are in their winter-retreat sites under sun-heated sandstone rocks (R. S., pers. obs.). The snakes actively select warmer-than-average retreat sites (Webb and Shine 1998b), and the availability of thermally suitable rocks has been severely reduced in many areas by anthropogenic processes such as rock theft (commercial collection for garden ornamentation), habitat destruction during illegal snake collecting for the pet trade (Webb and Shine 2000), and vegetation overgrowth (perhaps due to the cessation of traditional Aboriginal "fire-stick farming" practices; Pringle et al. 2003). Because restricted rock availability will concentrate all snakes within the population into a smaller number of retreat sites, such changes presumably modify mating systems (e.g., the ability of large males to monopolize groups of reproductive females) and, hence, attributes such as the incidence of multiple matings by females (a potential influence on offspring viability; see later in the chapter).

Courtship and mating induce habitat shifts, but the requirements for successful embryonic development typically will be more important for habitat selection (see Weatherhead and Madsen, Chapter 5). The rates and routes of embryonic differentiation and growth are highly sensitive to incubation conditions in snakes, as in other reptiles (Ji et al. 1997; Deeming 2004). Not only is hatching success decreased and occurrence of malformation increased when eggs are exposed to unfavorable levels of temperature or soil moisture, but the trajectories of embryonic development can be modified even by subtle shifts in conditions within that favorable range (Shine et al. 1996). During incubation, the thermal and hydric regimes experienced by the clutches influence offspring size, body shape, postnatal growth rate, and antipredator behaviors (Shine et al. 1996; Aubret et al. 2003; Weatherhead and Madsen, Chapter 5). This sensitivity imposes strong selection for the maternal ability to recognize and use nest sites that provide optimal incubation conditions for the eggs. Soil moisture levels determine offspring size in Common Keelbacks (*Tropidonophis mairii*) from tropical Australian floodplains, with dry soils restricting the uptake of yolk reserves and, thus, decreasing hatchling size (Webb et al. 2001). In turn, reduced hatchling size substantially decreases survival rates in the first year of life, so females in this population are under strong selection to oviposit in suitably moist nests (Brown and Shine 2004, 2005a). Hence, any anthropogenic manipulation that reduces the availability of these (spatially restricted) nesting conditions could negatively impact population viability in this taxon.

Embryos within the oviducts of viviparous snakes are buffered against environmental extremes to a much greater degree than are eggs in the nest, but they are not completely immune to such influences. First, gravid females may experience dehydration, because their basking regimes divert them from other activities such as foraging (Shine 1980). Second, even when moisture regimes are not a problem, maternal behavioral thermoregulation sometimes may be unable to maintain optimal developmental temperatures (e.g., Arnold and Peterson 2002; Shine et al. 2005). Annual variation in weather conditions during the gestation period (summer) accurately predicted the annual variation in offspring traits in a northern (cool-climate) population of *V. aspis* (Lourdais et al. 2004). Hence, even in viviparous species, females may need to select sites that facilitate precise thermoregulation. Reinert's (1984) careful multivariate analysis of radio-tracked American pitvipers revealed a strong pattern for gravid individuals to be found in more open (and thus warmer) microhabitats than was true for nongravid conspecifics.

Presumably reflecting the scarcity of thermally and hydrically suitable sites within the landscape, especially in areas with cool climates or thick forests, reproductive females of many snake species have been recorded as aggregating for oviposition or gestation (Graves and Duvall 1995). In extreme cases, all the females from a very wide area lay their eggs in the same communal nest site (*Demansia psammophis,* Covacevich and Limpus 1972; *Laticauda colubrina,* Voris and Voris 1995) or aggregate during gestation in small rookery areas (Shine 1980; Graves and Duvall 1995). Sometimes, more than one species may use the same communal oviposition site, clearly introducing massive vulnerability for local snake populations. One of us (X. B.) has recorded more than 120 newly hatched European Whipsnakes (*Hierophis [Coluber] viridiflavus*) and *Elaphe longissima* killed by local residents as the young snakes emerged from a communal oviposition site (a terrace) near a house. In different sites (situated in a 100-km radius), we have recorded communal oviposition by *Natrix natrix* and *N. maura* as well as the two species already noted. Thus, a high proportion of all the juvenile snakes within a large area may hatch at about the same time and in the same place.

Viviparous snakes do not require oviposition sites, but they nonetheless may cluster at especially favorable sites. Aggregation of gravid females in viviparous snakes has been recorded from a strikingly diverse array of phylogenetic lineages and geographic locations (e.g., elapids in Australia, Shine 1979; viperids in Africa, Stevens 1973; boids in North America, Dorcas and Peterson 1998). The result of such behavior is that an especially critical component of the snake population (gravid females) is concentrated in space and time, so that any disturbance to such areas that reduces maternal survival (or thermoregulatory ability) might be catastrophic for the local population. Accordingly, such sites need to be identified and given special protection. The same is true for communal oviposition sites, especially if females remain with their eggs and, hence, are vulnerable to disturbance during incubation (as in the Plains Threadsnake, *Leptotyphlops dulcis,* Hibbard 1964; and the Australian elapid, *Pseudonaja textilis,* Whitaker and Shine 2003). Female *Tropidonophis mairii* selectively oviposit beside existing eggs or eggshells (Brown and Shine 2005b), so that a small number of sites may be especially significant in terms of nesting intensity.

The most likely reason that specific habitats are important for reproduction involves the availability of suitable sites for incubation or gestation, but other pathways can be involved also. The strong ontogenetic (size-related) dietary shift in Reticulated Pythons (previously discussed) means that adult females (which grow much larger than adult males in this species) feed primarily on large mammalian prey (Shine et al. 1998a). Unlike the commensal rodents that form most of the diet of adult male pythons, the prey types taken by adult females are found mainly in relatively undisturbed areas of forests and swamps—and thus, such habitats are critical to maintaining high reproductive output in these heavily exploited systems (Shine et al. 1998a).

Intuition suggests that the critical issue is a simple one—to conserve the specific habitats that are crucial for reproduction; but other factors may also intrude (Weatherhead and Madsen, Chapter 5). It may also be important to maintain genetic separation among sympatric taxa, or even between adjacent but genetically differentiated local populations. Studies on seasnakes in Vanuatu (*Laticauda* spp.; Shine et al. 2002a) and gartersnakes on the Canadian prairies (*Thamnophis* spp.; Shine et al. 2004c) suggest that introgression between closely related taxa is prevented by species-specific male mate choice (which is, in turn, based on lipid-based pheromones on the female's skin; Mason 1993). A low level of introgression appears to be occurring in the gartersnake system, however, apparently because severely cold conditions at the extreme north of the species' range force all local gartersnakes (of both species) into the same communal overwinter dens. The usual temporal separation in emergence (and hence mating) times in spring is reduced by the short summer season (Shine et al. 2004c). Hence, reproductive males and females of the two species are brought into closer contact than would be the case in warmer climates elsewhere in their joint range. This situation provides obvious opportunities for habitat manipulation, perhaps for the establishment of artificial hibernacula (Burger and Zappalorti 1986), especially if the anthropogenic elimination of previously available overwintering sites has forced such taxa into close reproductive contact (see also Shoemaker et al., Chapter 8).

Temporally Variable Patterns of Vulnerability

Even in the tropics, most snake species display broadly seasonal reproductive cycles (Fitch 1982). Thus, important events in reproduction occur at predictable times within the year. Such seasonal shifts mean that the demographic impact of any threatening process (such as increased mortality risk for active animals due to predation or wildfire) will depend on the time of year at which that threat is experienced. However, reproduction-associated behaviors also can modify the link between risk and demographic impact. For example, intense predation by American Crows (*Corvus brachyrhynchos*) on Common Gartersnakes (*Thamnophis sirtalis*) in Canada occurs mostly in early spring, as the snakes begin to emerge from their long overwinter inactivity period (Shine et al. 2001). As in many seasonally active organisms, male gartersnakes tend to emerge from hibernation prior to females (Gregory 1974) and, thus, might be expected to experience higher mortality. In actuality, the reverse situation occurs—predation is concentrated on females rather than males (Shine et al. 2001). There appear to be two reasons for this unexpected pattern. First, females have greater energy reserves for future expenditure in reproduction, and their energy-rich livers make them a more attractive target for the crows. Second, female gartersnakes tend to emerge and disperse from the den later in the season but at times when the risk of crow predation is high, early in day and in very cold weather when few males are active. This risky behavior results from the high costs of sexual conflict, whereby females avoid intense (and potentially dangerous) courtship by males by dispersing when conditions are too cold for courtship (Shine et al. 2004b). They must then run the risk of predation. The end result of this complex interplay between predator preference and reproductive biology is that gartersnake mortality due to these predation events has more impact on recruitment rates to the next generation than might be expected from the number of animals killed because removing females from the population will have more effect than removing an equivalent number of males (Caughley 1977).

Similar complexities doubtless arise with threatening processes that overlap with other phases of the reproductive cycle. In many cool-climate snake species, vitellogenic and gravid females are the most conspicuous (and thereby vulnerable) component of the population during spring and midsummer because they tend to bask for longer periods than do non-gravid females, males, or juveniles (Charland and Gregory 1990; Bonnet and Naulleau 1996; Bonnet et al. 1999b). The composition of museum collections confirms that egg formation and pregnancy greatly increase the vulnerability of reproductive females in such situations (Shine 1981, 1994). No such effect is apparent in warmer climatic zones because optimal thermal regimes can be attained without risky exposure, and thus, gravid females may be virtually absent from museum collections of such species (Shine 1994).

The degree to which reproduction exposes females to direct killing by humans (and by other predators also; Bonnet et al. 1999b) depends on local climates. In cold areas, it may be of critical importance to prevent the direct persecution or collection of snakes during the times of year that gravid females are most vulnerable. The biases can be considerable—restricting collection to roads during springtime will probably generate a massively male-biased sample, whereas a similar collecting effort devoted to rocky habitats in midsummer may provide mostly gravid females (Shine 1981; Aldridge and Brown 1995; Bonnet and Naulleau 1996). Similarly, even nearby billabongs on tropical Australian floodplains can contain very different sex ratios of Arafura Filesnakes (*Acrochordus arafurae*: Houston and Shine 1993), so that concentrating collection in male-biased waterbodies could considerably reduce the harvest of reproductive females by local Aboriginal people. Such biases provide opportunities to allow harvesting (for subsistence, recreation, or commercial purposes) while minimizing the demographic impact on local populations (Fitzgerald and Painter 2000). Unfortunately, gravid females of many species are considered a delicacy by local hunters and, consequently, are heavily harvested for their eggs (see Shine 1986 for *A. arafurae*). In at least five species of water snakes from the Tonle Sap Lake in Cambodia that reproduce during the summer monsoon season (Saint Girons and Pfeffer 1971), authorities have calculated harvest rates during this season of up to 8500 snakes per day and a potential demand for snake carcasses (for crocodile food) of up to 18,900 kg/week (*Crocodile Specialist Group Newsletter* [2001] 20[3]:57–58 [online]). It is difficult to see how such a harvest could be sustainable.

The vulnerability of a population to some threatening process may vary from year to year as well as from month to month. In many snake species, adult females reproduce on a less-than-annual basis, reflecting the need for long foraging periods to accumulate sufficient reserves for offspring production (Fitch 1999). Because foraging success also varies among years, depending on factors such as prey densities and favorable weather conditions, we are likely to see some degree of synchrony in the times of clutch or litter production by females within a population. Many females may produce litters in the year immediately following a good feeding year, whereas few will reproduce in the year following a bad feeding year. For example, the proportion of adult-size female *A. arafurae* that were reproductive per year varied from 0 to 60% during a 10-year period (Madsen and Shine 2000b). Similarly, Martin (2002) noted a tendency for annual synchronicity in litter production by *Crotalus horridus* in North America. Given the vulnerability of gravid females to humans in both these species (Shine 1986; Brown 1993), it is clear that collecting is likely to have a more severe demographic impact in years when many females are reproductive than in other years, when predation will fall more evenly across sex and size classes. For intensively studied populations, we might also predict the circumstances in which collecting specific age classes will have little effect on longer-term population persistence. For example, hatchling *Liasis fuscus* have very low survival rates in years when their major prey species, the Dusky Rat (*Rattus colletti*), ceases breeding before the young pythons hatch (Madsen and Shine 2000a). Thus, the collection of hatchlings for the pet trade in such years would have minimal impact on the source population. More generally, temporal variation in the relationship between collecting effort and demographic impact has obvious implications for regulating harvesting and for concentrating hunting efforts to particular places and times.

Active Management of Snake Populations

Background Information and Habitat Maintenance

The effects of the limitations of our current knowledge of snake ecology on identifying conservation priorities also apply to field management. Even if information about the effects of reproduction on snake habitat selection, behavior, and movements are available to guide management practices, the flexibility of snakes makes it difficult to frame overall generalizations. It is difficult to translate the scattered available information into practical operations that can be used widely. But the flexibility of snakes is also a blessing for field management. Improvements of habitats for snake reproduction do not rely on precisely dimensioned building or laying sites. Rather, some simple general propositions can be made; they will not surprise any field herpetologist, but they nonetheless may be of use to resource managers.

1. As noted, snakes are especially vulnerable during reproduction. Usually, reproduction extends from spring to late summer in cold climates and is seasonal (but variable) in hotter regions. This annual schedule means that population surveys or harvesting control should be undertaken during specific time periods. In cold climates, capture rates may be highest on sunny days in early spring or after rain in summer, whereas in the tropics, wet days and at

night are likely to provide higher rates of encounters. Hot dry sunny days are unlikely to be profitable.

2. Most snakes need shelters that provide a range of thermoregulatory options as well as protection against predators. In any habitat, reproductive individuals of most species will benefit from predator-impenetrable cover (e.g., thick bushes, large logs, and piles of rocks). Therefore, it is essential to retain such objects within the landscape rather than removing them and to maintain a relatively open and heterogeneous habitat. Even without detailed ecological information, such habitat manipulations are likely to reduce the risk to this critical population component. The provision of suitable cover items (sheets of fibrocement, metal, and so forth) may provide gravid females with secure shelters that allow their access to optimal thermal regimes without having to bask (a behavior that exposes them to predation risk).

3. Snakes may face major difficulties when populations are fragmented, not only by roads but also by open cultivated fields or "cleaned" zones within a forest (e.g., in plantations). A large open area of only several hectares separating two populations can represent a serious obstacle during mate searching. In terms of snake ecology, an open area is a place where the snake cannot hide during its travels; hence, even a shaded tall tree plantation with clear ground is a risky zone for a snake. Building networks of connecting edges with rocks, grasses, and bushes can facilitate reproduction and dispersal.

Manipulating Reproductive Behavior of Free-Living Snakes

As noted, reproduction is risky for both sexes in many snake species. Determining the magnitude of mortality associated with reproductive behaviors can be relatively straightforward—we can simply count the number of dead animals, especially for roadkill victims. An analysis of the sexes and body sizes of snakes killed on roads in France revealed a strong seasonal peak in numbers of adult males run over by vehicles during the mating season (in springtime) and of females in early summer when the snakes migrated to or from oviposition sites (Bonnet et al. 1999b). Although juveniles constitute an important component of the population based on capture rates in shelter sites, animals in this age class are rarely found as roadkills (Bonnet et al. 1999b).

It may be possible to reduce the mortality rates of mate-searching male snakes by habitat manipulation, in particular by providing opportunities for them to avoid roads during mate-searching. In Manitoba, the roadkill mortality rates of *Thamnophis sirtalis* migrating to and from communal hibernacula were reduced by the construction of tunnels under a busy highway (Shine and Mason 2004). Drift fences funnel the snakes away from the road and through the tunnels; success in this case may have been facilitated by active road avoidance by mate-searching males of this species (Shine et al.

2004a). Such manipulations will probably prove too expensive to construct and maintain in most situations, however; and in any case, they are likely to be inefficient because in many cases the snakes do not cross the road at any particular point and hence may often miss the tunnel (Aresco 2005).

A logistically more attractive opportunity would be to remove the pressure on snakes to cross roads and hence reduce risks such as those associated with a female's migration from the usual home range to the nearest communal oviposition site (in egg layers) or gestational rookery (in livebearers). Current studies in central France showed that it is feasible to construct alternative egg-laying sites close to the snakes' usual habitats. By this simple manipulation, the distances moved by egg-laden females (and hence their probability of mortality) can be substantially decreased (X. B., pers. obs.). Adult males are also attracted because the oviposition site concentrates reproductive females. Neonates are also likely to benefit from such manipulations because they will hatch out in areas close to suitable habitat.

The artificial structures must be large enough to offer a wide range of thermal and hydric conditions, and they ideally should include a variety of construction materials differing in thermal and water-retaining qualities. The artificial oviposition sites that we have constructed in France consist of rock walls 1.5 m high enclosing an area approximately 15 m^2, large enough to provide stable thermal conditions for incubating eggs (Fig. 6.1). The area within the walls is filled with rocks, soil, and plant material, providing easy access to deep crevices. Plastic sheeting across the top (held down by rocks) maintains high levels of moisture within the oviposition site and prevents the growth of trees. The entire structure is surrounded by a wire fence, to prevent ingress by wild pigs, and extends to close contact with surrounding vegetation so that gravid female snakes can enter the structure without having to cross open terrain, where they would be vulnerable to predation. Within 4 years of construction, these sites were regularly used for oviposition by *Hierophis [Coluber] viridiflavus*, *Elaphe longissima*, and *Natrix natrix*; X. B., pers. obs.). The buffered thermal conditions of the artificial laying sites also provide suitable shelter during winter (Fig. 6.2). The three artificial laying sites were all built in a natural reserve and well-protected area (even though natural predators are abundant). All were successful in terms of being used by the snakes and their prey (small mammals, lizards, and amphibians; all of them reproducing in the laying sites). The next step will be to build similar edifices on both sides of roads where numerous roadkills have occurred. If the laying sites indeed divert reproductive females, adult males, and neonates from crossing the road, then we should observe a decrease in the incidence of roadkills at these sites. Because the network of roads is likely to increase in most places in the world, it is important to assess the efficiency of this simple and cheap method (the cost per site was 1500€). Kevin Shoemaker et al. (Chapter 8) state that although nesting sites

Fig. 6.1. Construction of an artificial site for winter retreat and egg-laying in Chizé Forest, France. (a) Construction process (b) Finished shelter site (with the upper sheeting pulled back to reveal construction details).

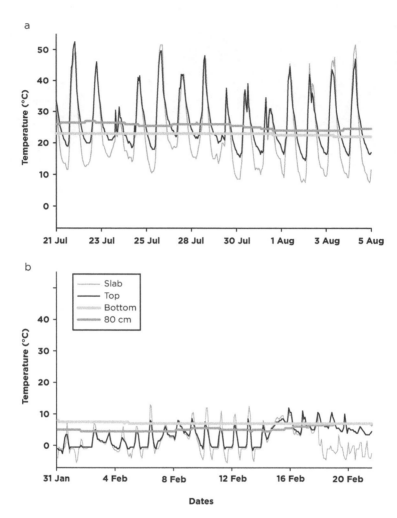

Fig. 6.2. Temperatures inside an artificial oviposition site in Chizé Forest, France. (a) During the time of the year when eggs are incubating (b) During winter, when snakes are inactive (January and February being the coldest months of the year). Temperatures were recorded using plastic models of snakes placed at 120-cm depth in the middle of the site (Bottom), under the plastic sheeting among the upper rocks (Top), at mid-depth (80 cm), and under a nearby concrete slab (Slab).

may be a limiting resource for many snake populations, the manipulation of nesting habitat is not common in snake conservation. We agree, but we also think that it may well play a key role in the future because it has great conservation potential in highly disturbed habitats, notably those fragmented by road networks or other features; in addition, such sites can have significant educational value.

Food Supply and Reproduction of Free-Living Snakes

The management strategies discussed so far revolve around enhancing reproductive output by increasing the probability of a female surviving long enough to produce offspring. An alternative approach is to directly boost reproductive output in the field (i.e., induce females to mature earlier, reproduce more frequently, or produce larger or more viable offspring). The most critical variable that drives reproductive rate presumably is the food supply; field studies show a strong correlation between prey abundance and snake reproductive rates (Shine and Madsen 1997; Bonnet et al. 2001), and laboratory studies confirm a strong causal link between these two variables (Seigel and Ford 1991). Recent field studies in which radiotracked Western Diamond-backed Rattlesnakes (*Crotalus atrox*) were given supplementary food (Taylor et al. 2005) provide further evidence for such a link, whereas earlier studies using field enclosures with *C. viridis* yielded less clear-cut results in this respect (Charland and Gregory 1989).

Would it be feasible to provide extra food to a natural population of snakes and, hence, increase reproductive rates? We are aware of only one attempt to do this in a conservation context, involving a remarkably dense population of endemic Shedao Island Pitvipers (*Gloydius shedaoensis*) on a small island in northeastern China (Li 1995; Shine et al. 2000). Adult snakes in this population feed almost entirely on migrating passerine birds and do so by scavenging dead birds (typically, killed by other snakes) as well as by ambushing live ones (Shine et al. 2000). Thus, feeding rates potentially could be enhanced by providing recently killed birds. In a year when bird numbers were low, Chinese conservation authorities scattered dead ducklings around the island to maintain snake feeding rates. Unfortunately, the effectiveness of that endeavor seems not to have been assessed.

The direct provision of additional food is unlikely to be feasible or effective in most snake populations, but habitat manipulations may have the same effect and be far easier to conduct. We can attempt to increase prey densities or to concentrate prey in areas where snakes are most capable of capturing them. Again, the pitvipers of Shedao provide the best example. Scientists have experimented with increasing the success rates of ambush-foraging snakes by attracting birds to specific places to facilitate snakes' foraging abilities. The methods to do this include trimming branches so that birds have to alight on twigs thick enough to also support snakes, constructing ponds as water sources to attract birds, and tying ears of grain to branches used as ambush sites by foraging snakes (see review by Shine et al. 2002b). It is not clear how such manipulations have affected snake reproduction, but the overall population densities of *G. shedaoensis* have increased during the period of these manipulations (Shine et al. 2002b). Presumably, simple techniques such as the construction or modification of bodies of water have considerable potential to enhance the availability of prey for anurophagous and piscivorous snakes.

Artificial Shelters

The reproductive rates of snakes are constrained not simply by prey availability but also by the rate at which the snake can capture and process that food. Thus, reproductive rates typically are higher in warm-climate populations than in cool-climate populations of the same species, reflecting greater opportunities during the year to forage, digest, and so forth (Adolph and Porter 1993). Thus, the provision of suitable (warm) microclimates—by means of adding sun-heated shelter items or trimming shading vegetation from preferred sites—might also enhance rates of reproduction. In practice, the provision of additional cover items may simultaneously increase prey abundance and provide thermal and hydric conditions for snake reproduction. Many snakes depend on prey that share their preference for secure retreat sites that are well hidden from predatory birds.

Creating artificial refuges can be a simple and efficient way to improve habitat quality for snakes (see also Shoemaker et al., Chapter 8). In the forest of Chizé in central France, we have set out more than 800 concrete slabs, which now support diverse communities of invertebrates, amphibians, lizards, small mammals, and four species of snakes that feed on these prey types. Interestingly, even though snakes regularly visit the slabs, all the small mammal species preyed on by the snakes actually build nests and raise their young under these shelters. Our surveys over more than 10 years have recorded potential prey (plus their nests) under virtually every cover item that we have distributed in this forest. The regularly spaced network of cover items also allows us to quantitatively survey the abundance of both prey and snakes, a task that would otherwise be almost impossible for logistical reasons. More than 100 snakes are captured (and released) under the slabs every year, whereas visual searches for active snakes (i.e., without slabs) in the same area over the same period typically produce only one-twentieth of this number; using this technique, the catchability of the snakes was greatly increased, not necessarily their density. The fact that the snakes intensively use the network of artificial shelters, however, suggests that an apparently minor management of their habitat (800 1-m² slabs scattered on more than 2000 ha) had a strong influence on their behavior. Whipsnakes fitted with transmitters and temperature dataloggers thermoregulated under slabs with the same efficiency and rates of thermal exchange as they did when using a mosaic of open and shaded areas for shuttling thermoregulation (X. B., pers. obs.). The slabs offer optimal thermal opportunities in the field without the risk of predation exposure. All five snakes species monitored in the area (N. *maura* is also studied nearby) used the slabs, suggesting that such artificial shelters are favorable for snakes in general. This is important because slabs allow reproductive females to avoid prolonged exposure while basking in sunlight during vitellogenesis and pregnancy.

The educational opportunities are significant also. For example, X. B. and Jean-Marie Ballouard (PhD student) constructed a small network of 250 slabs a few kilometers away from a primary school; the children at that school (8–12 years old) regularly inspect the slabs to study animal activity. We encourage managers to further test the applicability and the utility of this technique for conservation.

Combining Laboratory and Field Manipulations: A Role for Captive Husbandry

Although we are dubious about the feasibility of breeding snakes in captivity for reintroduction into the wild (at least for most species), there may be circumstances in which captive maintenance of animals can play a role (see Kingsbury and Attum, Chapter 7). It may be feasible to maintain animals only briefly in captivity, during critical phases of the reproductive cycle, and thereby minimize the sources of mortality that would otherwise be experienced either by the reproducing adults or their offspring. The idea behind this approach is straightforward, and the logistics much are less problematic than for long-term husbandry. Gravid females of many snake species rarely feed (Shine 1980), so the provision of prey items may be unnecessary. An additional benefit may be the availability of captive specimens for educational programs. For example, we again turn to our work with *Vipera aspis* in central France. Female vipers are at high risk of bird predation because of their extended basking during vitellogenesis and gestation, and they are at risk of starvation after parturition because they are typically emaciated at this time (Madsen and Shine 1993a). In a preliminary study, we captured 10 reproductive female vipers during late vitellogenesis (when they bask or shelter under slabs and thus are easily found) and kept them in captivity under favorable thermal and nutritional conditions during gestation. We released the offspring immediately after parturition, but retained the mothers until they had recovered body condition. The females were then released at their sites of initial capture in good condition after 6–12 months in captivity. Our results are encouraging; we recaptured four of these females 12–24 months later, all of them reproductive once again, and with larger body sizes than at the initial capture. Without such a period in captivity, it is unlikely that any of these females would have reproduced a second time (Bonnet et al. 1999c; Bonnet et al. 2002a).

A similar manipulation has been in place for many years in the African nation of Togo, where there is a major commercial enterprise based on hatchling Ball Pythons (*Python regius*) for the international pet trade. Expert snake-hunters locate and capture females before oviposition and then keep the females in captivity to collect their eggs. The eggs are incubated artificially, and the females are released. When the eggs hatch, 10% of the progeny are released to maintain local populations while 90% are sold into the pet trade (Aubret et al. 2003). This harvest relies on wild animals captured in the vicinity of Lome and has been sustained for more than 30 years without any notable decrease in the python population. One important caveat is that any program involving the capture and subsequent release of wild animals needs to ensure rigorous quarantine procedures to prevent the transfer of pathogens to wild populations, as may have occurred in North American gopher tortoise populations (Seigel et al. 2003).

Finally, it is worth noting that successful reproduction involves the quality and the quantity of offspring that are produced. Manipulations of food supply and thermal environments may enhance offspring viability as well as accelerate overall reproductive rates (Ford and Seigel 1989). In at least some cases, genetic quality also may be an issue (King, Chapter 3). Some good evidence comes from a small, genetically isolated population of European Adders (*Vipera berus*) in southern Sweden. Separated from any neighboring populations by agricultural modification of the landscape, these adders had very low genetic diversity. Rates of stillbirth were high, especially in females recorded to mate with only a single male during the mating season preceding their production of offspring (Madsen et al. 1992). Offspring viability increased in multiply-mated females, suggesting a genetic basis to this phenomenon (Madsen et al. 1992, 1995). To test this hypothesis (and, it was hoped, rescue the population from its long-term decline), males from other, genetically diverse adder populations were released at the Smygehuk site (Madsen et al. 1999). The results were dramatic, with an immediate increase in offspring viability and a strong population recovery to levels never recorded over the preceding decades of study (Madsen et al. 1999). This case history suggests that manipulating genetic diversity, as well as food supply and habitat quality, can enhance reproductive output and effectiveness, at least in some circumstances. Lack of genetic diversity is likely to be responsible for population decline in only a small minority of cases, however, and identifying and remediating the ecological factors responsible for decline (habitat destruction, disease, feral organisms, and so forth) generally will be a more productive avenue for conservation efforts (Caughley 1994). Perhaps the most important role for genetic studies in this context is to clarify the degree of gene flow among isolated populations and, hence, identify locally differentiated taxa of particular conservation significance, as well as guiding translocation efforts should these prove essential to facilitating population recovery (Ujvari et al. 2002).

Ecological Traps?

Several authors have suggested that artificial buildings supposedly favoring snakes might represent ecological traps (see details and references in

Shoemaker et al., Chapter 8). For example, wild animals may be unable to perform their normal behaviors in the novel sites. Even worse, artificial buildings could attract predators and increase the probability for disease transfer. We have advocated the use of artificial shelters and laying sites. Are these likely to introduce problems of these kinds? We doubt it, especially if the artificial shelters and laying sites are built in appropriate places (e.g., with close connection to surrounding snake habitats). First, the main snake predators are not likely to penetrate well-built sites (see Fig. 6.1), and the snakes are able to use well-protected routes when entering and leaving those structures. To avoid an overconcentration of snakes in the vicinity of artificial buildings, the solution is simple—multiple structures must be scattered about the landscape at a scale determined by the home range of the snakes. Significantly, this requirement necessitates research programs to identify a baseline for appropriate management. Second, artificial shelters combined with artificial ponds tend to increase food availability and diversity, as well as protecting against dehydration. We believe that well-fed snakes are likely to resist diseases, and thus the transfer of diseases and parasites, although potentially problematic, is unlikely to affect population health. Research on these issues will be of great value.

Future Directions and Research

Natural History of Snake Reproduction

The relatively conspicuous behaviors of snakes during reproduction offer a strong fulcrum for population survey, habitat management, and field observation as well as for engaging the public in conservation efforts. Unfortunately, for the vast majority of snake species, the phenology of reproduction is unknown. This ignorance is particularly regrettable for snakes in tropical countries, the places where these animals are most diverse and where conservation efforts are urgently needed. Similarly, the ecology of seasnakes is insufficiently documented, and coral reefs (one of their main habitats) are seriously threatened. Human demographics, the deleterious effects of climatic changes, and habitat destruction occur at a much higher rates in tropical Asia, Africa, and South America than in more intensively studied areas such as Europe, North America, and Australia (although major problems occur in these countries also). Most of the knowledge used to frame conservation actions (including all the chapters in the current book), however, originates from research conducted in the latter areas. We can redress this first bias by increasing our ecological knowledge on snake species from developing countries (especially in tropical biomes). We need to assess the extent to which strategies for field management can be extrapolated successfully across different systems, climates, and human cultures. Unfortunately, this is not likely to happen easily. In addition, the rates of publication of papers on reptile natural history are declining (McCallum and McCallum 2006). Given the low probability of an increase in the amount of field-based ecological research on tropical snakes, we need alternative approaches.

One technique for partially filling this void is to examine museum specimens. Even for abundant and widespread species such as night adders (*Causus* spp.) in Africa, the seasonal timing of vitellogenesis has been documented only recently (Ineich et al. 2006). Museum specimens can rapidly provide basic information on topics such as feeding ecology, community assemblages, and periods of vulnerability (e.g., Shine 1994). Ideally, before using such data for management, we need to combine it with first-hand field knowledge from experts familiar with specific localities, to identify conservation priorities and tactics. We also encourage field-based mark-recapture studies for tropical reptiles. Such studies are time consuming, but do not require complex or expensive equipment and, hence, may be feasible in developing countries.

Even in temperate countries, the level of ecological knowledge on snake ecology remains insufficient. The suggestions made here thus apply universally, albeit with less force in the temperate zone than in the tropics. We need to better document the life-history patterns of even well-known species (e.g., in Europe, the reproductive pattern of iconic snakes such as *Elaphe longissima* is still poorly known). In practice, the first priority is to document the timing of the main reproductive events: mating, vitellogenesis, oviposition, incubation, pregnancy, and parturition. The second priority is to determine the dietary composition of different age classes. The third is to document the spatial ecology, especially habitat use and movement patterns. Such information must be framed within a climatic and geographic context to be exploited for conservation purposes by field managers.

Artificial Oviposition Sites

The utility of artificial laying-hibernation sites should be tested in a broad range of geographic and climatic situations. The first step is to verify the attractiveness of the laying sites in relatively well-documented systems because the interpretation of the results will be easier. A simple procedure consists of recording the gradients of temperature (and humidity, if possible), as illustrated in Figure 6.2. If the artificial site provides a wide range of both stable and fluctuating thermal regimes, then it is likely to prove suitable for snakes of a range of body sizes, reproductive status, and species. Larger laying sites are likely to offer greater thermal and hydric buffering; small (<5 m³) structures might be a waste of time. The role of connective corridors and the density of artificial sites should be tested also. In the future, we can compare fecundity, growth rate, and survival between snake populations monitored in managed and nonmanaged habitats. This will allow us to better focus to our conservation efforts; the improvement of some habitats may be unnecessary.

Another obvious field for future inquiry involves manipulating snake behaviors to encourage the use of such artificial oviposition sites. Research on *Tropidonophis mairii* in tropical Australia suggests at least two such strategies. First, female keelbacks selectively oviposit beside existing eggshells (Brown and Shine 2005b), so priming the artificial site with old eggshells may be effective. Second, females return to the site they first encountered after hatching to lay their own eggs (Brown and Shine 2007), and thus translocation should be performed using eggs or hatchlings rather than older animals. The success of such efforts can easily be determined by experimental studies.

Habitat Conservation to Protect Snakes at a Large Geographic Scale

High priority should be given to protecting healthy ecosystems that currently contain abundant and diverse snake populations (see Beaupre and Douglas, Chapter 9). In practice, any areas where many living specimens or roadkills (especially reproductive individuals) are found may warrant protection, even if the species involved are regarded as common. Snake populations are decreasing too rapidly in many places for us to wait for the often slowly changing wildlife-protection rules. Many habitats suitable for snakes are regarded by managers (including farmers and foresters) as "rubbish" to be cleaned up. For example, fallen trees, heaps of debris, ruins of old buildings, and rubbish piles covered by brambles often provide important habitats for snakes and their prey. We do not encourage managers to develop rubbish dumps but, rather, to conserve habitats that shelter abundant snake populations.

Captive-Raised Snakes

We need more experiments into the fate of snakes that have been raised in captivity (either long or short term) and then released into the field. Snakes cannot cross long distances in fragmented habitats, especially those cut by large and intensively frequented roads. On the other hand, the restoration of many habitats is feasible, especially as agricultural regimes shift from broad-scale clearing to more environmentally sensitive practices (Mollard and Torre 2004; but see Tscharntke et al. 2005). The introduction of snakes to restored habitats should be considered. Prior to release, tests must be conducted to determine which category of snake is the most efficient to place in the field: neonates, juveniles, adults, or even eggs. Extreme nest-site philopatry suggests that, in at least some cases, we should release hatchlings rather than older animals if we wish to see translocated animals breeding at the release sites (Brown and Shine 2007). It is likely that some species can be released successfully and that others cannot; research can identify the most productive approaches.

Impacts of Climate Change

The sensitivity of snake reproduction to ambient weather conditions, especially thermal regimes, suggests that global climate change will have significant impacts. Warmer conditions in cool-climate regions (such as mountaintops) will shift the seasonal timing of reptilian reproduction and accelerate embryonic development. Such changes may enhance reproductive success—but at the same time, climate change may modify factors such as vegetation density (and thus basking opportunities), the existence of safe dispersal corridors, and the availability of critical edge habitat in fragmented landscapes (see Weatherhead and Madsen, Chapter 5). Other species may be affected also, with predators, parasites and competitors moving from lower elevations to interact with the snake species in question. Understanding the impacts of climate change on such abiotic and biotic factors and, hence, on the reproductive biology of snakes remains a major challenge for future research.

Acknowledgments

We thank D. Barré and numerous students for their help in the field and for the construction of the artificial laying sites in Chizé. Special thanks are due to the funding organizations (especially the Australian Research Council, the Centre National de la Recherche Scientifique, the Office National des Forêts-France and the Conseil General 79) that have supported our work for many years.

7

Conservation Strategies

Captive Rearing, Translocation, and Repatriation

BRUCE A. KINGSBURY AND OMAR ATTUM

The documented declines in snake populations necessitate identifying approaches for enhancing recruitment that go beyond habitat protection. Examples include habitat restorations and manipulations (see Shoemaker et al., Chapter 8), examining ways to enhance reproductive success (Shine and Bonnet, Chapter 6), and even manipulation of populations themselves. As we begin to succeed at restoring lost habitat to better support viable populations of snakes, we need the tools to replace species lost as a consequence of former degradation. Ideally, any such actions are conducted based on scientifically sound information and with a clear plan for success. These conservation approaches must be planned appropriately by considering habitat requirements, population dynamics, genetics, and evolutionary constraints or else the measures will fail despite good intentions (see King, Chapter 3; Jenkins et al., Chapter 4; Weatherhead and Madsen, Chapter 5). In this chapter, we review the practices of moving and captive-rearing snakes for later relocation for the purposes of population conservation, and we identify the reasons for the failure or success of such techniques when we endeavor to rescue snake populations. We then make recommendations for individuals planning such activities based on the lessons learned from past attempts.

Terminology

Any discussion of moving snakes around gets confusing unless we standardize terminology. Here, we adhere to the International Union for Conservation of Nature (IUCN 1998) definitions where possible. To begin, if a snake is captured in a given location and then returned, this action is a *replacement*.

Simple enough, and perhaps not even worth discussing. Yet replacements may be catastrophic for the host population if, in the interim, the specimen was housed with diseased animals and becomes itself a vector for a pathogen. The term *relocation* implies repositioning and, thus, not just a simple return or replacement. With the biology of snakes in mind, relocations made within the historical range of the animal are termed translocations; this term avoids the semantic ambiguity of the word *relocate*. The IUCN definition of *translocation* is the deliberate movement of wild individuals from one part of their distribution to another where the species historically occurred or is currently present. When relocations occur outside of the historical range of a species, we are speaking of *introductions*. Such manipulations are beyond the scope of this chapter, but discouraging such actions is supported by numerous worst-case scenarios for a variety of taxa, including snakes (e.g., *Boiga irregularis;* Savidge 1987; Rodda et al. 1999b).

On a more local scale, moving individual snakes from immediate danger and releasing them somewhere else to save the animal is also a translocation. Picking up a nuisance or threatening snake and moving it from one place to another is a common practice (Sealy 1997; Fischer and Lindenmayer 2000; Nowak et al. 2002; Butler et al. 2005) and is probably the most common form of relocation. Snakes may also be moved to save them from immediate danger, such as moving those in areas about to undergo development or those found on roads to undisturbed areas (Reinert and Rupert 1999).

If the effort is more substantive and planned, the intention may be to add snakes to a target-site population. Restoring a population by introducing additional individuals is most often referred to as a *repatriation,* which, for our purposes, is defined as the intentional release of animals into an area currently or formerly occupied by that species for the purposes of enhancing population viability. Deferring to IUCN (1998), when animals are released into an existing population, we might term such a translocation a supplementation or augmentation. We favor the term *augmentation* here to specifically refer to repatriations to areas where residents are still present. In contrast, *reintroductions* are repatriations to establish a population in an area where the species once occurred but is now locally extirpated or extinct.

The methods by which animals are released into previously occupied or new habitat can vary. If they are simply released into an area without any acclimation or experience with the site, we term that a *hard release.* On the other hand, if animals are released first into an enclosure for a period of time, or are otherwise acclimated to the release site, it is a *soft release.*

Snakes used in a repatriation effort have to come from somewhere. For one reason or another, snakes selected for a relocation effort may be rare or their removal may actually threaten the source population. Thus, we might need to generate a collection of individuals for relocation. Captive breeding of individuals from one or more source locations (which could, in fact, be the same population as the target location) is a possibility. Head-starting is an another approach, in which neonates are temporarily held in captivity and grown (perhaps) at rates exceeding those normally observed in nature (Aubret et al. 2004; King and Stanford 2006).

The Challenges of Repatriation

Moving snakes into other populations or saving an individual snake from impending death makes intuitive sense and provides a "feel good" response. Even if such relocated snakes survive, and they often do not, repatriations often fail when we apply a stringent definition of success, such as the establishment of a viable population (Fischer and Lindenmayer 2000). To achieve such success, released animals must first survive in their new environment by finding food and refugia and by avoiding predators, pathogens, and competitors. Survival alone does not necessarily determine success, however. Animals may survive repatriations but subsequently experience low fecundity (Lloyd and Powlesland 1994). Ultimately, success depends on the released animals reproducing and having their offspring reproduce to create a viable population (Dodd and Seigel 1991; IUCN 1998; Seigel and Dodd 2000).

Establishing viable populations through repatriation is challenging because repatriations are faced with all the limitations that small populations experience. Vulnerability to extinction is related to population size, with smaller populations more likely to go extinct as a consequence of Allee effects, demographic stochasticity, environmental stochastic events, disease, genetic drift, and reduced genetic variation (Caughley 1994). Given that repatriations often involve individuals from small or nonviable populations and that efforts to establish new populations are based on relatively few individuals, such efforts are already more likely to fail (Wolf et al. 1998; Fischer and Lindenmayer 2000). Also, animals moved from immediate danger into a new area often die in their new environment because they are not familiar with their new surroundings, spend more time moving, and thus are more vulnerable to predation or other dangers as a result (Fischer and Lindenmayer 2000; Plummer and Mills 2000; Sullivan et al. 2004). Given the challenges faced by any small population, the problems associated with releasing individuals into new environments, and the habitat limitations probably existing at many sites, it is easy to see why repatriation efforts are challenging at best. In addition, repatriations often occur as a last resort and, thus, may be too late because negatively compounding factors have already manifested a decline of the population (Griffith et al. 1989).

Constraints and Considerations

Repatriations can be expensive because of personnel needs, animal maintenance during captivity (including veterinary care and disease monitoring), costs of moving animals between sites, and postrelease monitoring (Fischer and Lindenmayer 2000). To maximize the chances for success of a repatriation attempt, we must consider the availability of and prospects for (1) alternative and more effective conservation measures; (2) ecological data that facilitate informed repatriation decisions (Dodd and Seigel 1991); (3) a suitable source of healthy animals, the removal of which will have a minimal effect on the viability of the source population; (4) postrelease monitoring; and (5) adequate funding to carry out all components of a repatriation effort (Kleiman 1989; IUCN 1998).

A thorough understanding of the ecology of the species to be repatriated should include the degree of genetic similarity between the source and target populations, so that all will possess the adaptations that will serve their needs at the release site (Kleiman 1989; Fischer and Lindenmayer 2000). If captive animals are used, medical screening should also be conducted to prevent disease transmission between populations, which can cause outbreaks and hamper recovery efforts (Jacobson 1994). A release site should be identified based on the availability of suitable habitat (Kleiman 1989; Shoemaker et al., Chapter 8). Also, local human communities should be consulted prior to the repatriation, especially in the case of venomous snakes (Kleiman 1989). Table 7.1 provides a checklist of steps to consider when planning and implementing a translocation. Table 7.2 provides a list of known studies of translocations in snakes.

Despite these challenges, there are several success stories, particularly involving birds and mammals (reviewed in Wolf et al. 1996, 1998). Documented successes, especially with snakes, are limited. We therefore concur with Dodd and Seigel (1991) that repatriation is not a proven recovery technique for snakes but, rather, an experimental conservation method that needs to be further tested and refined, perhaps on a species-specific basis. It may be more feasible to reverse population declines by simply protecting existing populations and improving the habitat (Reinert 1991; Fischer and Lindenmayer 2000; Seigel and Dodd 2000; Shoemaker et al., Chapter 8). That said, we feel that translocation holds promise and that many individuals and agencies will embark on translocation efforts with or without a general consensus on the utility of the technique. So, we hope to help them act in the most educated way possible.

Repatriation efforts are more likely to succeed if they are planned properly and issues relating to repatriation are addressed (Kleiman 1989; Fischer and Lindenmayer 2000). For example, prior to any repatriation, the researcher must mitigate the conditions that contributed to the species decline in the first place (Fischer and Lindenmayer 2000). Failure to do so addresses only the symptoms, rather than the cause, of the species decline and ultimately the repatriation will fail (Kleiman 1989; Hambler 1994; Seigel and Dodd 2000). Mitigation efforts may involve acquiring additional habitat, restoring the existing habitat, or preventing wildlife collection (Shoemaker et al.,

TABLE 7.1
A checklist of steps for successful translocation

Step	Challenges	Solutions
Establish goals for success	Lack of understanding of what constitutes success	Develop goals of project and determine how success will be established
	Inadequate funding or other commitments over time	Correctly identify all crucial project participants/stakeholders
		Obtain commitments early in process that are sufficient for achieving stated goals
Establish release site	Insufficient suitable habitat in terms of quality, size, or not all life history needs met (juvenile habitat, hibernacula, etc.)	Selection based on natural history
		Use of GIS modeling
	Historic challenges not addressed	Confirm historic threats ameliorated
	Stability of site uncertain	Use public property or secure permanent easements
		Identify all stakeholders and secure buy-in
Identify appropriate release life stage	Particular cohorts may be unlikely, or less likely, to establish themselves	Test for appropriate life history stage to release
Identify source population	Genetic suitability	Use stock from on-site or from same metapopulation, or otherwise from as nearby as possible
	Threats to source	Confirm source population sufficient to withstand withdrawal of transplants
Obtain transplants	Insufficient suitable transplants	Headstart wild-bred individuals to appropriate size
		Breed suitable stock
	Health of individuals	Screen for disease
		House in isolation from other reptiles
Release transplants	Excessive dispersal	Release most appropriate size/sex class
		Soft release into enclosure
		Release into hibernacula or into other situations forcing stasis to promote acclimation
Sustain the effort	Releases inadequate for success	Follow through on built-in commitment in order to proceed with multiple releases over time
Monitor the site	Inadequate monitoring may not capture emergent problems	Follow through on built-in commitment to conduct monitoring as part of project
	Inflexibility	Plan for capacity to respond to new information and challenges

Notes: For each step, the associated challenges are listed, along with potential solutions for those challenges. The table is organized by the steps involved, although, when the researcher is planning, challenges may need to be addressed earlier in process than when they arise in implementation. GIS, geographical information system.

TABLE 7.2.
Summary of studies of translocations of snakes

Species	Location	Stage	Source?	Outcome	Activity/Movement Patterns	Reference
Crotalus atrox	Southwest USA	A	WC	High mortality	Translocated snakes more active than normal	Nowak et al. 2002
Crotalus horridus	Northeast USA	A	WC	Low survival and release site fidelity	Unknown	Galligan and Dunson 1979
Crotalus horridus	Southeast USA	A, J	WC	Low survival and release site fidelity	Short-distance relocation did not cause excessive movements	Sealy 1997
Crotalus horridus	Northeast USA	A	WC	High mortality	Translocated snakes more active than normal; 1 year after release, movement patterns of translocated snakes start to resemble residents	Reinert and Rupert 1999
Drymarchon couperi	Southeast USA	A, J, H	WC, CB	Low survival	Unknown	Irwin et al. 2003
Heterodon platirhinos	Southeast USA	A	WC	Higher mortality than resident snakes	Translocated snakes more active than normal	Plummer and Mills 2000
Notechis scutatus	Australia	A	WC	Unknown	Translocated snakes more active than normal	Butler et al. 2005
Pituophis catenifer sayi	Northeast USA	A	WC	Low mortality	Translocated snakes more active than normal	Moriarty and Linck 1997
Sistrurus catenatus	Midwest USA	A, J	CB	Normal to high mortality, but depending on timing of release	Normal movement and activity patterns	King et al. 2004
Thamnophis radix	Midwest USA	J	CB	Normal mortality	Unknown	King and Stanford 2006
Thamnophis sirtalis parietalis	Canada	A	WC	Low combined release site fidelity/survival	Unknown	Macmillan 1995
Vipera berus	Sweden	A	WC	Augmentation increased genetic variation in host population	Unknown	Madsen et al. 1999

Notes: Format follows Dodd and Seigel (1991). A, adults; CB, captive bred; E, eggs; H, hatchlings or neonates; J, juveniles; U, unknown; WC, wild caught.

Chapter 8). For example, commercial wildlife collectors removed individuals from a Red-sided Gartersnake (*Thamnophis sirtalis parietalis*) population, thus disrupting a repatriation effort (Macmillan 1995).

Postrelease monitoring is vital to assessing repatriation success (Dodd and Seigel 1991; Fischer and Lindenmayer 2000). Studying the habitat use and movement patterns of repatriated snakes can provide valuable insights into their responses to their new environment. Monitoring allows the estimation of population size, mortality rates, causes of death, and reproductive rates. These variables can then be used in demographic models to determine which age class or sex will make the greatest contribution to recovery for future repatriations (Heppell et al. 1996; Sarrazin and Legendre 2000). In addition, the results of such studies can suggest ways to improve future repatriation techniques (Fischer and Lindenmayer 2000; Moehrenschlager and MacDonald 2003; Tuberville et al. 2005).

Specific details of repatriation efforts have been reported elsewhere (Kleiman 1989; Dodd and Seigel 1991; IUCN 1998), and here we summarize and expand on those findings. We also discuss the factors that contribute to repatriation failure, discuss ways to minimize these problems, and suggest future directions of research.

PostRelease Factors

Patterns of Dispersal

A typical response for animals experiencing a hard release is for them to disperse for long distances away from the release site, presumably looking for something familiar. Thus, these individuals have excessively large activity ranges and have linear or erratic movement patterns compared to resident animals (Bright and Morris 1994; Oldham and Humphries 2000; Moehrenschlager and MacDonald 2003; Tuberville et al. 2005). Such responses have been noted in snakes. Hard-released snakes have exhibited atypically large activity ranges and excessive dispersals from release sites (Moriarty and Linck 1997; Plummer and Mills 2000; Nowak et al. 2002). For example, translocated male Timber Rattlesnakes (*Crotalus horridus*) had a polygon activity ranges of 600 ha, the largest ever recorded for a terrestrial snake and ten times greater than those recorded for resident animals (Reinert and Rupert 1999). In addition, translocated *C. horridus* moved more frequently, traveled three times the total active-season distance, and moved almost four times the distance per day than resident snakes (Reinert and Rupert 1999). Movements in translocated Western Diamond-Backed Rattlesnakes (*C. atrox*) tended to be greater, and the activity ranges were approximately 30% larger than found in resident populations (Nowak et al. 2002). Excessive movement was also observed in hard-released Eastern Hognosed Snakes (*Heterodon platirhinos*), with translocated individuals having six times greater variance in the distance of daily movements and making more unidirectional movements of greater distances than resident snakes (Plummer and Mills 2000). This pattern was also observed in hard-released Bullsnakes (*Pituophis catenifer sayi*), whose activity ranged were four to six times the size recorded for other resident populations (Moriarty and Linck 1997). Nuisance Tiger Snakes (*Notechis scutatus*) that were hard-released traveled, on average, longer distances per month than resident snakes (Butler et al. 2005). The period immediately after release is the time when repatriated individuals are most likely to be lost to researchers, despite using radiotelemetry, because the animals move distances that exceed the receiver range (Moriarty and Linck 1997; Nowak et al. 2002).

Reasons for Postrelease Dispersal

The excessive movements of hard-released snakes suggest that they are searching for familiar environmental features or exploring and becoming familiar with their new surroundings (Reinert and Rupert 1999; Nowak et al. 2002). Hard-released animals are often disoriented in their new surroundings because they do not have attachments to particular areas, lack a mental map of refugia or hibernacula, and have no prior knowledge of areas with high prey density or where they may be especially vulnerable to predation (Bright and Morris 1994; Moehrenschlager and MacDonald 2003; Sullivan et al. 2004). Animals released at sites with poor habitat quality, or that have experienced anthropogenic disturbances, may also disperse widely to find more suitable habitat (Larkin et al. 2004). In contrast, resident snakes often exhibit movement patterns that indicate a familiarity and confinement to specific home range area (Reinert and Rupert 1999; Plummer and Mills 2000; Butler et al. 2005). Hard-released adults do not wander around indefinitely, however, and eventually they show more normal movement patterns with time (Moehrenschlager and MacDonald 2003; Sullivan et al. 2004; Tuberville et al. 2005). For example, 1 year after release and successful hibernation, the movement patterns of hard-released adult *C. horridus* decreased and resembled those of residents (Reinert and Rupert 1999).

Excessive and unidirectional movements in repatriated animals may represent a homing attempt toward the area from which they originated or were captured (Fritts et al. 1984; Oldham and Humphries 2000; Sullivan et al. 2004). Observations of homing behavior in various snake species usually involve subjects that either successfully returned to, or dispersed in the direction of, their capture site (Sealy 1997; Webb and Shine 1997a; Nowak et al. 2002; Shetty and Shine 2002). The snake homing ability appears to be a function of body size, with larger snakes being more likely to return to their original habitat (Fraker 1970; Weatherhead and Robertson 1990). An extreme case of such homing was exhibited by Burmese Pythons (*Python molurus bivittatus*), now established in Florida. Translocated snakes were tracked over 20 km back toward their previous areas of activity (M. Dorcas, pers. comm.).

Homing success is also inversely related to the distance that the animal is translocated. Snakes translocated short distances (2 km or less, depending on the species) from their capture site will often return to their original activity range and resume regular activity patterns, but they may avoid capture sites (Sealy 1997; Reinert and Rupert 1999; Nowak et al. 2002). Conversely, snakes released at distances of 8 km or more from their capture sites, typically do not return (Reinert and Rupert 1999). Distant translocations might thus spare the animal in the short term from any immediate anthropogenic disturbances, but with the trade-off that such snakes will have excessive movements on release, potentially decreasing their probability of survival (Reinert and Rupert 1999; Nowak et al. 2002). Taken together, these studies suggest that, although distant translocations might spare an animal in the short term from any immediate anthropogenic disturbances, translocations involving large-bodied nuisance snakes with high site fidelity might not be successful, in part because the homing efforts of the snakes makes contact with humans or altered habitat unavoidable (Reinert and Rupert 1999; Nowak et al. 2002; Shetty and Shine 2002; Butler et al. 2005).

When researchers use translocation to augment the populations of territorial species, the resident individuals may force the released animals out of the area and cause excessive movement of released individuals (Fritts et al. 1984). The effects of resident snakes on repatriated individuals are unknown, although it is probably minor, given the relatively low intensity of territorial defense in most snake species (Madsen 1984; Rivas and Burghardt 2005). In some snake species, resident individuals may actually indirectly assist translocated snakes to find shared resources. For example, translocated *C. horridus* probably learned of hibernacula locations by following or trailing resident animals (Reinert and Rupert 1999).

Consequences of Excessive Dispersal

Hard-released animals generally experience higher mortality than resident animals (Blanchard and Knight 1995; Mullen and Ross 1997; Brown and Day 2002). Higher mortality was observed in translocated *C. horridus* (Reinert and Rupert 1999) and *C. atrox* (Nowak et al. 2002) than in resident snakes. An early (1992) translocation of adult Massasaugas (*Sistrurus catenatus*) resulted in 100% mortality (R. Johnson, pers. comm.). Although resident and translocated *H. platirhinos* had similar overall mortality rates, translocated individuals averaged one-third the survival duration (number of days alive) of resident snakes (Plummer and Mills 2000). Not surprisingly, then, mortality associated with postrelease dispersal is a principal factor behind repatriation failures (Moehrenschlager and MacDonald 2003).

Such mortality that involves adults may particularly hinder recovery efforts because adult mortality is especially damaging to the recovery of small populations (Larkin et al. 2004).

Mortality associated with greater movements by translocated animals may be the result of vulnerability to predation or anthropogenic disturbance (Eastridge and Clark 2001; Butler et al. 2005; Roe et al. 2006). Snakes in natural populations are already susceptible to road mortality (Andrews and Gibbons 2005), and the encounter rates on roads are frequently high (Bonnet et al. 1999b; Roe et al. 2006). Road mortality has been observed in repatriated *Pituophis catenifer sayi* (Moriarty and Linck 1997), and the human persecution and anthropogenic disturbances experienced by translocated *C. atrox* increased their mortality rates (Nowak et al. 2002).

There are consequences to excessive dispersal beyond the direct effects of mortality. Excessive dispersal or poor fidelity to the release site is undesirable for conservation purposes because animals that have excessive dispersal may have little or no range overlap with other animals (Fritts et al. 1984; Oldham and Humphries 2000; Larkin et al. 2004). Even if individual survival is high, the establishment of new, or the augmentation of small, populations will be limited by the inability to find conspecifics with which to breed. Any animals that leave the population can also be considered "dead" with respect to that population (Seigel and Dodd 2000; McKinstry and Anderson 2002; Nowak et al. 2002). Thus, repatriation success may be highly dependent on release-site fidelity.

Release-site fidelity is also desirable because the new areas are often chosen with respect to specific features in the available habitat (e.g., high prey density and reduced predator abundance and anthropogenic disturbances) that would presumably increase fitness and survival. Suppose a snake is released into a nature reserve large enough to contain the activity ranges of individuals representing a healthy population, but the reserve is surrounded by roads and residential areas. If the snake leaves the reserve because of a lack of release-site fidelity, the animal is presumably more susceptible to anthropogenic disturbances. In addition, release-site fidelity and minimizing excessive movements is desirable because animals that move more have higher energetic costs (Sullivan et al. 2004). No negative relationships between movement and body condition have been observed in translocated *C. atrox* (Nowak et al. 2002), *C. horridus* (Reinert and Rupert 1999), or *H. platirhinos* (Plummer and Mills 2000), however, which suggests that translocated snakes can forage successfully as long as the habitat of the release site is adequate.

Model simulations showed that the extinction probability of a repatriated population decreases with higher site fidelity (Seigel and Dodd 2000), a relationship that was supported by studies on other vertebrate species (McKinstry and Anderson 2002; Moehrenschlager and MacDonald 2003). Thus, direct mortality and excessive dispersal from the release site both reduce the chances of repatriation success. Among snakes, combined release-site fidelity rates and survival were low: 29% in translocated *C. atrox* (Nowak et al. 2002), 9% in *C. horridus* (Galligan and Dunson 1979), and approximately 13% in the Eastern Indigo Snake (*Drymarchon couperi;* Irwin et al. 2003). This suggests that multiple releases, consisting of numerous animals, are required to offset the low survival and poor release-site fidelity. Data on long-term site fidelity will be crucial, however, because short-term data can be misleading (Ashton and Burke 2007).

Mortality often decreases once a translocated individual has survived a critical acclimation period (Jones and Witham 1990; Brown and Day 2002). The highest mortality should occur within the first year of release and decrease over time once the individual demonstrates activity patterns similar to residents (Haskell et al. 1996; Mullen and Ross 1997; Reinert and Rupert 1999). Repatriated animals, however, do not always experience higher mortality than resident populations (Oldham and Humphries 2000). For example, in contrast to the 100% mortality observed in R. Johnson's (pers. comm.) translocation efforts previously mentioned, King et al. (2004) reports that repatriated head-started *S. catenatus* in Wisconsin experienced a mortality rate similar to resident individuals. If mortality can be minimized, the chance of repatriation success increases greatly, even when a small number of animals are released (S. Taylor et al. 2005).

Once the snakes have settled into their release site, they may still be vulnerable to mortality at particular seasonal or life-cycle stages. A critical bottleneck in areas with cold winters may be hibernation and subsequent successful initiation of the next activity season. We (N. Bieser and B. A. K.) are monitoring a cohort of head-started *S. catenatus* in Michigan that had approximately 90% survival through an entire activity season. At the end of the season, however, they did not move in a manner similar to the residents, toward known hibernation areas; instead, the head-started snakes persisted in the summer activity areas. Ultimately, many positioned themselves in areas with other snakes (resident *S. catenatus* and *Thamnophis sirtalis*) before snowfall, but others do not appear to have located suitable areas. In other studies, snakes have found hibernacula and even survived the winter, only to perish in the spring (B. Johnson, pers. comm.; A. Lentini, pers. comm.). In areas where winters are not severe enough to threaten translocated snakes, they may have a longer period during which to establish themselves.

Increasing Release-Site Fidelity

Conservation efforts should use and test a variety of methods to increase the release-site fidelity and survival of repatriated individuals because these are among the main factors that limit repatriation success (Fischer and Lindenmayer 2000; Griffin et al. 2000). One method that is increasingly used to reduce dispersal is the soft release of repatriated wildlife. Soft releases allow individuals to acclimatize, undergo prerelease conditioning, or develop fidelity to the release site by the individuals' being placed in outdoor enclosures on-site prior to their release (Pedrono and Sarovy 2000; Lockwood et al. 2005; Tuberville et al. 2005). Through this process, individuals acclimate to the environmental conditions (e.g., temperature and photoperiod) and develop references to landmarks surrounding the release site (Brown and Day 2002). Soft releases also facilitate the monitoring of individual behavior and health, and provide a time for recovering from any handling stress prior to release.

The variables of interest in soft releases, such as length of time in an enclosure or amount of supplemental food and refuge, can each be established at a particular level (Bright and Morris 1994; Biggins et al. 1998; Truett et al. 2001). Generally, as time in a soft release enclosure increases, the dispersal distance from the release site decreases, survivorship increases, and release-site fidelity increases (Lohoefener and Lohmeier, 1986; Lockwood et al. 2005; Tuberville et al. 2005). To our knowledge, the effects of soft releases have not been tested in snakes. When retained within a small (50-m circumference) enclosure for a brief (2-week) period, Massasaugas stopped patrolling the perimeter of the enclosure after approximately a week, which suggests that a relatively short acclimation period may at least blunt the initial drive to disperse (N. Bieser and B. A. K., pers. obs.). One-quarter of the snakes returned to the release enclosure area, even entering it; three ended the season there. The soft release apparently assisted those individuals in becoming somewhat anchored to the release site. In our case, we positioned the enclosure in suitable summer habitat. The propensity to home back to the enclosure suggests future placements should include a hibernaculum.

The main disadvantage of soft-release enclosures is that they are expensive to build, may require extensive maintenance, and might not always be logistically feasible in certain habitats or situations (Moehrenschlager and Macdonald 2003). Soft-release enclosures are presumably more challenging for larger or arboreal species, but relatively low-budget or short-term soft releases are possible for snakes (Nishimura 1999). Although the enclosure size for soft release needs to be thoroughly investigated, we suspect that the enclosed space does not need to emulate even a minimum seasonal range size. Instead, enclosure size should be driven by other considerations: capacity to support some snake mobility, safety of the enclosed (= exposed?) snakes, and cost. The intent is not to have one or more individuals associate the enclosure with a home range but to establish it as a familiar place from which to explore.

Animals often show fidelity to hibernacula by using the same site from year to year (Macmillan 1995). Releasing animals late in the year, just prior to when animals normally enter hibernation, and releasing animals in a hibernation state directly into hibernacula are methods that might reduce dispersal from the release sites and increase survival in translocation efforts

(Eastridge and Clark 2001). Head-started Massasaugas released in autumn, just prior to entering hibernacula, had smaller activity ranges and higher mortality than animals released in the spring (King et al. 2004). The higher mortality in this case, however, might have been due to inadequate recovery from radiotransmitter implantation surgery prior to hibernation (King et al. 2004). In another study, translocated *C. horridus* that successfully over-wintered exhibited a resident-like affinity to their hibernacula by returning those sites at the end of the activity season (Reinert and Rupert 1999). Translocated snakes do not always develop fidelity to new hibernacula—other studies found that less than 10% of translocated *C. horridus* (Galligan and Dunson 1979) and 11% of *T. sirtalis parietalis* (Macmillan 1995) were captured in the spring following their release at hibernacula sites.

Captive Handling

Head-starting

Head-starting is the practice of raising animals in captivity with the intent of releasing them into the wild after they reach a certain size or age class. In addition to avoiding the risk of predation during maturation, these individuals are then closer to reproductive maturity at release. The source of individuals used in a head-starting effort can be a captive breeding program or, as is more often in the case of snakes, neonates obtained from temporarily holding wild-caught gravid females in captivity (King et al. 2004; King and Stanford 2006). Head-starting has been used extensively in game fish, for which the focus has traditionally been on releasing large numbers of animals and maximizing productivity (Brown and Day 2002). In contrast, head-starting efforts involving snakes should focus on increasing the survivorship of young snakes (King and Stanford 2006), thereby enhancing recruitment into the population (Shine and Bonnet, Chapter 6).

Snakes make good candidates for head-starting because, unlike other taxa, snakes are relatively inexpensive to keep in captivity and there is no evidence of imprinting (Weatherhead et al. 1998; Davis and Stamps 2004). In addition, some snake species share redeeming features with other head-started taxa, such as relative lack of parental care, high fecundity, and high juvenile survival (in captivity), that make them ideal candidates for head-starting (King 1986; Bloxam and Tonge 1995; Heppell et al. 1996; Marsh and Trenham 2001). Head-starting can be a useful snake conservation tool because individual survivorship typically increases with age and size due to reduced risk of predation (King 1986; Stanford and King 2004).

An animal's growth rate is determined by its genotype and nutritional input (Madsen and Shine 2000b; Aubret et al. 2004). Captive conditions have the potential to affect snake hatching mortality, neonate size, and the individual's survival on release (Gutzke and Packard 1987; Burger and

Zappalorti 1988a). Therefore, breeding and raising snakes in captivity should focus on producing offspring that are morphologically, behaviorally, and genetically similar to wild cohorts (Brown and Day 2002). Given the cost and space requirements of housing snakes, and the goal of increasing population recruitment, a primary objective of head-starting is to quickly get individuals as large as possible. Consequently, head-starting schedules commonly exclude hibernation to extend the growing season (King and Stanford 2006), producing individuals that are larger than resident animals of similar ages (Elsey et al. 1992).

Complications due to the retention of gravid females in captivity are possible. For example, gravid female DeKay's Brownsnakes (*Storeria dekayi*) exhibit a negative correlation between length of captivity and neonate body size and body condition (King 1993c). To avoid this sort of impact, small enclosures can be used to confine gravid females in the field until parturition (Pilgrim 2001). Collection too soon may also interfere with courtship, mating, and fertilization. Early collection and retention may be unavoidable, however. We obtain gravid *Nerodia sipedon* when we can find them (late spring) as opposed to when they are closer to parturition (mid-summer; S. Gibson and B. A. K., pers. obs.). Conversely, collecting gravid *Sistrurus catenatus* is possible right up to parturition (N. Bieser and B. A. K., pers. obs.)—the later into the summer it is, the easier they are to find. Unfortunately for this species, gravid females are susceptible to collection by poachers as well. Similar challenges will probably exist for many species.

Faster growth rates during early life history can also confer a "silver-spoon" effect on head-started individuals (Madsen and Shine 2000b). Not only are larger individuals more likely to survive than smaller ones (Haskell et al. 1996; O'Brien et al. 2005), but they can experience other long-term effects that may enhance their fitness long after their release. For example, Water Pythons (*Liasis fuscus*) with faster growth in their first year (because of supplemented food supply) continued to experience greater than average growth than smaller cohorts later in life (Madsen and Shine 2000b). Other advantages to snakes having larger body sizes include higher mating success in males, higher female fecundity, and larger size of offspring produced by those females (King 1986; Weatherhead et al. 1995; Stanford and King 2004). Although the silver-spoon phenomenon has been observed in head-started alligators (Elsey et al. 1992), it has not been quantified in the few published studies of head-started snakes (Aubret et al. 2004; King and Stanford 2006). For instance, on reaching a certain body size, some head-started snakes experience similar survivorship to that experienced by resident individuals of the same cohort (Aubret et al. 2004; King and Stanford 2006).

If individuals released as part of a head-starting program consist of the age class(es) with the highest reproductive output, then wild-born offspring should be produced sooner and population growth rates should increase.

Stochastic models have shown that long-lived species with high adult survival rates experience reduced extinction probability when reproductively active adults are released, as opposed to juveniles or nonbreeding adults (Sarrazin and Legendre 2000). Conservation efforts for long-lived species with delayed sexual maturity should focus on protecting subadults and adults because changes in hatchling mortality have limited effects on populations compared to changes in adult mortality (Heppell et al. 1996; Heppell 1998; Seigel and Dodd 2000). This concept also applies to snakes. Conservation efforts for long-lived snakes with delayed maturity, such as the Broad-headed Snake (*Hoplocephalus bungaroides*) are more sensitive to protecting juveniles and adult classes than neonates (Webb et al. 2002b). In contrast, populations of earlier-maturing Small-eyed Snakes (*Cryptophis nigrescens*) will respond more positively to protecting the younger age classes (Webb et al. 2002b).

One way to offset expected mortality of head-started adults after release is to arrange multiple releases of a large number of animals over several years and perhaps longer. Such an effort involving *Sistrurus catenatus* in Ontario, Canada, was intentionally designed to release head-started subjects over multiple seasons to avoid catastrophic losses (A. Lentini, pers. comm.). Political and practical considerations forced a single release, however, and an unusually high winter water table appeared to contribute to a loss of all of the snakes. The manipulation of sex ratios within each released cohort to maximize population growth is another possibility. Any uncertainty regarding sex ratios in head-started hatchlings or juveniles increases the probability of extirpation (Sarrazin and Legendre 2000). Fortunately, ascertaining the sex of young snakes is generally straightforward (Fitch 1987a). Should sex ratio manipulations be contemplated, both short- and long-term impacts on the population must be considered (Wedekind 2002). Studies of how sex ratio manipulations may affect snake populations are lacking.

Releasing juveniles or subadults may also have potential benefits. Juveniles or subadults may be less affected by long-term captivity and often disperse shorter distances from release sites than adults (Pedrono and Sarovy 2000; Moehrenschlager and Macdonald 2003; Larkin et al. 2004). Another potential benefit to releasing these age classes is that, in some species, these cohorts may have higher survival in repatriations than adults (Stanley Price 1989; Moehrenschlager and Macdonald 2003). The trade-off is that there will be lower initial recruitment into the population until the younger age classes reach sexual maturity and reproduce (Heppell et al. 1996). Among studies involving snakes, neonate or juvenile *Crotalus horridus* can make better candidates for repatriations than adults because younger snakes have less fidelity to previous hibernacula or activity ranges (Reinert and Rupert 1999).

Captive Rearing and Environmental Enrichment

Enriching captive conditions has been used as a prerelease method to prepare head-started animals for their new environment; enriched captive conditions have possibly been responsible for higher postrelease survival rates (Biggins et al. 1998). In contrast, rearing animals in simple artificial environments could lead to individuals who are unprepared for life in the wild (Brown and Day 2002). Although the positive effects of enrichment have been well documented for mammals and birds (Biggins et al. 1998; Griffin et al. 2000), little is known about the effects of enrichment on snakes. Work by Almli and Burghardt (2006) indicates that snakes will, in fact, benefit from enriched housing.

Enrichment studies on captive-bred fish (reviewed in Brown and Day 2002), which are often maintained in simple enclosures (similar to snakes), can be used to infer the potential benefits of enrichment prior to repatriating a captive snake population. The impact of captive enrichment on postrelease survival rates is perhaps best documented in hatchery and head-started fish (Brown and Day 2002). Hatchery fish that do not experience any enrichment are both physically and behaviorally underdeveloped. They have less stamina, are less wary of predators, and can be more prone to risk-taking once released (often motivated by starvation). These factors are believed to be among the reasons why only 1–5% of hatchery fish survive to adulthood, with most of the mortality occurring within a few days of release.

Enrichment can increase the fitness of released animals by helping them develop hunting skills (a trial-and-error process in snakes) and enhancing their problem-solving skills (Biggins et al. 1998; Brown and Day 2002; Swaisgood and Shepherdson 2005). Even a single exposure to a predator can increase wariness and antipredator responses in future encounters (Brown and Day 2002). Small cage size and cage homogeneity often do not provide the mechanical stimuli necessary for long bone development in mammals and contribute to poor fitness (Wisely et al. 2005); it is possible that analogous effects are experienced by snakes.

Cages can be enriched simply through increasing enclosure complexity by adding substrates such as dirt, leaf litter, and vegetation. Climbing structures also allow more efficient use of space and provide exercise opportunities, shade, hiding spaces, and temperature gradients. Concealing food or scents can elicit foraging and investigating behaviors (Brown and Day 2002; Swaisgood and Shepherdson 2005).

Outdoor enclosures provide the highest form of enrichment for captive animals because they minimize human-animal interactions, provide animals ample opportunity to hunt prey, and increase their conditioning to microorganisms (Biggins et al. 1998). When equipped with an enriched environment, outdoor enclosures used in head-starting snakes have the potential to facilitate the success of a future soft release by developing release-site fidelity.

Such enclosures might also provide opportunities to educate local communities that might be impacted by the release of a head-started population (Burghardt et al., Chapter 10). Despite the possible benefits of larger, enriched enclosures reported for other vertebrate taxa, the few published studies of head-started snakes involved raising them in small simple cages. Once released, however, head-started individuals appear to experience mortality rates similar to wild cohorts (King et al. 2004; King and Stanford 2006).

Captive Breeding and Genetics

Individuals used in repatriations or head-starting programs should originate from areas as close as possible to the release site to maintain locally adapted genotypes (IUCN 1998; Ficetola and Bernardi 2005). Molecular techniques provide a means to fulfill this requirement even when the origin of breed stock is not certain. For example, mitochondrial and nuclear microsatellite markers were used to determine the diversity and natural population structure of captive Jamaican Boa (*Epicrates subflavus*), which will help optimize captive breeding efforts prior to any reintroduction (Tzika et al. 2008a, 2009). Translocating individuals that are adapted to a certain habitat or that originate from an evolutionarily distinct deme into a different habitat are more likely to fail because of the lack of adaptations for their new environment. Mixing ecologically or genetically different populations could result in outbreeding depression and the loss of unique alleles and local adaptations, which can result in a reduction of fitness (Ficetola and Bernardi 2005; King, Chapter 3). The occurrence of outbreeding depression may be temporary, however, because selection may alleviate this phenomenon. In situations in which there are no resident individuals at the repatriation site, repatriations may succeed despite the repatriated animals' originating from distant or unknown localities (reviewed in Stanley Price 1989; Dunham 1997).

   If a large number of animals die or leave the study site, repatriated populations may also experience more reductions in genetic variability than would otherwise be faced by small populations. Therefore, the founder population will be smaller than the actual number of animals released. Small founder populations are more susceptible to genetic drift and reduced genetic variation, which can eventually lead to lower fecundity and survival (Bodkin et al. 1999; Madsen et al. 1999). The genetic diversity among animals within a repatriated population can be increased by combining different population sources of founders, releasing large number of individuals, or reducing translocation mortality (Bodkin et al. 1999). Establishing a viable population in the short term, however, may be a higher priority than maintaining genetic variation of a repatriated population, especially if the species is at extreme risk of extinction (Stanley Price 1989). Nevertheless, genetic variation among repatriated founders ultimately has the potential to affect the long-term viability of that population (Madsen et al. 1999).

Future Directions

Recipes for Success?

Repatriations are challenging and should not be the first resort for rescuing a declining population of snakes. If the decision is made to carry out such an effort, however, successful repatriations will have several common attributes (see Tables 7.1 and 7.2). The factors associated with the original extirpation, such as habitat degradation and persecution, must be reversed to avoid a similar fate for the repatriated animals (Smith and Clark 1994; Wolf et al. 1996, 1998; Seigel and Dodd 2000; Dunham 2001). Successful repatriations generally will require a long-term commitment of time and resources by both land managers and researchers (Beck et al. 1994). Monitoring long-term repatriation projects provides opportunities to correct mistakes, thereby improving the likelihood of success and the likelihood that future repatriation efforts will also succeed. A large number of individuals may have to be released multiple times over a long duration to offset the higher mortality and extinction rates inherent in smaller populations (Beck et al. 1994; Smith and Clark 1994; Wolf et al. 1996, 1998; Fischer and Lindenmayer 2000). The number of individuals released, however, is dependent on species fecundity and survivorship (Stanley Price 1989; Wolf et al. 1998; Seigel and Dodd 2000) because the potential for success often increases as the number and clutch size increases for a species (Griffith et al. 1989).

Repatriation success is also associated with the habitat quality of the release site (Griffith et al. 1989; Smith and Clark 1994; Wolf et al. 1998). Releasing animals into poor habitat will probably fail regardless of other precautions taken or the adaptability of the species (Griffith et al. 1989). Repatriations are also more likely to be successful if they occur into prime habitat in the core of the species' historical range versus the periphery (Griffith et al. 1989; Smith and Clark 1994; Wolf et al. 1996, 1998). Release sites should also be large enough to encompass the typical patterns of dispersal for the species of interest from release site (Moriarty and Linck 1997).

The Future: Learning from Past Mistakes

For better or worse, translocation of individual snakes is a common practice. Interest in moving snakes for the purposes of conservation is growing, and resource managers are often eager to try it. From the research summarized in this chapter, however, it is evident that many of these efforts are not reported to a more general audience or lack scientific rigor. As a result, we are not learning as quickly as we would like about how to effectively translocate snakes. We strongly encourage the publication of any such efforts, even if the studies are opportunistic, anecdotal, have small sample sizes, or result in failure. We understand the hesitation many managers may feel about publishing the results of their efforts given the controversy surrounding these methods (Burke 1991; Dodd and Seigel 1991; Reinert 1991). Managers or researchers may be hesitant to publish because they may be cited in journals as repatriation failures if the majority of animals die during the translocation and criticized for making mistakes or for not following general guidelines. The emotional response and accountability both increase if the species is rare or endangered. Despite potential consequences, we still encourage the publication of these translocation failures and mistakes. These publications will raise awareness about attempted repatriations and, we hope, prevent other conservationists from making similar mistakes.

Snake repatriation needs more experimental study. Attempted repatriations should be conducted with clear objectives and greater scientific rigor. Few translocation studies have adequate sample sizes for hypothesis testing or are experimental in their design, and thus, they lack the power needed to test the effectiveness of this conservation technique. Increasing sample sizes may be easier said than done because of the lack of funding and the limited sample availability when working with nuisance animals or imperiled species. The former challenge may be addressed as funding entities accept the need for better information regarding translocation. The latter problem reflects the notion of "doing something" to rescue animals without clear regard for the consequences of the translocation. It also leads to the suggestion of using common species for preplanned studies that test the effectiveness of repatriations as a conservation tool. For example, we are engaged in an exploration of head-starting and translocation approaches for the recovery of the Copper-bellied Watersnake (*Nerodia erythrogaster neglecta*, federally recognized as threatened, and we are initially using the Northern Watersnake (*N. sipedon sipedon*) as a surrogate to develop approaches. As we become confident of our capacity to translocate watersnakes effectively, we can graduate to the imperiled species. If we fail, we can adjust our approach or abandon it as untenable before negatively impacting the species that is already in dire straits.

The remaining uncertainties pertaining to translocation, repatriation, and captive rearing provide many directions for future research. Some information gaps go well beyond the actual practice of moving snakes around. For example, we lack a thorough understanding of when a site is good enough for a population rather than just for an individual, such that a translocation might actually succeed in repatriating a population. Examined from a different perspective, we might not know which factors in the landscape led a population beyond the tipping point that causes local extirpation. For snake translocations, we need more attention to be directed to the question of which life stages to release. Adults may make a more immediate reproductive contribution but only if they persist within the new habitat. Alternatively, juvenile mortality may be so great that an inadequate number of individuals reaches maturity. Should we release more females to get an early reproductive bump? With respect to captive rearing, we know how to raise snakes for the pet industry, but we still need to know more about how such snakes do in the natural environment.

Future studies should test techniques that improve orientation and decrease the adjustment period of translocated animals in their new environment. These studies could include testing the effects of captive conditions and of soft versus hard releases on postrelease dispersal and survival. Soft release techniques appear to be an important topic on which to focus our energy. The use of enclosures, at least temporarily, to hold translocated snakes should be helpful, although larger enclosures may be quite expensive.

To conclude, many translocation projects are already underway. We want to conduct these sorts of projects correctly, yet we lack a thorough understanding of how to make snake repatriations effective. Acquiring this knowledge will require the consideration a number of key steps and the resolution of challenges during the planning and implementation phases. Planned, hypothesis-driven studies, with follow-through in the form of reporting and publication, are greatly needed. For many snakes that play integral roles in their respective communities, these strategies may be their only hope.

8

Habitat Manipulation as a Viable Conservation Strategy

KEVIN T. SHOEMAKER, GLENN JOHNSON, AND KENT A. PRIOR

Fifteen years ago, in a volume otherwise focusing on snake ecology and behavior, Seigel and Collins (1993) saw fit to include a chapter on snake conservation (Dodd 1993b). In his chapter, Dodd bemoaned the unquestioned acceptance of habitat manipulation practices such as conservation corridors and road-crossing structures: "there is an urgent need to evaluate what are rapidly becoming accepted...management techniques" (1993b: 384). In an effort to provide managers, planners, and field practitioners with a framework for making informed habitat management decisions, in this chapter we review the use of habitat manipulation in snake conservation, evaluate the extent to which habitat manipulation has been successful in achieving conservation goals, and make recommendations regarding the use of habitat manipulation in future snake conservation endeavors. In so doing, we highlight knowledge gaps and profitable applied research opportunities, the investigation of which should lead to improved conservation practices.

Habitat manipulation is often uncritically embraced as a practical management "fix." Its particular appeal may lie in the hope that habitat functions (e.g., the ability to support viable snake populations) might simply be restored through direct manipulation of habitat remnants—no net loss, everybody wins; ecological costs and benefits apparently optimized without harmful social or economic consequences. The recent conversion of Wisconsin farmland to support a population of the state-listed Butler's Gartersnake (*Thamnophis butleri*), displaced to make way for a Target store, serves as a case in point (Wisconsin Department of Natural Resources 2007). But the question remains: Is habitat manipulation effective in improving the conservation status of snakes, or would scarce financial resources be better spent protecting existing habitat? Habitat manipulation projects may not function as managers intend; artificial or novel habitat elements may introduce threats that snakes are unequipped to detect, functioning as evolutionary traps (Kolbe and Janzen 2002; Schlaepfer et al. 2002). Even habitat manipulation projects that are ecologically benign can potentially drain resources from more effective conservation strategies.

We consider here three broad categories of habitat manipulation: (1) manipulation of targeted habitat features (e.g., basking sites and hibernacula), (2) manipulation of the seral stage of natural communities (e.g., prescribed fire), and (3) manipulation of ecological landscapes (e.g., linear corridors). Published studies in these categories were identified by searching the ISI Web of Knowledge database and the online meta-search engine Google Scholar for all references using the words *snake* or *reptile* along with terms such as *habitat management, artificial hibernacula,* or *prescribed fire* (the complete list of search terms is available on request) and by searching the bibliographies of relevant publications. We also solicited the assistance of colleagues by posting requests on relevant electronic mailing lists, contacting state-employed herpetologists and nongame wildlife specialists (in the United States), contacting university scientists currently conducting snake conservation research. Published studies that quantitatively evaluated the response of snakes to habitat manipulation were evaluated for overall strength of evidence (rigor of experimental design and strength of response), and key results from each study were compiled in a table. Due to variation in methodology and types of data reported, a meta-analysis of the rates of management success or factors influencing management success (Gates 2002) was deemed impractical.

We were able to locate 33 published studies relevant to our three broad categories of habitat manipulation (Table 8.1). The majority of studies ($n = 22$) investigated snake response to vegetation management (e.g., logging or prescribed fire). Several studies evaluated the use of herbivore-exclusion fencing and road-crossing structures in snake conservation. The response of snakes to the manipulation of targeted habitat features (e.g., artificial hibernacula, retreat sites, and basking sites) was not well-documented in the literature.

Manipulation of Targeted Habitat Features

Habitat management can target specific habitat needs of snakes, including basking, retreat, hibernation and estivation, feeding, and nesting.

Basking Sites and Gestation Areas

Thermoregulation is closely associated with physiological function (e.g., shedding, digestion, locomotion, and gestation) and serves as a fundamental

TABLE 8.1
Summary of published documentation on habitat manipulation in snake conservation

Reference	Location	Focal or Dominant Snake Species	Total Number of Snakes in Study	Premanipulation Monitoring?	Postmanipulation Monitoring?	Long-Term Monitoring? (≥ 3 Years)	Treatments Replicated?[a]	Response of Individual Snakes to Habitat Manipulation	Change in Abundance after Manipulation	Change in Fertility, Mortality, or Other Fitness Indicator	Overall Strength of Evidence[b]
A. Manipulation of targeted habitat features											
Artificial basking/gestation sites											
Webb et al. 2005	Southeastern Australia	*Hoplocephalus bungaroides*	2	Y	Y	Y	Y	+++	N/A	N/A	+++
Artificial hibernacula											
Zappalorti and Reinert 1994	New Jersey	*Pituophis melanoleucus,* other species	139	N	N	N	N	++	N/A	N/A	+
B. Manipulation of the seral stage of natural communities											
Prescribed fire											
Jones et al. 2000	Oklahoma	*Virginia striatula,* other species	48	N	Y	N	Y?	+	N/A	N/A	+
McLeod and Gates 1998	Maryland	*Carphophis amoenus, Coluber constrictor,* other species	156	N	Y	N	N	M	N/A	N/A	U

TABLE 8.1—continued

Reference	Location	Focal or Dominant Snake Species	Total Number of Snakes in Study	Premanipulation Monitoring?	Postmanipulation Monitoring?	Long-Term Monitoring? (≥3 Years)	Treatments Replicated[a]	Response of Individual Snakes to Habitat Manipulation	Change in Abundance after Manipulation	Change in Fertility, Mortality, or Other Fitness Indicator	Overall Strength of Evidence[b]
Wilgers and Horne 2006	Kansas	Diadophis punctatus, Tropidoclonion lineatum, other species	U	N	Y	N	N?	M	N/A	N/A	U
Kilpatrick et al. 2004	South Carolina	Tantilla coronata, Carphophis amoenus	U	N	Y	N	Y	U	N/A	N/A	U
Masters 1996	Northern Territory, Australia	Ramphotyphlops endoterus, other species	24	N	Y	Y	N	U	U	N/A	U
Cavitt 2000	Kansas	Coluber constrictor, Thamnophis sirtalis, other species	550	Y	Y	N	N	–	U	N/A	–
Litt et al. 2001	Florida	Tantilla coronata	U	N	Y	N	Y?	–	N/A	N/A	–
Setser and Cavitt 2003	Kansas	Coluber constrictor, Thamnophis sirtalis, other species	92	N	Y	N	N	– –	N/A	N/A	–
Mechanical clearing Johnson 1995; Johnson and Leopold 1998	New York State	Sistrurus c. catenatus	U	N	Y	Y	Y	+	N/A	N/A	+

Clear-cutting/site preparation

Reference	Location	Species									
Enge and Marion 1986	Florida	*Coluber constrictor, Cemophora coccinea,* other species	280	N	Y	N	N	++	U	N/A	+
Greenberg et al. 1994	Florida	*Tantilla relicta,* other species	U	N	Y	N	N	+	N/A	N/A	+
Kavanagh and Stanton 2005	New South Wales, Australia	*Demansia psammophis,* other species	U	N	Y	N	Y?	N/A	N/A	N/A	+
Crosswhite et al. 2004	Arkansas	*Agkistrodon contortrix, Coluber constrictor*	66	N	Y	N	N	U / +	U / +	N/A	U
Perison et al. 1997	South Carolina	*Diadophis punctatus,* other species	21	N	Y	N	N	U / −	N/A	N/A	U
Renken et al. 2004	Missouri	*Storeria occipitomacu-lata, Virginia valeriae*	U	Y	Y	Y	Y	U	U	N/A	U
Russell et al. 2002	South Carolina	*Carphophis amoenus, Coluber constrictor*	U	Y	Y	N	Y?	−	N/A	N/A	U

Group selection harvesting/salvage logging

Reference	Location	Species									
Greenberg 2001 (natural canopy gaps)	North Carolina	*Carphophis amoenus, Diadophis punctatus,* other species	108	N	Y	N	Y?	+	N/A	N/A	+
McLeod and Gates 1998	Maryland	*Pantherophis alleghanien-sis,* other species	226	N	Y	N	N	++	N/A	N/A	+
Ross et al. 2000	Pennsylvania	*Thamnophis sirtalis, Diadophis punctatus,* other species	347	N	Y	N	Y?	+	N/A	N/A	+
Cromer et al. 2002	South Carolina	*Diadophis punctatus, Nerodia erythrogaster, Storeria dekayi*	387	N	Y	N	Y?	+	N/A	N/A	U
Goldingay et al. 1996	New South Wales, Australia	Various species	13	N	Y	Y	Y	U	U	N/A	U

TABLE 8.1—continued

Reference	Location	Focal or Dominant Snake Species	Total Number of Snakes in Study	Premanipulation Monitoring?	Postmanipulation Monitoring?	Long-Term Monitoring? (≥ 3 Years)	Treatments Replicated[a]	Response of Individual Snakes to Habitat Manipulation	Change in Abundance after Manipulation	Change in Fertility, Mortality, or Other Fitness Indicator	Overall Strength of Evidence[b]
Right-of-way management											
Yahner et al. 2001a	Pennsylvania	Storeria occipito-maculata, Diadophis punctatus, other species	50	N	Y	N	N	+++	N/A	N/A	+
Herbivore exclusion/restoration of riparian habitat											
Homyack and Giuliano 2002	Pennsylvania	Nerodia sipedon, Regina septemvittata, Thamnophis sirtalis	~500	N	Y	N	Y?	++	N/A	N/A	+
Szaro et al. 1985	New Mexico	Thamnophis elegans vagrans	~25	N	Y	N	N	++	N/A	N/A	+
Leynaud and Bucher 2005	Salta, Argentina	Bothrops neuwiedii, Phimophis vittatus, other species	153	N	Y	N	N	U	N/A	N/A	U
Bowers et al. 2000	South Carolina	Nerodia fasciata, Storeria occipitomaculata, other species	626	N	Y	N	N	U	N/A	N/A	U

C. Alteration of ecological landscapes

Road crossing structures

Reference	Location	Species									
Dodd et al. 2004; Smith and Dodd 2003	Florida	*Pantherophis alleghaniensis*, other species	772	Y	Y	N	N/A	U	N/A	+++	++
Shine and Mason 2001	Manitoba, Canada	*Thamnophis sirtalis parietalis*	U	N	N	N	N/A	(+)?	N/A	(+)?	+
Woodhouse et al. 2002	Ontario, Canada	*Sistrurus c. catenatus*	U	Y	Y	Y	N/A	– –	N/A	N/A	U
Aresco 2005	Florida	*Nerodia fasciata, Coluber constrictor*, other species	363	Y	Y	N	N/A	U	N/A	U	U
Rodriguez et al. 1996	Central Spain	*Elaphe scalaris, Malpolon monspessulanus, Vipera latastei*	U	N	Y	N	Y	(+)?	N/A	N/A	U

Notes: The magnitude and direction of snake responses to habitat manipulation are summarized using the following rating system: +++, strong positive; ++, positive; +, weak positive; –, weak negative; – –, negative; M, mixed; N, no; N/A, study not designed to provide this information; U, unclear; Y, yes.
[a] Adequate number, no pseudoreplication.
[b] + indicates for manipulation; – indicates against manipulation.

driver of habitat selection for many snake populations (Reinert 1993; Shine and Madsen 1996; Weatherhead and Madsen, Chapter 5). Open-canopy basking habitat is critically important for many snakes, especially large-bodied species (Stevenson 1985). Gravid females of many viviparous species spend the gestation season basking within open-canopy gestation habitat (Brown 1993; Parker and Prior 1999; see also Weatherhead and Madsen, Chapter 5; Shine and Bonnet, Chapter 6), often forfeiting opportunities for feeding or other nonbasking behaviors (Keenlyne and Beer 1973; Seigel and Ford 1987; Seigel et al. 1987). In addition, many snakes bask extensively on emergence from hibernation. This behavior may be instrumental for the completion of spermatogenesis in some species (Gregory 1982).

Snakes often favor basking sites that provide low-cost access to a wide thermal gradient (Spellerberg 1975, 1988). For instance, Gray Ratsnakes (*Pantherophis* [*Elaphe*] *spiloides*) apparently prefer forest-field ecotones for thermoregulation (Blouin-Demers and Weatherhead 2001b). Although access to direct sunlight can be critical for thermoregulation, snakes often bask in or near some form of shaded cover or below-ground retreat (Burger and Zappalorti 1988b; Nilson et al. 1999). Rocky outcrops and talus slopes often provide abundant thermoregulation and retreat opportunities, and these are consequently used by many snake species (e.g., Brown et al. 1982; Parker and Prior 1999).

Declines and even extirpations of snake populations may be linked to the loss of basking habitat. Anecdotal records suggest that an endangered population of Massasaugas (*Sistrurus catenatus*) in New York has declined as basking habitat has reverted to a closed-canopy state after an 1892 fire (Johnson and Breisch 1993). One of us (K. T. S.) recently assessed the case for habitat manipulation (canopy removal) at this site. Massasaugas generally selected the warmest available microhabitats for basking, but the average temperatures at these sites were substantially lower (by approximately 3 °C) than selected basking sites at an open-canopy reference location (Shoemaker 2007). Habitat management to improve basking habitat is therefore likely to prevent a further decline of this endangered snake population.

Artificial basking habitat may consist of simple clear-cut patches or artificial tree-fall gaps within a forested matrix (Schmidt and Lenz 2001; Gregory 2007). Many reptiles are well-suited to take advantage of small canopy gaps (Vitt et al. 1998) and may even follow tiny sun-flecks across a forest floor (Huey 1982). In addition, snakes are not known to be territorial; individuals of many species bask communally (Gillingham 1987; Gregory et al. 1987). Efforts to create or improve the basking habitat for snakes should therefore focus on quality and strategic location of basking sites rather than on the size of manipulated areas.

In some cases, vegetation removal may be insufficient to create optimal basking habitat. As already noted, habitat heterogeneity can be important for effective thermoregulation; homogenization of basking habitat is probably undesirable in most cases. Poorly planned vegetation management may increase the visibility of sedentary species that rely on crypsis as a predator-avoidance mechanism (e.g., Graves 1989). The potential costs to snakes of vegetation removal, such as increased predation rate and decreased crypsis, should be evaluated as part of any study of managed basking habitat.

Retreat Sites

Retreat sites function primarily to shelter snakes from potential predators (Webb and Whiting 2005) and from extreme temperatures (Huey et al. 1989); they are used extensively by many species (Whiting et al. 1997; Whitaker and Shine 2003; Pearson et al. 2005; Sherbrooke 2006; Shine and Bonnet, Chapter 6). In lieu of basking, some snakes access solar energy by selecting large flat rocks in open, sunny places (colloquially termed snake rocks) as retreat sites (Huey et al. 1989; Webb and Shine 2000). Snakes also use retreat sites to protect themselves from desiccation (Clark 1970; Whiles and Grubaugh 1993), to lay eggs (Henderson et al. 1980), and to forage (Webb and Shine 2000; Russell et al. 2004; Shine and Bonnet, Chapter 6). Because snakes spend much of their time stationary within retreat sites, retreat-site selection can influence fitness substantially (Webb and Shine 1998a; Kearney 2002; Pringle et al. 2003; Webb et al. 2004).

Populations of the endangered skink Adelaide Pygmy Bluetongue (*Tiliqua adelaidensis*) in Australia increased in size in response to the establishment of artificial burrows (Milne and Bull 2000; Souter et al. 2004). Similar examples, however, are difficult to find in the snake literature (see Table 8.1). Artificial retreat sites (Webb and Shine 2000) and the selective removal of canopy vegetation from formerly occupied retreat habitat (Webb et al. 2005b) may improve the conservation status of the endangered Broad-headed Snake (*Hoplocephalus bungaroides*). The endangered Concho Watersnake (*Nerodia paucimaculata*) in Texas makes extensive use of retreat sites within rocky shoreline habitat, which can be in short supply when water levels are high; the creation of elevated rocky shoreline habitat may improve the conservation status of this snake (Whiting et al. 1997). Riprap (large stones and boulders used to stabilize waterways and prevent erosion) is used by snakes for basking and retreat in some locations (Herrington 1988; Perry et al. 1996; Wylie et al. 2002), suggesting that construction of rock piles may provide retreat habitat for many snakes, especially those adapted to talus slopes (Herrington 1988; Schmidt and Lenz 2001). Strategically placed brush or rock piles may similarly serve to create valuable retreat habitat for snakes (Frier and Zappalorti 1983; Seymour and King 2003).

Hibernation Sites

Hibernacula are a special and particularly important type of retreat site for snakes in temperate climates. Winter kill and other hibernation-related losses represent some of the most important documented sources of mortality for snakes (Gregory 1982; Shine and Mason 2004), suggesting that hibernation habitat is a limiting resource for many temperate-zone populations. Choosing a proper hibernaculum is critical because a poor choice is almost certainly fatal (Reinert 1993). A suitable hibernacula must (1) provide protection from freezing temperatures (Bailey 1949), (2) maintain relatively cool temperatures to reduce wasteful metabolic expenditures (Goris 1971), (3) provide protection from desiccation (Costanzo 1989), (4) provide protection from predation (Burger et al. 1992), (5) provide access to an adequate supply of oxygen (Gillingham and Carpenter 1978; Shine and Mason 2004), and (6) remain free of molds and other pathogens (Goris 1971).

For communally hibernating species, hibernation site improvement may be a relatively cheap, simple, and effective management strategy (Shine and Mason 2004). For example, the Red-sided Gartersnake (*Thamnophis sirtalis parietalis*) hibernates communally by the thousands within limestone caverns in central Canada. Observing high overwinter mortality at some hibernation sites, researchers suggested that levee banks could be erected to protect hibernacula against flooding and that insulation of hibernacula might be used to protect snakes against freezing temperatures (Shine and Mason 2004).

Many snake species will use human-made structures such as building foundations and sewer lines as hibernacula (Zappalorti and Reinert 1994; Seymour and King 2003), raising the intriguing possibility that artificial hibernacula could benefit wild snakes. In reality, attempts at creating artificial hibernacula often fail due to one or more critical violations of the criteria we have listed (Bailey 1949; Goris 1971). Although failures documented in the literature generally involve captive snakes forced to use created hibernacula (see Goris 1971), wild snakes may also be threatened by poorly designed structures.

Artificial snake hibernacula have promise as a management technique (Shine and Bonnet, Chapter 6). Human-made hibernacula effectively decreased overwinter mortality from nearly 100% to approximately 10% at a commercial snake farm in Japan (Goris 1971). At this site, hibernacula were created by filling shallow holes with gravel for drainage, boulders to provide hibernation cavities, and packed soil for insulation (Goris 1971). Similarly, researchers in Oklahoma constructed a snake hibernaculum by filling a concrete-lined underground chamber with stacked concrete blocks. After a pump was installed to improve drainage within the hibernaculum (high mortality was observed in the first year of the study due to flooding), overwinter mortality fell to a relatively low 15% (Gillingham and Carpenter

1978). Artificial hibernacula similar to these were constructed for wild Northern Pinesnakes (*Pituophis m. melanoleucus*) in New Jersey. Many pine snakes, as well as individuals of other species, have been documented using these structures (Zappalorti and Reinert 1994). The development of artificial hibernacula was not accompanied by population monitoring (Zappalorti and Reinert 1994), and the conservation success of these structures remains unclear.

The effectiveness of artificial hibernacula in snake conservation has not yet been demonstrated in the literature (see Table 8.1). Until the effectiveness of this technique is firmly established, artificial hibernacula should generally be restricted to captive management studies and experimental field studies (Shine and Bonnet, Chapter 6). We caution against the use of artificial hibernacula as *quid pro quo* for the destruction of known hibernacula.

Foraging Sites

Improvements to foraging habitat may be an effective conservation strategy for prey-limited populations. In Australia, prey availability may be the primary determinant of population size for *H. bungaroides,* at least at the site level (Shine et al. 1998b). Artificial cover objects have been shown to elicit a positive response from Lesueur's Velvet Geckos (*Oedura lesueurii*)—the primary prey of the *H. bungaroides*—and may indirectly benefit Broadheaded Snakes by augmenting prey populations (Webb and Shine 2000). In central New York, plots cleared of woody vegetation in an effort to benefit *S. catenatus* showed higher Massasauga prey (small mammal) abundance and diversity following vegetation removal (Johnson 1995).

The manipulation of habitat structure may improve foraging success without changing prey density. In China, trimming tree branches improved foraging success for Shedao Island Pitvipers (*Gloydius shedaoensis*) by forcing birds to alight on branches strong enough to support snakes (Shine et al. 2002c). Mesocosm experiments with ratsnakes (*Pantherophis*) have shown that the structural complexity of foraging habitat can influence foraging success; for this species, intermediate levels of habitat complexity (density of woody vegetation) maximized foraging success (Mullin et al. 1998).

Nesting Habitat

Nesting ecology is poorly understood for most egg-laying snake taxa. Nesting habitat should supply embryos with the warmth necessary for proper development (see Weatherhead and Madsen, Chapter 5); adequate moisture; and protection from predators, pathogens, and plant roots (Burger and Zappalorti 1991; Burger et al. 1992; Shine and Bonnet, Chapter 6). Although nesting sites may be a limiting resource for many snake populations, the manipulation of nesting habitat is not common in snake conservation.

Some snake species commonly use human-altered habitat for nesting. In a study of Northern Pinesnakes in New Jersey, all observed nesting events occurred in areas recently disturbed by humans (Burger and Zappalorti 1988b). Ratsnakes use human-made leaf piles and compost piles as nest sites (Weatherhead and Madsen, Chapter 5). Rock walls are used for nesting by some species (Shine and Bonnet, Chapter 6). Artificial nesting sites (carefully constructed rock piles) have been used by several snake species in France (Shine and Bonnet, Chapter 6). European snakes (*Natrix* spp.) frequently nest in manure piles, where bacterial decomposition supplies supplemental warmth (Madsen 1984; Spellerberg 1988). In fact, northern populations of *N. natrix* may be dependent on the presence of manure piles for oviposition (Shine et al. 2002a). In Germany, wildlife managers placed manure piles next to preferred Dice Snake (*N. tessellata*) foraging habitat to facilitate nesting, but the success of this management intervention remains to be seen (Herzberg and Schmidt 2001).

Manipulation of Habitat Features as a Viable Conservation Strategy

We found virtually no published studies evaluating the manipulation of targeted habitat features in snake conservation; thus, the efficacy of these methods cannot be assumed. Manipulation of habitat features (especially hibernacula) should be accompanied by well-planned research and monitoring efforts. Nonetheless, small scale and specificity can make the manipulation of targeted habitat features an attractive option compared with the manipulation of entire communities.

Manipulating the Seral Stage of Natural Communities

Reversing Vegetative Succession

Many reptiles of forested regions depend on early-successional habitat. Nearly 75% of snake and lizard species in the southern United States require open-canopy habitats within their range, and more than half of these species are primarily associated with early-successional habitat (Trani 2002b). In a Pennsylvania study, powerline right-of-ways (ROWs) maintained in an early-successional state supported a greater diversity and abundance of snakes than the surrounding forested habitat (Yahner et al. 2001a, 2001b; Yahner 2004). Some regions (e.g., the northeastern United States) are undergoing extensive reforestation as agricultural areas are abandoned (Motzkin and Foster 2002), with major implications for all early-successional species (Litvaitis 1993; Brawn et al. 2001). Natural vegetation succession can pose a threat to populations of snakes adapted to early-successional habitat (Johnson and Leopold 1998; Kingsbury 2002; Smith and Stephens 2003;

Webb et al. 2005b). Natural succession has been implicated in the presumptive loss of four native snake species and the decline of eight others from the Fitch Natural History Reservation in Kansas (Fitch 2006). Forest regrowth has also been implicated in the extirpation of several Aspic Viper (*Vipera aspis*) populations in Switzerland (Jäggi and Baur 1999).

Before European colonization, a large portion of central North America consisted of grasslands (prairies) maintained by drought, fire, and grazing ungulates (Vickery et al. 1999). As prairies were converted to agriculture, however, prairie natives were often relegated to low-quality remnant habitat patches. Included among the native North American prairie fauna are several snake species including the Plains Hog-nosed Snake (*Heterodon nasicus*; Wright and Didiuk 1998) and the Eastern and Desert Massasauga (*S. c. catenatus* and *S. catenatus edwardsii*; Mackessy 2005), some of which are locally or nationally threatened. The coastal pine communities of the southeastern United States are also adapted to frequent fire disturbance (Ford et al. 1999; Greenberg 2000; Kilpatrick et al. 2004). Resident snake species of southeastern fire-adapted pine communities include the Eastern Indigo Snake (*Drymarchon couperi*), crowned snakes (*Tantilla relicta* and *T. coronata*), the Short-tailed Snake (*Lampropeltis extenuata*), and pinesnakes (*P. melanoleucus*; Greenberg 2000).

Conserving populations of early-successional reptiles may be accomplished by simulating or harnessing natural processes such as fire and grazing (Howe 1994). Prescribed fire has been used to maintain open-canopy habitat for the prairie-adapted Eastern Massasauga in several locations (Johnson and Leopold 1998; Wilson and Mauger 1999; Johnson et al. 2000). Evidence for the success of prescribed fire in snake conservation is mixed (see Table 8.1). A Kansas study suggested that prescribed burns and wildfire may increase long-term viability for Eastern Racers (*Coluber constrictor*), despite an apparent negative short-term response (Cavitt 2000). In a Florida study, prescribed fire successfully altered herpetofaunal community composition to closely match that of a reference site with a "natural" fire-disturbance history. However, fire management appeared to elicit a negative response from the only snake included in this study, the Peninsula Crowned Snake (Litt et al. 2001). A similar study in Maryland documented a herpetofaunal community shift in response to fire, with some snake species (notably *C. constrictor*) responding positively and others (*Storeria dekayi, Thamnophis sirtalis, Lampropeltis getula,* and *Carphophis amoenus*) responding negatively (McLeod and Gates 1998).

In cases in which prescribed burning is impractical, herbicide application or mechanical brush clearing may be used to discourage woody vegetation (Wigley et al. 2000). Herbicides can be particularly useful for clearing small areas of woody vegetation (Wigley et al. 2000) and so may be an effective means of creating basking habitat for snakes (see Johnson and Breisch 1999). Herbicide application may be toxic or otherwise harmful to snake species and other nontarget organisms, and it must therefore be used only after expert consultation and a thorough impact evaluation. The use of Round-Up® (Monsanto, Inc.) is especially discouraged because of reported detrimental effects on larval amphibians (Relyea 2005).

Mechanical brush clearing, using either hand tools or heavy machinery, can be used to create or maintain early-successional habitat. In New York, Massasaugas responded favorably to mechanically cleared treatment areas; 10% of above-ground radiolocations occurred in or around mechanically cut treatments that constituted only 2.5% of the total core habitat at the study site (Johnson 1995). Management success, however, was apparently short-lived (Johnson and Breisch 1999).

The response of herpetofauna to commercial forestry operations has been fairly well studied (reviewed in Russell et al. 2004). Some snake populations may benefit from the mosaic of seral stages resulting from logging activities (Greenberg et al. 1994; Ross et al. 2000; Crosswhite et al. 2004; Shipman et al. 2004; Loehle et al. 2005). In a post hoc analysis of Australian faunal survey records, researchers noted that three of seven snake species occurred more often on sites that have been clear-cut than on undisturbed forested sites; only one species occurred primarily on undisturbed forested sites (Kavanagh and Stanton 2005). In Pennsylvania, a reduction in tree basal area due to logging was positively correlated with snake abundance (mostly *T. sirtalis;* Ross et al. 2000). Tropical heliothermic snakes may be attracted to canopy openings created by logging (Vitt et al. 1998; Fredericksen and Fredericksen 2002). Still, much remains to be learned about the response of snakes to logging and other forestry practices (Goldingay et al. 1996; see Table 8.1).

The relative merits of prescribed fire, mechanical cutting, and herbicide application in achieving snake conservation goals cannot be addressed adequately using available evidence (see Table 8.1). In a study of powerline ROWs in Pennsylvania, researchers found that snake abundance and diversity was generally higher in ROWs cleared with herbicides than those cleared by mechanical means (Yahner et al. 2001a). Combinations of fire, herbicide, and mechanical clearing may be more effective in achieving conservation goals than any of these management tools used alone. An Oklahoma study found that snakes were most abundant on plots that had been treated with herbicide and subsequently burned than on plots treated with herbicide alone (Jones et al. 2000). Whenever possible, pilot management studies should assess the relative effectiveness of alternative management options.

Prescribed fire and other forms of vegetation management may injure or kill snakes directly. Studies investigating the direct mortality of snakes after prescribed fires generally indicate that mortality rates are low enough to be of little concern at the population level (Erwin and Stasiak 1977; Floyd et al. 2002). For reasons that are unclear, shedding snakes may be most vulnerable to direct mortality (Means and Campbell 1980). Every attempt should be made to limit direct snake mortality due to high-impact management activities by conducting these activities during times and seasons when snakes are least likely to be active (Dalrymple 1984; Johnson et al. 2000).

Perhaps of greater conservation concern, vegetation management may indirectly increase the mortality rates for target snake species. In a Kansas tallgrass prairie, researchers recorded increased predation of large-bodied snakes by raptors on fire-managed plots (Wilgers and Horne 2006). At the same Kansas study site, small earthworm-eating snakes tended to be healthier and more abundant on unburned plots than on regularly burned prairie plots (Wilgers and Horne 2006). Researchers in Australia showed that low-intensity fire may degrade Southwestern Carpet Python (*Morelia spilota imbricata*) habitat by eliminating favored retreat sites (Pearson et al. 2005). To minimize the risks inherent in high-impact management activities such as prescribed fire, unmanaged plots should always be interspersed with managed plots as refugia from direct mortality, predation, and habitat degradation (Setser and Cavitt 2003).

Promoting Vegetative Succession: Herbivore Exclusion

Herbivore exclusion functions to jump-start vegetative succession. Just as the introduction of grazing mammals can be used to maintain early-successional habitat, the exclusion of grazing mammals can be used to improve the quality of habitat devoid of vegetative cover (Szaro et al. 1985; Leynaud and Bucher 2005). Some researchers have documented a positive response of snakes to herbivore-exclusion plots (see Table 8.1). The exclusion of grazers from riparian areas has been correlated with increased snake abundance in Pennsylvania (Homyack and Giuliano 2002) and New Mexico (Szaro et al. 1985). The results of a similar domestic herbivore exclusion study in Argentina were inconclusive (Leynaud and Bucher 2005).

Manipulation of Vegetation Communities as a Viable Conservation Strategy

On the one hand, according to most published accounts, snakes respond favorably to anthropogenic canopy disturbance (see Table 8.1). On the other hand, none of the published studies we reviewed documented improved snake conservation status in response to habitat manipulation. Moreover, given the abundance of literature on the effects of prescribed fire on herpetofaunal communities (Russell et al. 1999), surprisingly few studies demonstrated a benefit to snakes. In general, long-term population-level studies of the response of snake populations to prescribed fire and other large anthropogenic disturbances are sorely needed.

Habitat Manipulation from a Landscape Perspective

The number, size, and distribution of habitat patches within a landscape can have important ecological consequences (Harris 1984; Turner et al. 2001; Haila 2002). Dispersal rates, mortality rates, and other ecological processes of direct relevance to population viability may be related to landscape composition and configuration (Turner 2005; Jenkins et al., Chapter 4). To effectively manipulate habitat at the landscape scale, detailed site-specific knowledge is often required (Roe et al. 2003; Fischer et al. 2004; Roe and Georges 2007). Therefore, any review of this topic should be interpreted with some caution; management success at one site may not translate into success at other sites.

Dispersal Habitat

Dispersing snakes typically experience higher mortality rates than snakes engaged in sedentary behaviors (Bonnet et al. 1999b; Kingsbury and Attum, Chapter 7). This generalization may be especially relevant for snakes inhabiting areas with a high road density (Andrews and Gibbons 2005; Roe et al. 2006). Carefully placed artificial hibernacula and other critical habitat elements may reduce the need for dispersal, thus limiting dispersal-related losses (Shine and Bonnet, Chapter 6). That said, dispersal movements play an important role in the life cycle and evolution of many snake species (Gregory et al. 1987; Roe et al. 2006) and should not necessarily be discouraged. For example, species that hibernate communally, such as the Red-sided Gartersnake (*T. sirtalis parietalis*) and the Timber Rattlesnake (*Crotalus horridus*), engage in semi-annual migrations to and from communal hibernacula. Such migrations are probably important in maintaining gene flow among den sites (Gregory 1982). Increasingly, conservation professionals are implementing measures to restore landscape connectivity such as road-crossing structures and dispersal corridors.

Anthropogenic development and associated habitat losses often result in the isolation of populations that were formerly able to exchange genetic information. In such cases, wildlife managers may wish to create linear corridors (e.g., riparian buffer zones) to improve landscape connectivity (Harris 1984). The use of movement corridors has yet to be refined as a management strategy for snakes. For example, an Australian study showed that many reptiles were functionally isolated despite the existence of linear forest remnants (Driscoll 2004). In heavily forested regions, powerline ROWs and other linear features maintained in an early-successional state may serve as effective movement corridors for early-successional snake species.

In some cases, manipulation of matrix habitat may facilitate snake dispersal in the absence of habitat corridors. Snakes tend to avoid habitat areas lacking protective cover (Shine et al. 2004a; Andrews and Gibbons 2005).

As part of the restoration of the Guadiamar River in Spain following a damaging toxic discharge, researchers used artificial cover objects to successfully promote recolonization by snakes and other reptiles through an otherwise unfavorable matrix (J. M. Pleguezuelos and R. Márquez, pers. comm.). Similarly, research in Indiana indicated that the use of partially submerged brush and debris piles may facilitate the colonization of artificial wetlands by the state-listed Copper-bellied Watersnake (Lacki et al. 2005).

Road-Crossing Structures

Of all human-built structures, roads are perhaps the most harmful to snake populations. Countless snake populations are threatened by existing roads or by proposed road-construction projects (Weatherhead and Madsen, Chapter 5). Roads can serve as a source of direct mortality or as a dispersal barrier (Forman and Alexander 1998). The ecology and behavior of snake species affect the type and magnitude of the threat posed by roads (Andrews and Gibbons 2005). Although few studies have demonstrated a detrimental effect of roads on snake populations (e.g., Row et al. 2007), it is likely that many populations have already been severely impacted. After relatively few generations, highway construction apparently contributed to genetic differentiation among occupants of different ratsnake hibernacula (Prior et al. 1997). Recently, roads have been shown to effectively impede gene flow among Timber Rattlesnake den sites (R. Clark, pers. comm.; King, Chapter 3).

Harmful road impacts can be mitigated by crossing structures, often consisting of a roadside barrier directing animals toward a culvert or overpass (Forman and Alexander 1998; Dodd et al. 2004; Aresco 2005). Such structures have been shown to reduce road mortality (and presumably increase dispersal rates) for herpetofauna, especially turtles (Aresco 2005) and amphibians (Langton 1989). Road-crossing structures may be most beneficial in cases in which snakes engage in cyclic mass movements or in which roads cut through snake hotspots (Smith and Dodd 2003; Dodd et al. 2004; Aresco 2005). In Manitoba, Canada, road-crossing structures have reduced road mortality for Red-sided Gartersnakes traveling to and from communal hibernacula (Shine and Mason 2001; Shine and Bonnet, Chapter 6). In cases in which snakes are reluctant to use road-crossing structures, structures may be made more attractive to snakes; well-placed cover objects may facilitate the use of road-crossing structures by snakes and other vertebrates (Rodriguez et al. 1996). In one case, researchers used pheromones to entice migrating Red-sided Gartersnakes to use road-crossing culverts (Shine and Mason 2001).

Road-crossing structures carry no guarantee of conservation success. Culverts may flood, causing them to lose their value for terrestrial animals. Improperly designed crossing structures may strand animals in a highway median strip or lead them into unsuitable habitat (J. Brown, pers. comm.).

Fence and culvert systems must be regularly maintained to ensure long-term effectiveness (Dodd et al. 2004). Finally, in many cases, snakes simply do not make use of road-crossing structures (see Wright 2006). Unfortunately, such management failures are rarely documented and disseminated to the conservation community.

Because money is limited, road-crossing structures should be carefully placed where they will have the greatest conservation benefit. Behavioral experiments (e.g., Andrews and Gibbons 2005) can be paired with geographical information system (GIS) simulations of snake movement patterns to aid managers in maximizing the positive impacts of road-crossing structures. Carefully designed experiments and observational studies can identify factors (adjacent habitat types, diameter of culvert, length of culvert, temperature, light within culvert, etc.) influencing the use of crossing structure by snakes (Yanes et al. 1995; Rodriguez et al. 1996).

Managing Land-Cover Diversity

Some snake species require multiple habitat types within their range, either to fulfill basic needs or because of phenological shifts in habitat selection. For these species, habitat may be managed for land-cover diversity (Spellerberg 1988; Smith and Stephens 2003). Creating and maintaining land-cover diversity need not be expensive or time consuming; a well-designed prescribed fire regime should result in a mosaic of seral stages, which can be favorable for reptile communities (Masters 1996; Litt et al. 2001; Smith and Stephens 2003). Maintaining a mosaic of patch types may also function to increase ecosystem resilience. An Australian study indicated that, although the diversity and abundance of reptiles (mostly lizards) was low on recently burned plots, these patches were likely to serve as fire-breaks, benefiting the integrity and resilience of the reptile community as a whole (Masters 1996).

Semi-aquatic snakes may benefit from created wetlands or the creation of a mosaic of wetland and upland habitats. In an Ohio study, snakes were frequently associated with mine-reclamation wetlands (Lacki et al. 1992). Constructed wetlands were used readily by the state-recognized endangered Copper-bellied Watersnake (*Nerodia erythrogaster neglecta*) in Indiana (Lacki et al. 2005). In central California, wetlands were created on former agricultural land to benefit the Giant Gartersnake (*T. gigas*); although Giant Gartersnakes have used the created wetlands, population-level management success has not yet been demonstrated (Wylie et al. 2002). In Germany, the management of *Natrix tessellata* habitat featured the restoration of a mosaic of natural habitat features (such as inlets and fluvial islands) to a heavily disturbed riverine system. Anecdotal evidence suggests that this project has been successful (Lenz and Schmidt 2002), but scientific documentation of management success is not yet available.

Managing Patch Size

Island biogeography theory (MacArthur and Wilson 1967) continues to influence landscape ecologists and land-use planners. Within this conceptual framework, island (patch) size is generally considered a key determinant of species richness (Harris 1984). In a New Hampshire study, snake diversity and abundance was generally higher on large habitat patches (> 10 ha) than on small patches (< 1 ha); thus, increasing the size of existing patches may benefit snake communities more than the creation or preservation of isolated habitat patches (Kjoss and Litvaitis 2001). In contrast, the Copper-bellied Watersnake may benefit from an increase in the number of habitat patches (shallow ponds and wetlands) rather than an increase in the size of existing patches. Because the Copper-bellied Watersnake specializes on an ephemeral food resource (amphibians), increasing the number and heterogeneity of ponds theoretically increases the likelihood that at least one pond will contain high prey densities at any given time (Roe et al. 2004). Although it is clear that the size and arrangement of habitat patches influences habitat quality for snakes, further research is necessary to understand the conservation implications of alternative landscape configurations on snake populations and communities.

Future Research

Based on the available literature, few (if any) studies have demonstrated that habitat manipulation has resulted in improved conservation status for a snake taxon or population. Few studies explicitly measured the response of snake populations to habitat manipulation (see Table 8.1); the response variable most commonly measured was the number of animals captured in pitfall traps, funnel traps, or under cover objects. In addition, the inference power of many of the studies we reviewed was compromised by pseudo-replication (Hurlbert 1984) and a lack of temporal or spatial controls. Pseudo-replication arises when experimental units treated as replicates are predisposed to respond to experimental treatments in similar ways. Pseudo-replication can increase the likelihood that a study will report a significant response when management in reality had no effect (see Hurlbert 1984). Admittedly, independent replicates of experimental treatments are difficult to achieve in habitat management studies; replicate plots can be spatially autocorrelated even when separated by hundreds or even thousands of meters. Until the safety and effectiveness of habitat manipulation is firmly established in snake conservation, the success of habitat manipulation projects should be monitored experimentally—using proper controls and replication wherever possible.

Managers should exercise extreme caution when using habitat manipulation to mitigate the impacts of proposed development projects (e.g., the replacement of a drained wetland with a created wetland; see Perry et al. 1996). Increasingly, artificial snake hibernacula and gestation sites are being incorporated into construction projects to ease permitting restrictions (Kelly and Hodge 1996). Highway departments and conservation agencies are erecting road-crossing structures to reduce the harmful effects of road-building projects on snakes and other wildlife. The results of this review indicate that such mitigation efforts, although perhaps beneficial, cannot be justified as *quid pro quo* for the loss of natural habitat.

Moving toward Evidence-Based Conservation

Responding to a paradigm shift in the medical profession, many conservation professionals have called for a shift toward "evidence-based conservation" (Smallwood et al. 1999; Fazey et al. 2004; Sutherland et al. 2004). Essentially, the evidence-based conservation paradigm calls for the development of testable management hypotheses based on a systematic review of published and "gray" literature (Smallwood et al. 1999). Then well-designed monitoring programs provide raw evidence for or against alternative management hypotheses (Smallwood et al. 1999; Nichols and Williams 2006). Finally, the timely publication and dissemination of all findings completes the cycle by making information available for future management efforts (Box 8.1).

Developing an Evidence-Based Plan for Habitat Manipulation

The evidence-based conservation paradigm requires that key management questions (e.g., Which of various management alternatives has historically been most successful in improving population viability for this snake species?) are evaluated through a systematic review of published and unpublished literature before drafting a management plan (Gates 2002; Fazey et al. 2004; Sutherland et al. 2004). Where possible, meta-analysis should be used to gain a rigorous understanding of the information content and implications of previous research (Fazey et al. 2004; Sutherland et al. 2004). Admittedly, conducting systematic reviews can be difficult for land managers who may lack access to scientific databases or lack the necessary expertise. We strongly encourage collaboration in this effort between natural resource managers and academic or consulting ecologists. At the very least, systematic reviews will inform conservation professionals of promising new habitat manipulation techniques and specific areas in need of further research (Fazey et al. 2004; Sutherland et al. 2004).

In cases in which the ecology, behavior, and management needs of a target species is undocumented, "expert" knowledge (generally the experience-

---

BOX 8.1 An evidence-based approach to snake habitat manipulation

Step 1. Once a potential habitat-related problem is recognized, review all information relevant to the habitat and management needs of the target species and to the potential consequences of alternative management regimes.

Step 2. Establish plausible management hypotheses (e.g., prescribed fire will result in increased mammal densities, leading to increased snake densities).

Step 3. Develop a plan for habitat manipulation that explicitly addresses the management hypotheses (e.g., replicate management treatments, use appropriate controls, and monitor response variables). Submit the plan to external review.

Step 4. Implement a monitoring program before the initiation of any habitat manipulation. *Note:* This is an ongoing process that continues throughout the initial and any subsequent habitat manipulations.

Step 5. Implement the management plan.

Step 6. Evaluate the weight of the evidence supporting each alternative management hypothesis. Document and disseminate all important findings. Adaptive management (repeat steps 1–6): Revise the management hypotheses as necessary based on the findings and further review of outside sources of information.

---

based opinions of other wildlife management professionals) is regularly used in the drafting of management plans (Sutherland et al. 2004). As a rule, however, "expert" knowledge is not based on sound science and should be considered weak evidence (Smallwood et al. 1999; Sutherland et al. 2004). The use of anecdotal sources appears to be common in snake habitat management; for instance, "expert" opinion was used to justify the inclusion of rice fields as key Giant Gartersnake habitat in a U.S. Fish and Wildlife Service Habitat Conservation Plan and regionwide recovery plan even though there was no documented evidence to support this decision (Smallwood et al. 1999). Ideally, managers faced with a lack of information should conduct pilot experiments or observational studies (perhaps working with local academic or consulting ecologists) to generate the evidence needed to develop informed management hypotheses.

After reviewing the available sources of information, management professionals should articulate one or more testable hypotheses related to their management goals. For example, the literature suggests that *Coluber constrictor* may avoid prescribed burn units immediately after a fire yet prefer the same habitat several months postfire (see McLeod and Gates 1998; Cavitt 2000). A further literature review might suggest several plausible hypotheses to explain this phenomenon: (1) *C. constrictor* is responding to postfire population cycles of primary prey, (2) *C. constrictor* requires more protective cover than is afforded immediately after a fire, (3) *C. constrictor* requires a more heterogeneous thermal regime than is available immediately after a fire, and (4) *C. constrictor* is responding to a complex interaction of one or more of these factors. To test these alternative hypotheses, a management plan for this species may be developed that monitors the relationships among snake densities, prey densities, protective cover, and thermal regimes after a prescribed fire.

Although it is not yet standard procedure, habitat management plans for the conservation of at-risk snake populations should be subjected to external review before implementation (Smallwood et al. 1999). Just as peer review underpins the integrity of academic publications, external review of management proposals ensures that the best available evidence is used in practice (Smallwood et al. 1999).

Monitoring the Success of Habitat Manipulation

Well-designed monitoring efforts ultimately provide the evidence for or against alternative management hypotheses. To generate a baseline against which to gauge the effects of any subsequent management action (Smallwood et al. 1999), monitoring programs should generally be implemented several years before habitat manipulation is initiated (Gibbs et al. 1999; Renken et al. 2004). Note that exceptions can and should be made in cases in which a population appears to be in immediate danger of extirpation (e.g., Daltry et al. 2001). With foresight, unfocused baseline monitoring can be replaced by data collection efforts addressing the management hypotheses (Nichols and Williams 2006).

To assess the effects of management on snake population viability, monitoring protocols should be able to detect changes in population-level characteristics such as increased reproductive ability, reduced mortality rate, increased adult survival, and increased population size (Seigel et al. 1998; Renken et al. 2004; Dorcas and Willson, Chapter 1). Long-term capture-recapture studies can be a powerful means of evaluating trends in population size, age structure, coarse movement patterns, mortality rates, and more (White and Burnham 1999). For long-lived snake species, population-size indicators may have a prohibitively long response time (Dorcas and Willson, Chapter 1). For such species, it may be more appropriate to focus on monitoring indicators of fertility or mortality.

Adaptive Management

The probability of success for any habitat manipulation project is in part a function of site-specific criteria (Fazey et al. 2004). Management decisions should therefore be based on a combination of outside information (e.g., systematic literature reviews) and site-specific information gained as part of the adaptive management process (Fazey et al. 2004). In the adaptive management paradigm, subsequent management actions are informed by previous monitoring efforts in an iterative feedback process (Gibbs et al. 1999) (see Box 8.1). Adaptive management is not a trial-and-error process but a hypothesis-driven process—and therefore it has a place within the evidence-based conservation framework. Unfortunately, adaptive management is often misused and rarely functions as intended (Gibbs et al. 1999).

Integrating habitat manipulation and experimental research will not only benefit conservation efforts but will enhance our knowledge of population ecology and the response of animal populations to environmental change. We challenge conservation biologists and natural resource managers to combine their skill sets, share their successes and mistakes, and make evidence-based conservation for snakes a reality.

Acknowledgments

We are grateful to everyone who took the time to respond to our requests for information—including (in alphabetical order) C. Anderson, K. Andrews, T. Anton, R. Baker, R. Christoffel, P. deMaynadier, J. Edwards, S. Friet, S. Gillingwater, P. Gregory, S. Harp, M. Howery, J. Kapfer, B. Kingsbury, C. Kostrzewski, J. Lamb, J. MacGregor, J. Moore, J. Pleguezuelos, J. Powers, D. Roberts, R. Sadjak, B. Shacham, W. Sherbrooke, G. Sorrell, D. Sparks, K. Stanford, O. Thornton, and B. Zappalorti. This chapter benefited from thoughtful reviews by the editors of this volume and by J. Gibbs.

9

Snakes as Indicators and Monitors of Ecosystem Properties

STEVEN J. BEAUPRE AND LARA E. DOUGLAS

Efforts to conserve snake species inevitably occur in the context of conservation plans related to other organisms, typically those organisms prioritized by managers and the public. In comparison with concerns about the populations of organisms deemed important for recreational, economic, or aesthetic reasons, snakes in need of protection may be overlooked or deprioritized. In some cases, snakes have been vilified as harmful to humans or to at-risk species that might be potential snake prey items (Dodd 1987; Scott and Seigel 1992). Concerns about snake species are difficult to address under policies that view snake conservation as antagonistic or, at best, irrelevant to other conservation priorities. While there are clearly exceptions (see examples in King, Chapter 3; Kingsbury and Attum, Chapter 7; Shoemaker et al., Chapter 8), typical snake conservation efforts require persuasion of skeptical managers, landowners, farmers, and other stakeholders of the value of snakes in ecosystems and the importance of preserving these unique organisms. We propose that conservation efforts directed at the improvement of ecosystem properties and processes could benefit by viewing snakes as indicator organisms whose physiological, ecological, and population characteristics yield information about the status of the system as a whole.

The Importance of Ecosystem Monitoring

The growing importance of management at the ecosystem level, often for purposes of conservation, forestry, or fish and wildlife population maintenance, has necessitated the development of methods for monitoring and assessing ecosystems. Initial assessments of ecosystems are required for devising management plans (Carpenter 1996). Adaptive management plans use flexible strategies that can respond to observed changes in ecosystems as they occur, and they necessitate more consistent monitoring (Heissenbuttel 1996; Ringold et al. 1996; Lindenmayer et al. 2000). Some ecosystem properties can be monitored using remote-sensing methods (Coppin et al. 2004). But the inclusion of organismal, population, and community characteristics provides an advantage over the monitoring of physical environmental traits alone because many critical ecosystem properties are, at least in part, biological (NRC 1986).

Despite a growing trend toward the oversight of ecosystems rather than individual species, the aspects of ecosystems that merit monitoring and management remain a source of disagreement. Some managers still prefer the practicality of conducting manipulations for single species and have had difficulties adapting to the often abstract goals of ecosystem monitoring (Clark 1999). Properties of ecosystems that have been used for ecosystem monitoring include biodiversity, sustainability, health, integrity, services, keystone species, structure, absence of disease, resilience, stress, and state of human interaction (Callicott and Mumford 1997; De Leo and Levin 1997; Callicott et al. 1999; Clark 1999; Costanza and Mageau 1999; Rapport et al. 1999; Lackey 2001). Of these concepts, the dominant goals that have emerged for ecosystem-level monitoring are ecosystem health and integrity (Callicott and Mumford 1997; De Leo and Levin 1997; Callicott et al. 1999; Costanza and Mageau 1999).

Both ecosystem health and integrity have been ambiguously defined and remain poorly understood. Most proposed definitions have been criticized because the terms *health* and *integrity* imply the existence of an objective ideal state for the system and obscure the problems faced in areas that have experienced intermediate degrees of disturbance (Schaeffer et al. 1988; Wicklum and Davies 1995; Callicott et al. 2000; Calow 2000; Hunter 2000). A recent operational definition of ecosystem health provides indicators that are informative, but do not necessarily require reference to a particular successional stage. Costanza and Mageau defined a *healthy system* as "one that can develop an efficient diversity of components and exchange pathways (high organization) while maintaining some redundancy or resilience as insurance against stress, and substantial vigor to quickly recover or utilize stress in a positive manner" (1999: 109). The vigor, organization, and resilience components encompass desirable factors in previous definitions of both ecosystem health and integrity (Rapport et al. 1999). The use of these indicators may be complicated, particularly in systems undergoing long-term succession or in comparisons among systems in which these components are measured in different ways (Hunter 2000; Lackey 2001). Applying indicators of ecosystem health, therefore, may often require a specific examination of the function and composition of a system and the selection of a reference system for comparison and evaluation (Schaeffer et al. 1988; Lackey 2001).

Monitoring efforts are further complicated by a lack of consistent standards for how ecosystem health components can best be monitored. Several types of measurements and concepts are poorly defined in the literature, with no consistent agreement among managers of which variables should be monitored or which conditions are desirable (Grumbine 1994; Carpenter 1996). Relevant physical boundaries for monitoring efforts are difficult to distinguish because ecosystems are never closed and because officially recognized reserve boundaries may not mirror biologically relevant boundaries (Christensen et al. 1996). A committee report to the Ecological Society of America identified three primary problems associated with limitations to monitoring of ecological systems (Christensen et al. 1996): problems with data quality, inadequate understanding of ecological principles and inaccurate ecological models, and uncertainty fostered by the possibility of novel perturbations.

In this chapter, we investigate these problematic aspects of ecosystem monitoring in the context of using snakes as indicators for monitoring ecosystem patterns and processes. We critically evaluate the quality of data available in many studies of disturbances and address failures to establish comprehensible and universally applicable definitions and standards for ecosystem management. We also examine the possible use of mechanistic ecology to improve the predictive power of models for use in novel situations, as well as the special features of snakes that make them useful as bio-indicators. Finally, we make some recommendations as to future directions for research in this area.

Use of Organisms in Ecosystem Monitoring

Because of funding limitations, time constraints, and the proclivity to produce a simple assessment strategy for the ecosystem of interest, many studies focus on a single species as an ecosystem indicator (Thomas 1982; Verner et al. 1986; Soule 1991; Lindenmayer et al. 2000). Landres et al. defined an *indicator species* as "an organism whose characteristics (e.g., presence or absence, population density, dispersion, and reproductive success) are used as an index of attributes too difficult, inconvenient, or expensive to measure for other species or environmental conditions of interest" (1988:317). Plant species are frequently used as indicators of the physical conditions at a given location; the presence of plant species with known requirements for moisture, temperature, and nutrient concentrations provides evidence that at least minimum requirements for that species are met. Plant species presence has frequently been used (often erroneously) as evidence for the presence of an entire plant and animal community (Whittaker 1970).

Animals may also be used as indicator species. Bird species are commonly used because they are both sensitive to changes and easily observed (Morrison 1986). Comparative study suggests, however, that reserves designed to encompass bird species distributions would protect fewer species than reserves designed around distributions of other taxa (Moore et al. 2003). In some cases, the presence or absence of indicator species has been used to predict properties of related taxa; for instance, certain groups of bird or butterfly species could be used to predict combined bird and butterfly species richness (Fleishman et al. 2005). Often, species are selected and numerical responses are measured according to the interests of managers whose goals may be conservation, game populations, recreation, or forestry (Landres et al. 1988). In such situations, the species selected for monitoring may have little or no relationship to important ecosystem properties. Monitoring a selected species may not even be conducive to the goal of protecting the monitored organism itself, because a consideration of community and ecosystem features may be necessary to understand changes observed even in a single species (Grumbine 1994).

Other studies have chosen to monitor a species believed to be ecologically important in the system in question, although ecological importance has been evaluated differently in many studies. Species whose presence can be associated with species richness in a particular landscape type have sometimes been used as indicators of biodiversity (Chase et al. 2000). Some studies have selected abundant species or those that occur at a majority of habitat sites for monitoring (Dufrene and Legendre 1997; Lindenmayer et al. 2000; Lindstedt 2005).

Sensitivity to environmental change can also play a role in the selection of species. Some studies have used amphibians for monitoring due to their sensitivity to perturbation (Lindstedt 2005). Lungless salamanders (plethodontids), have been used in monitoring studies due to their tight linkage to ecosystem processes such as succession or changes in microclimate (Welsh and Droege 2001). Organisms with high sensitivity to contamination have been repeatedly used as indicators when toxicity is a primary concern of managers (Landres et al. 1988; Lindenmayer et al. 2000).

In other cases, organisms have been identified as important in determining food-web dynamics in the system (Lindenmayer et al. 2000; Lindstedt 2005). Such identifications can be problematic because of the often poorly understood nature of a "keystone species" (Grumbine 1994); species that are identified as important in an ecosystem for other reasons may be incorrectly assumed to have a disproportionate impact on trophic dynamics. A few studies have used top predators as indicators in an ecosystem. Top predators can be useful in examining ecosystem processes because of their dependence on other organisms and tendency to influence trophic dynamics (Matthews et al. 2002). An overall index combining physiological and ecological responses in marine predators has been used to categorize functional responses between predators and prey (Boyd and Murray 2001). Species with important impacts on trophic dynamics fulfill a functional role in their communities that can make them useful indicators.

Indicator organisms are often used for monitoring the population trends of species other than the indicator organism. Correlations among guild members are often assumed, so that numerical changes in the population of one organism are used as a proxy for population changes in other organisms in the same guild (Mannan et al. 1984). Given the likelihood of possible interactions (such as competition or predation) among guild members, such an assumption is problematic (Polis and McCormick 1987). Organisms may also be grouped according to habitat preferences, with population size trends in one species assumed to apply to others with similar habitat requirements (Thomas 1982). Assumptions that these organisms respond in the same ways are erroneous; even organisms in the same guild may not use resources or habitat in the same ways and may have differing needs and behaviors that render their responses very different (Mannan et al. 1984; Block et al. 1986; Landres et al. 1988; Lindenmayer et al. 2000). The failure of most indicator studies to address any of the intra- or interspecific mechanisms that may interact to yield a particular change makes the use of one organism's population as a proxy for other organisms difficult, if not impossible (Landres et al. 1988). Even in studies in which population sizes of multiple species respond to a change in the same way, some individual species will act as exceptions due to biologically relevant factors such as resource responses or interactions with other species that may be obscured by numerical analyses (James 2003).

Many managers and biologists also use organisms as indicators of habitat quality. Habitat quality measurements for some species are already provided for under the Habitat Evaluation Procedures mandated by the U.S. Fish and Wildlife Service or the Wildlife and Fish Habitat Relationships Program of the U.S. Department of Agriculture (USDA) Forest Service. If conclusions from habitat quality studies could be applied to multiple species, management would become much simpler (Nelson and Salwasser 1982; Thomas 1982). Some studies have found indicator species or assemblages whose presence or absence successfully predicts diversity, stability, or presence of a threatened species within specific habitat types (Orrock and Pagels 2003; Wilson and McCranie 2003; Ilmonen and Paasiverta 2005). In other cases, characteristics that may be assumed to divide animals into groups according to habitat use are actually more closely associated with phylogeny than ecology (Price 1982). Organisms with large home ranges are often chosen as umbrella species under the assumption that protecting the home ranges of those species will protect other organisms occupying them (Ozaki et al. 2006). If the umbrella species that are chosen can successfully adjust to habitat modifications, however, those species' home ranges may not encompass high-quality habitat for other species (Ozaki et al. 2006).

Habitat indicator studies typically assume that the population density for a species can act as an index of habitat quality for that species, as well as for other organisms occupying similar habitat (e.g., members of the same feeding guild) (Mannan et al. 1984; Landres et al. 1988). The use of population density as a variable to represent habitat quality may exclude seasonal variation in habitat or habitat use, variability in resource availability, and interactions with other species (Landres et al. 1988). High population densities may also be associated with density-dependent reductions in fecundity or fitness, for example, due to territoriality or agonistic interactions, as in some eagles (Ferrer and Donzar 1996). The use of one species to represent effects on numerous other species can also obscure the mechanisms responsible for observed changes in population levels.

Despite the nonmechanistic nature of using numerical responses (which we define as changes in abundance) of indicator organisms as proxies for other ecosystem variables, indicator species can still be useful, particularly in cases in which the mechanisms and processes of interest are difficult to observe or measure. Ideally, the performance of candidate indicator species in predicting ecosystem properties should be established and validated prior to the use of monitoring for management and prediction purposes. In cases in which the use of indicator species to monitor ecosystems remains desirable, the indicator organism should be ecologically relevant rather than chosen based solely on the interest of resource managers for commercial, recreational, or conservation purposes. Indicator species should further be chosen as species that are sensitive to change (Odum 1971; Wilson and Mc-Cranie 2003) and will display responses that vary according to the processes of interest for monitoring (Landres et al. 1988). In particular, such species tend to be those with large body size, specific habitat requirements, permanent residency in the area of interest, slow reproduction, sit-and-wait foraging mode, and large home range sizes We propose that snakes have a largely untapped potential to act as excellent indicator organisms. In addition to frequently possessing many of the properties deemed desirable in indicator species, snakes can, in many cases, avoid the problems associated with a poor mechanistic understanding that impede the effectiveness of many indicator organisms.

Snakes as Bioassessment Tools

While snakes have seen some use as indicator organisms by some biologists (Stafford et al. 1977; Matthews et al. 2002; Lind et al. 2005), their use by resource managers has been almost nonexistent, largely because the priorities of managers are typically set according to nonbiological considerations associated with public opinion and economic harvest (Scott and Seigel 1992). Snakes have been identified as a group having a high correlation with performance in other organisms and a good predictive ability for sites of conservation importance (Moore et al. 2003).

Snakes in general, and pitvipers (Crotalinae) in particular, tend to have life-history characteristics that make them vulnerable to population declines, such as long life spans, late sexual maturity, relatively high annual survival in undisturbed populations, low reproductive frequency, site fidelity, and high mortality among neonates and juveniles (Scott and Seigel 1992; Shetty and Shine 2002). There are differences in these respects among snake species or populations (Reed and Shine 2002; Webb et al. 2002b). Vulnerability to change associated with habitat destruction, climate change, or other long-term trends affecting ecosystems makes these organisms useful as indicators of processes affecting their ecosystems (Weatherhead and Madsen, Chapter 5). A comparative study in Honduras identified 103 species of particularly vulnerable reptiles and amphibians; of these, 21 were snakes (Wilson and Mc-Cranie 2003). Most snake home ranges shift relatively little over the course of their lifetimes; therefore, any changes observed in snakes can be linked to changes in local environment (Bauerle et al. 1975).

Many snake species have additional features that make them excellent model organisms in detailed field studies. Snakes have thermally dependent physiological functions (Huey 1982), making them excellent models for studies of environmental effects on growth and reproduction. Many snake species can be easily captured during spring or fall aggregations and tracked for behavioral observations, physiological studies, and routine mark-recapture (Diller and Wallace 2002). For snakes with low diet variation, foraging and digestion models can be constructed to aid in an understanding of snake behavior and time and energy budgets (Beaupre 2002). Behaviors can also be identified based on the location and body posture of an animal and, in many species, a behavior may be continuous for hours if not days (Beaupre 2008). Therefore, episodes of activity can be assigned to functional categories with duration of these episodes serving as a variable responding to environmental change (King and Duvall 1990).

As ectotherms, snakes specialize in surviving in low-energy environments; therefore, they may be slower to respond numerically to changes in food availability than terrestrial endotherms (Scott and Seigel 1992). Low-energy specialization tightly couples snakes to their resource environment (Beaupre and Duvall 1998a) and allows prolonged periods of inactivity during winter months in temperate climates, or during dry seasons in the tropics, with a population's activity following relatively stable patterns among years (Scott and Seigel 1992). We propose that this relative stability may act as an advantage in using snakes as indicators. Surveys taken at the same time period can be compared relatively easily, although caution to account for changes in detectability resulting from annual variations in environmental conditions is still necessary. Small, short-term fluctuations that might occur in populations of endotherms responding to periodic droughts or shifts in food availability typically do not occur in reptile populations. Snakes can be excellent indicators of longer-term trends, making them useful in studies of ecosystem-scale processes that might occur over several years.

The high biomass conversion efficiency in many snakes also contributes to a tight linkage between variation in resource availability or the physical environment and properties measured in the snakes (Beaupre and Duvall 1998a; Lind et al. 2005). Snakes are excellent model organisms for mechanistic investigations of the impacts of ecosystem-scale changes, particularly in studies that attempt to use physiological changes in organisms for assessment of community or population changes (Beaupre and Duvall 1998a).

The use of a variety of habitats during periods of hibernation, foraging, mate searching, and gestation or oviposition also makes snakes useful indicators of habitat quality (Scott and Seigel 1992). A manager concerned about protecting a sufficient variety of habitats to maintain snake populations would, by necessity, maintain a larger variety of total habitats than a manager concerned about a species with a smaller home range or narrower habitat proclivities, making snakes useful as potential umbrella species in conservation (Landres et al. 1988).

Many snake species are large enough for the implantation of temperature-sensing radiotransmitters (Beaupre and Duvall 1998a). Telemetry enables the repeated location of the same individuals, which allows the collection of useful information for indicator organisms, such as data about their movements, mortality, and habitat use (Beaupre and Duvall 1998a). Temperature-sensing radiotelemetry is uniquely beneficial because it allows an examination of the thermal environment, which enables the estimation of the thermal impacts of the environment on ectotherm physiological processes, nearly all of which are temperature-dependent (Huey 1982).

Chemical Monitoring

Snake ecology has been used in a number of studies examining the effects of contamination. As carnivores with relatively sedentary lifestyles compared to birds (another organism commonly used as a pollution indicator), snakes are likely to experience bioaccumulation that can be associated with particular localities (Bauerle et al. 1975). Snake tissues have been analyzed for the presence of dichlorodiphenyldichloroethylene (DDE), dichlorodiphenyltrichloroethane (DDT), beta-benzene hexachloride, heavy metals, and other chlorinated hydrocarbons in toxicity studies (Bauerle et al. 1975; Hopkins et al. 1999, 2001; Campbell et al. 2005). One series of studies that tested for residues of organochlorine insecticides in snake fat bodies associated changes in organochlorine residues over time with changes in the presence of pesticides in the ecosystem (Fleet et al. 1972; Stafford et al. 1977; Fleet and Plapp 1978).

Without an understanding of the mechanisms that relate chemical loads to individual responses (such as growth or reproduction), it is difficult to attribute any observed change or lack of change to the chemical factor of concern. For example, initial evidence that a mollusk could act as an indicator for heavy metal contamination was later contradicted by evidence that mortality in the mollusk never reached levels high enough to adequately indicate the degraded state of the system (Lindenmayer et al. 2000). Use of indicator organisms for detecting the presence of toxicants or pollutants may confuse remediation efforts unless the mechanisms for shifts in mortality, behavior or physiology are understood.

Numerical Monitoring

Numeric trends in snake populations can be useful, particularly in observing long-term variability (Lind et al. 2005). Although accurate population estimates in snakes have long been considered difficult to obtain due to problems associated with low detectability and biased sampling methods (Turner 1977; Dorcas and Willson, Chapter 1), mark-recapture studies using a variety of capture methods are often used as a baseline for conservation assessments of a snake population (Savidge 1991; Ota 2000; Wylie et al. 2002; Winne et al. 2005). Numerous studies have also observed long-term or short-term trends in snake populations (e.g., Fitch, 1999). Many broad observations about populations have been used to support decisions about snake conservation. Evidence for declines is typically associated with large-scale concerns affecting many snake species simultaneously (Gibbons et al. 2000; Zhou and Jiang 2004). Several population studies have associated numerical trends with changes in the ecosystem (Parker and Brown 1973; Larsen and Gregory 1989; Diller and Wallace 2002) and, particularly, with changes in prey populations (Madsen and Shine 2000a; Madsen et al. 2006).

Trends in population numbers should be carefully examined by creating accurate demographic models so that sampling problems do not lead to spurious conclusions (Lind et al. 2005). Numerical responses of two distinct ratsnake populations had been suggested to parallel one another in annual changes in long-term studies (Weatherhead et al. 2002). The two populations monitored did not parallel one another in age structure or in overall long-term trend, however, implying that different processes produced parallel changes (Weatherhead et al. 2002). In this case, although the two populations changed numerically in similar ways, the predictive value of these results is unclear because the mechanisms underlying parallel changes were unlikely to be identical. Similarly, two populations of gartersnakes from similar dens surveyed in different studies exhibited different patterns of population change, implying different mechanisms can act in different locations (Larsen and Gregory 1989).

The lack of mechanistic understanding is most likely a problem with most numerical monitoring endeavors. Population size or density data simply describe populations, and changes in their values do not imply a specific mechanism. Therefore, drawing conclusions about biological processes acting among populations requires additional assumptions about the underlying causes of population size changes. In ratsnakes, different mechanisms could be deduced because of differences in population structure between the two monitored populations (Weatherhead et al. 2002). A study using one species as an indicator for other unmonitored species would be ineffective at determining whether the mechanism behind a population change in the indicator species reflects anything about the mechanisms governing population changes in unmonitored organisms.

One benefit of using snake populations for monitoring efforts is their trophic position. As obligate predators, snake populations necessarily reflect the populations of their prey items, among other factors. If food availability is insufficient, snakes eventually starve (Beaupre 2002, 2007; Matthews et al. 2002). But because snakes are low-energy specialists, they may not respond numerically to prey population changes as rapidly as the prey populations themselves change (Scott and Seigel 1992). Snake populations do change over longer time periods, however, and remain useful for detecting long-term trends in prey populations. For example, snakes have been used in detecting declines in amphibian prey populations associated with the introduction of nonnative trout (Matthews et al. 2002). Snake populations have also been shown to respond numerically to the indirect effect of a prey base increase in response to the removal of large herbivores from plots in an African savanna (McCauley et al. 2006). Other information collected from snakes, such as physiological body condition data, could reflect shorter-term changes in prey populations (Beaupre 2002, 2007).

In addition, snake abundance, prey preferences, distribution, habitat specialization, population structure, body size, foraging mode, and reproductive frequency all influence the applicability of information about a particular snake species to other species (Fitch 1999; Reed and Shine 2002; Phillips et al. 2003). For example, evidence suggesting that numerical trends in diurnal reptile species closely paralleled other taxa, whereas trends in a nocturnal species did not (James 2003), supports the notion that choosing the correct indicator species is important. Because of their unique properties, snakes can respond to resource availability at an intermediate time scale. Monitoring efforts using snake numerical responses can avoid some of the spurious conclusions that might result from monitoring populations that change too rapidly, but monitoring snake population sizes alone still produces an incomplete mechanistic understanding of observed changes.

Ecological Monitoring

Aspects of snake ecology have occasionally been used to monitor ecosystems. Some studies have attempted to associate changes in community composition with changes in ecosystem properties. For example, Ishwar et al. (2001) described relationships between altitude and both reptile community structure and species richness. If an understanding of the reasons for community change could be gleaned from such studies, it could be possible to identify groups of organisms useful for monitoring purposes.

Predation by snakes can have clear impacts on prey species. Although prey populations are typically unaffected (Fitch 1987b; Reynolds and Scott 1987), the alteration of the foraging behaviors and preferences of prey species is possible (Kotler et al. 1993; Bouskila 1995; Patten and Bolger 2003). In one obvious exception, the introduction of an invasive snake predator (such as the Brown Treesnake) to an area with unadapted prey species, prey populations themselves can be decimated (Savidge 1987). Rodents in areas subjected to predation from both snakes and owls have been shown to exhibit different foraging behaviors and activity times compared to rodents in areas subjected to only owl predation (Bouskila 1995). Similarly, correlation between snake population size and prey bird nest failure rates varied with habitat characteristics and bird species (Patten and Bolger 2003). An understanding of the ecological relationships between predator and prey species could therefore enable the use of snakes as ecological indicators even when the direct application of snake population estimates is of limited utility.

Physiological Monitoring

Monitoring of physiological variables in snakes has been conducted repeatedly (Brown 1991; Beaupre 1995a, 1995b, 1996, 2002, 2007; Seebacher 2005; E. Taylor et al. 2005). Some snake species search available habitat for small mammal activity and select foraging sites accordingly (Duvall et al. 1990; Clark 2004), with observable physiological changes in the animals over relatively short periods (Beaupre 2008). Even short-term changes in resource environments (prey species populations) can result in physiological changes in snakes.

The application of physiological monitoring to the use of snakes as indicator species has been relatively unexplored. As indicated earlier, observed changes in a population or community offer little insight regarding the underlying mechanisms. Discerning the mechanisms behind an observed effect can be best achieved by examining the effects and processes at lower levels of organization (Levin 1992; Dunham and Beaupre 1998). An understanding of population ecology can be augmented by knowledge about the processes impacting individual births, deaths, and immigrations; in aggregate, these processes dictate changes in population sizes (Dunham et al. 1989; DeAngelis et al. 1991; Beaupre 2002). Studies focusing on physiological processes in individuals can therefore provide important early-warning mechanistic information about processes observed at the population or community scales (Beaupre 2002).

Dunham and colleagues (Congdon et al. 1982; Dunham et al. 1989; Dunham 1993; Dunham and Overall 1994; O'Connor et al. 2006) have developed a theoretical basis for understanding the evolution of reptile life histories in response to characteristics of local operative environments (e.g., biophysical, resource, predation and disease, competitive, social, and demographic). This body of theory (e.g., Fig. 9.1) explicitly represents the mechanistic connections among operative environments; behavioral (time use) and physiological (energy and mass) allocations; and individual patterns of growth, reproduction, and mortality. We note that, in addition to aiding an understanding of life-history evolution, Dunham's approach explicitly represents these dependencies in ecological time. The relevance of this approach to conservation and restoration biology or environmental change is direct (Dunham 1993; O'Connor et al. 2006). Changes in operative environments (due to natural or anthropogenic factors) directly influence individual time and mass-energy allocations, which in turn influence population growth rates and persistence. Each behavior-time, or mass-energy, allocation decision carries with it a risk of mortality. Individual organisms can be considered as integrators of their local operative environments. Changes in the environment are manifested first by effects on behavior (time allocations) and mass-energy budgets of individuals, and later by shifts in growth, reproduction, mortality, and ultimately, population density. The initial effects of changes in local operative environments can be directly measured in many snakes as changes in daily time and energy allocations, as well as changes in fundamental bioenergetic variables (e.g., feeding rates, body condition, growth rates, and field metabolic rates; Beaupre 2008).

Many bioenergetic variables are easy to measure and can be obtained during routine mark-recapture and telemetry studies. Both techniques have enjoyed extensive use in snake ecology. For example, although somewhat technically and financially demanding, the measurement of field metabolic rate by the doubly-labeled water method is well-developed (Lifson and McClintock 1966; Nagy 1980; Nagy and Costa 1980; Speakman 1997) and provides an integrative perspective on the total metabolic expenditure on maintenance, biochemical activity (e.g., synthesis), and physical activity (e.g., locomotion and work). Because field metabolic rate (FMR) integrates the maintenance and activity portions of the energy budget (Fig. 9.1), its magnitude rises and falls with changes in physical and biochemical activity. Differences in FMR between conditions of high and low food availability have been documented in Timber Rattlesnakes (*Crotalus horridus;* Beaupre 2008).

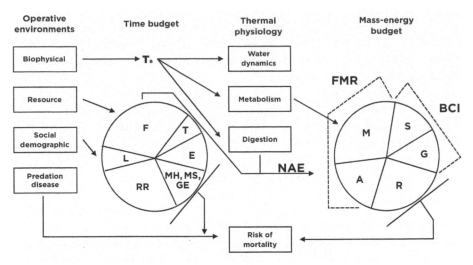

Fig. 9.1. An explicit representation of the relationships among operative environments, time and mass-energy allocations, physiological responses, and population dynamics. Operative temperatures on the left simultaneously influence tandem behavioral-time and mass-energy budgets. Operative environments determine constraints on time and energy allocations as well as apportionments of available time to competing behaviors (foraging, moving, resting or in retreat, mate searching, mate handling, gestating, ecdysis, and thermoregulation) and available mass-energy to competing functions (maintenance, activity, growth, reproduction, and storage). Each behavior-time or mass-energy allocation decision carries with it the risk of mortality and a fitness pay-off. Critical responses such as FMR and BCI provide insights regarding changes in time and mass-energy allocations. These variables reflect environmental influences on the life history of the organism. A, activity (physical and biochemical); BCI, body condition index; E, ecdysis; F, foraging; FMR, field metabolic rate; G, growth; GE, gestation; L, locomotion; M, maintenance; MH, mate handling; MS, mate search; NAE, net assimilated energy; R, reproduction; RR, resting or in retreat; S, storage; T, thermoregulation; T_b, body temperature. (Adapted from Dunham et al. 1989; Dunham 1993; O'Connor et al. 2006)

A more readily measured bioenergetic variable, body condition, can be easily obtained from routine measurements upon capture and recapture. Recently, body condition has been identified as a useful variable for monitoring and understanding organismal function in the context of conservation (Stevenson and Woods 2006). Like FMR, body condition integrates two important aspects of mass-energy allocation: growth and storage (see Fig. 9.1). As such, rises and declines in body condition are direct indicators of the mass-energy status of individuals. When food resource conditions are poor, body condition is expected to decline; when food resources are abundant, body condition should improve. Physiological variables, such as body condition and FMR may be especially useful in understanding the responses of uncommon, cryptic, or long-lived, infrequently reproducing species to environmental change. Such species may be less amenable to traditional numerical population analysis.

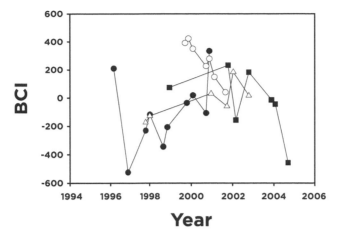

Fig. 9.2. Body condition index (BCI) of four representative adult male radiotagged Timber Rattlesnakes monitored from 1995 to 2006 in Madison Co., Arkansas. These individuals were chosen because of the long time frames over which each was monitored.

As an example of the rapid response and potential utility of physiological monitoring, consider the long-term studies of a Timber Rattlesnake population that has been under study in northwest Arkansas since 1995 (Wills and Beaupre 2000; Beaupre and Zaidan 2001; Cundall and Beaupre 2001; Zaidan and Beaupre 2003; Browning et al. 2005; Beaupre 2008). During routine mark-recapture and radiotelemetry studies, captured snakes are weighed and measured (snout vent length, SVL). In some cases, radiotagged animals are measured multiple times per year, over several years. A body condition index (BCI) is derived for each capture event. We define BCI as the deviation in actual mass from predicted mass based on a nonlinear regression relating body mass to SVL. The nonlinear regression is fit to the population length-mass data set, in this case $Weight(g) = 21.62 + ([1.62 \times 10^{-5}][SVL(cm)]^{3.887})$ ($P < 0.0001$; $R^2 = 0.92$; $N = 220$). A positive deviation indicates an animal that is heavier than the population mean at a particular SVL; a negative deviation indicates an animal that is lighter than the population mean. Body condition in Timber Rattlesnakes is a direct indicator of recent foraging success and can vary, sometimes dramatically, from year to year in this system (Beaupre 2008).

We present BCI data in two ways, first, as long-term traces of body condition for radiotagged adult males recaptured several times over the 12-year study (Fig. 9.2) and second, as mean body condition values for all adults (SVL > 60 cm) captured during each year of the study (Fig. 9.3). In the case of the long-term data on individual radiotagged snakes (Fig. 9.2), it is clear that there is substantial annual variation and also variations within years in body condition. Because of the relatively large size of ingested meals (of which

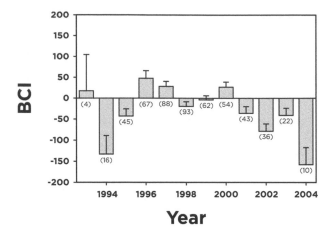

Fig. 9.3. Mean body condition index (BCI) for all adult (snout vent length > 60 cm) Timber Rattlesnakes captured during each year of a 12-year study in Madison Co., Arkansas. Numbers in parentheses indicate sample sizes.

approximately 90% of mass is absorbed), snake body condition can fluctuate dramatically even over short time scales (e.g., from June to August). The annual mean BCI for *C. horridus* in northwestern Arkansas also varies significantly from year to year (Fig. 9.3; one-way ANOVA; $F = 8.17$; $P < 0.0001$; $N = 540$), reflecting variation in prey abundance. Tukey HSD multiple comparisons suggest that 1996, 2004, and 2006 exhibited reduced body condition and that 1998 and 1999 exhibited improved body condition (Fig. 9.3).

Body condition is an easily measured response variable that serves as a barometer for feeding rates in snakes (Jayne and Bennett 1990; Bonnet et al. 1999c; Bonnet et al. 2002a; Beaupre 2008). As such, BCI is a likely candidate for use as a monitoring variable that reflects increases and declines in feeding by snakes, which in turn are related to fluctuations in available prey base. In the Ozark system, these prey-abundance fluctuations are probably due to variations in acorn mast crops that directly support small mammal populations (Wolff 1996). Thus, BCI in Timber Rattlesnakes reflects the integrated response of multiple trophic levels (oak trees, small mammals, and snakes).

The monitoring of physiological responses of snakes may be particularly useful in degraded systems in which snakes serve as near top predators. For example, forests of the Ozark Mountains were clear-cut during the late nineteenth and early twentieth centuries and have been allowed to grow back with little or no coordinated management under a fire-suppression policy. The resulting forests typically consist of even-age closed-canopy environments that exhibit relatively little ground-level productivity, low oak recruitment, and strong coupling to acorn mast crops (Spetich 2002). In an effort to restore these degraded forests to a higher-quality ecosystem for wildlife, the Arkansas Game and Fish Commission plans to use thinning, clear cuts, and controlled burns to manipulate forest structure and enhance seed production on the ground. These manipulations should directly impact both the thermal properties of the manipulated plots and the resource base for small mammals and larger wildlife. With the diversification of ground-level resources, we expect a decoupling between small mammal populations and acorn mast production. With higher density and lower variance in small mammal populations, there should be increased mean BCI and decreased BCI variance among snakes that use the manipulated plots (both in comparison with the premanipulation values and in comparison with control plots). Thus, Timber Rattlesnake BCI may serve as an indicator of changes in primary productivity and trophic interactions that result from large-scale habitat manipulations for specific conservation or restoration goals.

Future Research

Snake species possess numerous features that make them ideal as ecological indicator species. In particular, large-bodied snakes support telemetry studies that enhance monitoring and data collection, and they facilitate a high degree of mechanistic understanding. Unlike many other long-lived or large-bodied species, snakes exhibit limited migratory behavior and few long-distance movements to complicate the association of observed responses with a particular landscape. In addition, the home-range sizes of snakes are large enough that snakes can be useful for integrating the effects of environmental change (e.g., climate, habitat structure, or restoration). Furthermore, snake behavior and physiology are tightly coupled with environmental variation. We are only beginning to explore the utility of this relationship.

Snakes have been used as indicators in a variety of contexts, including in studies of environmental toxicity resulting from contamination and in population studies as proxies for other organisms that are presumed to change in parallel. We caution, however, that in both types of studies the mechanisms responsible for any changes (or any lack of change) observed in snakes are not determinable from population or mortality data alone. Therefore, it becomes necessary to assume that the species being monitored by proxy undergo either the same (unknown) processes recorded in snake populations or parallel processes that cause equivalent changes in other species of interest. Parallel population changes in the absence of knowledge regarding parallel mechanisms may generate misleading conclusions and inappropriate management strategies. Such errors may be particularly egregious when novel environmental change disrupts historically operating mechanisms. Novel environmental change presents tremendous challenges to managers, who typically rely on past experience to make decisions. Ironically, situations in which nonmechanistic monitoring is most likely to lead to flawed conclusions are the very situations where managers are most in need of indicator organisms to provide them with accurate information.

The behavioral and physiological properties of snakes can also be used as indicators of ecosystem function. Snakes respond relatively quickly to environmental change, with physiological changes rapid enough to garner information about very short-term fluctuations but with population sizes stable enough to reflect long-term effects. This property makes snakes useful as monitors, albeit in a different way than shorter-lived organisms in which long-term changes can easily be lost in the "noise" of annual population fluctuations. In the case of physiological monitoring, snakes are used less as indicators for other species and more as monitors for species with which they are known to interact. Observed changes in well-chosen snake species that have well-understood interactions with other organisms or their biophysical environment can be used to predict changes in multiple species.

Although mechanistic monitoring efforts are doubtless more time-consuming and research-intensive than surveys of populations, they are much more useful in novel circumstances. In cases in which rare, cryptic, long-lived, or infrequently reproducing species are involved, physiological monitoring may be the only viable option. A focus on mechanisms not only improves our ability to interpret results and relate perturbations to particular responses, but also contributes simultaneously to the broader goals of understanding ecosystem function and developing best practices for conservation.

A clear understanding of the mechanisms that affect physiology and behavior is required before monitoring of individual responses can become a useful assessment tool. To this end, ecologists must shift their focus from phenomenological approaches (e.g., simple trends in density or diversity) to more mechanistic approaches. Many of the common data collected in conservation-oriented studies, such as information about the thermal environment, prey base, weight and length, movement, and habitat choice, could be used in developing a more mechanistic understanding of the system. Mechanistic approaches should be designed to not only assess organismal response but also critically test the mechanisms that underlie the response.

In our view, there are four critical areas of research in need of attention and funding. First, continued research must establish functional relationships between various aspects of environmental change (natural or anthropogenic changes in food availability, thermal environment, and hydric environment) and the physiological and behavioral responses of individuals. Second, research must relate individual responses to longer-term population trends (e.g., birth rates, death rates, and density). Third, efforts should be expended to examine the interactions among critical mechanisms associated with operative environments (e.g., thermal balance, energy balance, and water balance). Such efforts should yield an improved understanding of the relative importance of candidate mechanisms in specific systems and improve our ability to select appropriate candidate species for monitoring (i.e., species that respond to the most appropriate variables for monitoring needs). Finally, a clear understanding of the candidate species' role in its community or ecosystem is required to draw strong conclusions about higher levels of function based on individual responses. The more we understand about mechanism, the greater will be our ability to anticipate and predict the responses of individuals, communities, and ecosystems to novel environmental change.

Acknowledgments

All research described herein was conducted with the approval of and all relevant permits from the University of Arkansas Institutional Animal Care and Use Committee, the Arkansas Game and Fish Commission, the Arkansas Natural Heritage Commission, and the Ozark Natural Science Center. The chapter was significantly improved by comments from J. Agugliaro, M. McCue, S. Mullin, R. Seigel, M. Smith, J. Van Dyke, R. Wittenberg, and two anonymous reviewers. The Ozark Natural Science Center, the Arkansas Natural Heritage Commission, and the Arkansas Game and Fish Commission graciously allowed permission to work on their property. The study of Arkansas Timber Rattlesnakes was supported by funds from the University of Arkansas Research Incentive Fund, the Arkansas Science and Technology Authority (Grant #97-B-06), and the National Science Foundation (Grants #IBN-9728470, #IBN-0130633, and #IBN-0641117).

10

Combating Ophiophobia

Origins, Treatment, Education, and Conservation Tools

GORDON M. BURGHARDT, JAMES B. MURPHY,
DAVID CHISZAR, AND MICHAEL HUTCHINS

The viper is by many taken for an image of malice and cruelty;
but in reality she is guilty of no such thing.

M. CHARAS (1677:3)

An Enduring Challenge

Human attitudes toward snakes have ranged from fascination, awe, and
worship to fear and loathing (Aymar 1956; Morris and Morris 1965). Certainly, the latter attitude is prevalent in many countries today, even in those
viewed as civilized, educated, and environmentally enlightened. Popular enter-
tainment continues to play on fears of snakes, as well as spiders, bats, and
other animals. But is fear and antipathy toward snakes universal and inevitable? The theme of this chapter is that the conservation of snakes is
more difficult than for other vertebrate groups owing to the general bad
reputation that snakes have in many regions of the world. They are loathed
in ways that render rational discourse insufficient for their conservation.
Drivers, for example, have been shown to go out of their way to run over
snakes on highways (Langley et al. 1989). The consequences can be tragic,
and not just for the snakes. In April 2007, a couple in Croatia was hiking
with their 18-month-old baby in a carriage, along with some friends. A male
friend saw a Long-nosed Viper (*Vipera ammodytes*) on the side of the path
and, concerned it would attack them, kicked the snake high into the air like
a soccer ball. Unfortunately, the snake landed in the carriage and bit the
baby on the chin. The infant died in spite of prompt medical care (Zoran
Tadic, pers. comm.; translated from newspaper accounts).

Snakes present profound psychological mysteries. We review the ways snakes have been perceived by different cultures over time, summarize scientific studies on fear of snakes, present observations on reptile exhibits in zoos, and discuss some possible solutions that may aid the conservation of snakes in the twenty-first century.

Cultural, Religious, and Historical Issues

Although various institutions (religious, governmental, sporting, health, and so forth) have contributed to the lack of respect for snakes as valued components of ecosystems, it is also probable that evolutionary predispositions underlie the widespread fear of snakes (Murray and Foote 1979; Quammen 2000).

The Mixed Legacy of Myth and Religion

Bierlein's (1994) analysis of universal themes in creation and other myths worldwide contains more entries on serpents than all other animals combined. For example, he recounts creation stories such as the Greek myth of Eurynome and Ophion. In this myth, Ophion, the great serpent of the waters, mated with Eurynome, the goddess of all things. Eurynome took on the appearance of a bird and laid a giant egg; Ophion coiled around and incubated the egg until it hatched, producing all living creatures. The two gods lived together for a time, but Ophion's bragging about his role in creation grew tiresome to Eurynome, so she "bruised his head with her heel" and threw him out of Mount Olympus and into the "dark regions of the earth" (Bierlien 1994:46). Similarities with the Biblical story of Genesis are probably no accident.

The Bible, viewed by many as justifying ill treatment of snakes, is more ambivalent about serpents than most people realize. For example, in Proverbs we find: "Three things are too wonderful for me; four I do not understand: the way of an eagle in the sky, the way of a snake on a rock, the way of a ship on the high seas, and the way of a man with a girl" (Proverbs 30:18–19, New Revised Standard Version, NRSV). Furthermore, Jesus is quoted as instructing his disciples to "be wise as serpents" (Matthew 10:16, NRSV). This attribution is probably derived from the serpent in the Garden of Eden, considered to be the most cunning of all animals (Genesis 3:1, NRSV).

Actually, because of their mysterious and powerful abilities, snakes figured in religious practices well before the time of the ancient Israelites, including Old Kingdom Egypt (Piccione 1990), and were revered by many cultures. A concise, well-referenced review by S. A. Cook explains why snakes may have received more attention than other animals.

Its gliding motion suggested the winding river. Biting its tail it symbolized the earth surrounded by the world-river. Its patient watchfulness, the fascination it exerted over its victims, the easy domestication of some species, and the deadliness of others have always impressed primitive minds. Its swift and deadly dart was likened to the lightning; equally marvelous seemed its fatal power. It is little wonder that men who could tame and handle the reptiles gained esteem and influence. Sometimes the long life of the serpent and its habit of changing the skin suggested ideas of immortality and resurrection. (1911:677)

Respect and wonder toward snakes is one thing; worshiping them is something else. Christian, Jewish, and Islamic monotheism viciously suppressed snake worship along with all other types of paganism and pantheism, beginning in Europe and the ancient Middle East. How better to oppose the worship of snakes than to portray them as degenerate brutes with malevolent tempers, deadly venom, and evil cleverness. Some nonvenomous snakes were perceived as dangerous as well. For example, large constrictors have commonly been implicated in human and livestock deaths (Murphy and Henderson 1997). We revisit this widespread role of snakes in religious practices later.

Ambivalence toward Snakes and the Rise of Scientific Natural History

By the sixteenth century, scholars were anxious to document credible knowledge about all aspects of nature, and snakes were no exception. Herpetologists from this period, such as Abbatius (1589), Severini (1651), Redi (1675), and Charas (1677), were typically respectful of snakes, even though their primary focus was on understanding the animals' venomous nature. They typically reviewed the work of ancient Greek and Latin authors and covered the depiction of snakes in culture, art, coinage, and religion prior to their more scientific and naturalistic treatments. They had little use, however, for ancient prejudices. Edward Topsell (1658), a parson, wrote a compendium of the entire animal kingdom and humbly proclaimed that his book on snakes was more extensive than any ever written. He noted, following traditional Christian dogma, that God created all creatures good and suggested that only heretics do not accept this. Thus, concerning snakes he wrote: "if we can be brought to acknowledge a difference betwixt our shallow capacity, and the deep wisdom of God, it may necessarily follow by an unavoidable sequel, that their uses and ends were good, although in the barrenness of our understanding, we cannot conceive or learn them" (Topsell 1658:591). Today, we are more confident in the role of science in revealing such understanding, but most biologists share Topsell's contempt for hatred expressed toward the existence of any organism.

Later scholars often demurred from the view that all God's creation was good. Carolus Linnaeus (1758) expressed an obvious contempt for snakes, suggesting, "These foul and loathsome creatures are abhorrent because of their cold body, pale color, cartilaginous skeleton, filthy skin, fierce aspect, calculating eye, offensive smell, harsh voice, squalid habitation, and terrible venom. And so their Creator has not exerted his powers (to Make) many of them." At the same time, he relied on Albertus Seba's magnificent natural history folio volumes (Seba 1734–1735), which emphasized snakes. Seba apparently found more beauty in snakes than in almost any other group of animals, as shown by the quantity and esthetics of the illustrations devoted to them.

Charles Owen's (1742) book on serpents was composed of three parts. The first two covered biology and behavior; the last consisted of six essays on the role of serpents in the Bible along with three essays on snake worship. The title of the last essay reflected the author's attitude: "Upon the Adoration of Different Kinds of Beasts by the Egyptians, with Instances of the Same Stupidity in Other Nations." Still, Owen accepted the view that God created all animals good and naturally perfect. Even snakes had important roles to play in nature; if animals hurt one another or people, it was due to "the Effect of moral Evil" (Owen 1742:36), a rather Augustinian concept. On the same page, he claimed that "serpents, tho' venomous, are of special Use to Mankind, as they are Part of the *Materia Medica*." He noted the use of snakes in a host of remedies, which are still prevalent in much traditional Asian medicine (Zhou and Jiang 2004). The use of snake venom in modern medicine and medical research is also flourishing (e.g., Huang and Ouyang 1984; Markland et al. 2001). In fact, over 70 medical uses of snake venom for diagnostic and therapeutic purposes have been documented (Russell 1980).

Charles Owen's book is an instructive example of the problems and contradictions seen in contemporary attitudes toward snakes; he accepts the biological, medical, and even ecological role of snakes while still being repulsed by them as "Instruments of divine and human vengeance" (Owen 1742:44). Over 100 years later, another amateur naturalist, Arthur Nicols (1883), reviewed many aspects of the lives of snakes much less credulously than Owen. Nevertheless, he was not enamored with them, claiming they are surprisingly stupid for animals so otherwise advanced on the vertebrate "scale." Discussing fossil snakes, he noted that the coldness of northern Europe prevented constrictors from living there. His final sentence reflected his general attitude toward snakes: "Possibly that rigorous cold which locked these islands in the grasp of the ice exterminated a race in every respect undesirable as joint possessors with man of that which he deems his special inheritance—the earth" (Nicols 1883:58). God's declaration in Genesis of the eternal enmity of snakes and humans remained robust.

About the same time, however, one of the most accurate, balanced, and sympathetic popular books ever published on snakes appeared, authored by

Catherine Hopley (1882) and dedicated to Sir Richard Owen. Still useful in its descriptions of snake behavior, Hopley's book anticipated later books (including Greene 1997): "those who can look at a snake with unprejudiced eyes and study its habits, find continual reason to wonder at and admire the extraordinary features which exhibit themselves in its organization.... But apart from science there is a glamour of poetry, romance, and mystery about snakes, and not without reason" (Hopley 1882:2). Can the views of Hopley and current ophiologists prevail in fostering snake conservation before it is too late?

The Ontogeny of Ophiophobia

Many predatory animals present stimuli that can be used by other species in avoiding them (Hirsch and Boles 1980; Isbell 2006). Some animals avoid and flee from snake odors; for human and nonhuman primates, visual cues eliciting avoidance are more relevant. Malagasy lemurs, evolving in the absence of highly venomous snakes (viperids and elapids) or large boids, show little or no fear of them compared to most Old and New World monkeys (Mitchell and Pocock 1907). Nonetheless, some colubrids do prey on lemurs (A. Mori, pers. comm.).

Human historical experience with venomous snakes has probably shaped our responses to them. In Africa, where hominids evolved, venomous snakes are common and there are no simple rules for visually discriminating harmless from truly dangerous species. Thus, detecting and indiscriminately avoiding *all* snakes was probably favored by natural selection. This much seems uncontroversial. Based on neurological data, Isbell (2006) suggested that the detection and avoidance of predatory and venomous snakes might have played a pivotal role in the evolution of the primate brain. Given these evolutionary forces, the primary scientific question in ophiophobia becomes the role of innate or instinctive snake aversion versus readily learned, but not inevitable, aversion, versus such aversions lacking any biological basis at all.

The history of research on primate ophiophobia goes back almost 200 years, when Broderip (1835) presented a python in a basket to a young chimpanzee. Initially exhibiting the curiosity typical of young primates and peeking into the container, the ape responded with "terror" on detection of the snake. This observation set the stage for a variety of follow-up studies, including one by Charles Darwin (1872), finding much the same result. Although virtually all the early work was anecdotal by today's standards, Darwin at least included a control stimulus, a turtle, and observed a stronger fear reaction toward the snake than toward the turtle. Subsequent researchers followed in Darwin's footsteps but found alarm reactions in primates on presentation of a range of stimuli and then argued that it was no longer possible to claim that snakes had exclusive ownership of this domain. Instead, snakes merely possessed several attributes that characterized fear-eliciting objects: unfamiliarity, movement, abruptness, rapidity of change and visual (chromatic) intensity. Instead of positing an innate recognition of snakes per se, theory shifted toward a polyvalent feature detection (PFD) concept in which the combination of fear-eliciting cues was critical (Mitchell 1922; Kohler 1925; Yerkes and Yerkes 1936; Haslerud 1938). This shift occurred slowly and some writers continued to adhere to the theory of innate snake recognition; but this notion was essentially considered passé by the middle of the twentieth century, when innate or instinctive components of virtually all aspects of human behavior were discounted in mainstream psychology. Rachman (2002) provides a brief history of the rise and fall of the conditioning paradigm to the study of fears and aversions.

Morris and Morris (1965) reviewed this research and at times seemed to accept the then-prevailing PFD concept, but at other times they balked, particularly when discussing some of their own experiments at the London Zoo. In the end, Morris and Morris (1965:214) spoke of fear of snakes in the "narrow sense" when they wanted to connote recognition of snakes per se and fear of snakes in the "broad sense" when they wanted to refer to the PFD concept. Instead of holding the two theories to be alternative and mutually exclusive points of view, Morris and Morris suggested that both types of cognition might be necessary to explain the full spectrum of findings. Furthermore, ophiophobia in either sense can be overcome by repeated positive experiences with these organisms, at least in some people. Still, the extreme emotional and physiological arousal seen in many phobic individuals, which, for some, is resistant to extinction, suggests that ancient emotional brain centers, not just cognitive processes, are involved.

Related to the question of snake recognition is the difficult issue of whether the reactions to snakes by nonhuman primates are innate or learned. Some writers argue for a widespread innate basis for ophiophobia among nonhuman primates, suggesting that humans could have inherited this trait from their ancestors (Mitchell and Pocock 1907; Mitchell 1922). Others take the position that ophiophobia is acquired through experience (Jones and Jones 1928). This difference of opinion is most dramatic when fear of snakes in the narrow sense is assumed to be the only basis for ophiophobia. In this case, an emotionally charged image of snakes is assumed to be either learned or innately present as the mechanism mediating the fear reactions. On the other hand, the difference of opinion fades into trivia when the broader PFD view is considered. This view presumes that both innate and experiential factors participate in the cognitive and affective components of a person's response to snakes. Thus, the PFD concept is consistent with most existing data and is therefore able to defuse theoretical conflicts, or so it seemed.

PFD was the prevailing view among most ethologists and psychologists post–World War II and continued to be the prevailing view (Wolin et al.

1963; Joslin et al. 1964) until it was discovered that Vervet Monkeys (*Cercopithecus aethiops*) produced distinctly different alarm calls to leopards, eagles, and snakes (Struhsaker 1967). Debate centered initially on whether these three calls referred to the three classes of predators or to the defensive behaviors that were appropriate in each case (climbing trees in the case of the leopard call, looking up and heading for dense brush or trees in the case of the eagle call, and standing up and scanning the ground in the case of the snake call). If the calls denoted the classes of predators, then the calls could be said to have external referential meaning; if the calls denoted the antipredator behaviors that the caller would soon exhibit or the caller's current emotional state, then the calls had egocentric meaning in that they reflected the caller's mood or intentions. This difference is subtle but theoretically important (Smith 1977). It should be noted, however, that the calls would be adaptive in either case, provided that listeners responded appropriately. From the perspective of natural selection, the effectiveness of the escape responses is vital, whereas the cognitive mechanisms mediating those responses are meaningful only if one class of mechanisms gives rise to more effective escape responses than another class of mechanisms. At present, there are no data that address this latter issue.

Cheney and Seyfarth (1990) showed that vervet alarm calls convey multiple types of information to conspecifics. The calls denoted both the signaler's intentions and the signaler's perceptions. Moreover, naïve neonatal vervets emit the three types of calls under appropriate circumstances without any prior experience (Seyfarth and Cheney 1980; Seyfarth et al. 1980). Therefore, we can conclude that vervet monkeys have an innate capacity to recognize dangerous snakes and to behave appropriately in their presence, including the ability to warn conspecifics about the danger. The snake alarm call (termed the "chutter" by vervet researchers) is given not only to pythons, which are predators of vervets, but also to venomous snakes such as mambas and cobras that are too small to devour a vervet but are potentially dangerous. The presence of harmless snakes or lizards does not result in the chutter response. Hence, chutter does not mean "snake" in the same broad categorical sense that we use the word. In spite of this slight semantic ambiguity, it appears that snake recognition in nonhuman primates, in both the narrow and the broad senses, is supported by scientific evidence. The vervet research leaves no doubt about the existence of the narrow (snake-specific) category, and research summarized earlier implicates the broad (general snakelike stimuli) category. Whether both types occur in humans has yet to be established, but it is fairly clear that at least the PFD or broad type of ophiophobia occurs in *Homo sapiens*. Further research needs to be conducted on the recognition of specific dangerous versus harmless snakes.

About the same time that the Vervet Monkey studies were documenting specific reactions to dangerous snakes, other studies were resurrecting the critical importance of age and experience. Wild-caught monkeys seemed more fearful of snakes and models than lab-reared Rhesus Monkeys (*Macaca mulatta*) (Mineka et al. 1980), and laboratory experiments supported the view that infants and juveniles learned to avoid snakes by observing the reactions of adults (Mineka et al. 1984; Cook and Mineka 1990). An experimental field study of the related Bonnet Macaque (*Macaca radiata*) in India, however, found that juveniles and subadults did not differ from adults in the latency to respond to models of venomous or harmless nonvenomous snakes or large dangerous pythons (Ramakrishnan et al. 2005). Observations of infants and mothers led to the conclusion that both innate recognition and maternal restriction of curiosity about snakes might be the two mechanisms at work rather than observational conditioning based on the fearful responses of other troop members. In fact, there is now an extensive literature on fears in humans and nonhuman primates in which reactions to snakes (or in most cases models and pictures) are measured in rather sophisticated and complex sets of experiments that cannot be reviewed here (see references in LoBue and DeLouche 2008).

The most recent position taken by many researchers concerning snake phobia in humans is that while humans are not necessarily born with a fear of snakes, they are salient stimuli and humans can rapidly acquire a fear of them. This view is based on the now-famous Mineka lab experiments and others that seemed to discount instinctive recognition of snakes and emphasized conditioning, especially of a social nature (cf. Öhman and Mineka 2001). This reaction could even be mediated through second-hand experiences such as scary stories (Wilson 1984, 1994). Wilson did not distinguish between the broad and narrow concepts of snake fear outlined by Morris and Morris (1965), but he asserted that there is probably a genetic foundation for the attitudes that we end up with and for the fact that we can acquire these attitudes very quickly. Morris and Morris (1965) wrote that "snake hatred" reached its peak in children at age six and then declined gradually thereafter; they reported data for children only through 14 years of age. Curiously, most research on snake phobia has been focused on college students, who should exhibit lower levels of "snake hatred." Before returning to some recent studies on the nature of snake fears, phobias, and aversions, it is useful to examine a series of studies from a clinical perspective that tried to overcome such responses, regardless of their origin.

The earliest experimental study of human avoidance behavior required a subject to enter a room containing the feared object and then to attempt to approach, touch, and handle that object. Typical measures included the subject's final proximity to the object, various physiological responses (e.g., increased pulse rate and palmar sweating as indicated by electrical resistance) and self-reported fear, usually on a Likert-type scale (Lang and Lazovik 1963; Geer 1965; Lang 1969). Levis (1969) introduced a new method in which the subject sat at one end of a 3.35-m track with a Plexiglas box containing the feared stimulus at the opposite end of the track. The subject could press a button to advance the feared object 30 cm closer to him or her per press. Thus, one dependent variable was the number of presses (= final proximity of object), and another was the latency before each successive press of the button. Because the subject was seated, a variety of physiological measurements could be taken continuously so that the subject's physiological responses and self-reports could be compared at each step. The use of the phobic test apparatus (PTA) became popular, and comparisons of data obtained with PTA to data obtained with traditional methods were reviewed (Borkovec and Craighead 1971). The results indicated that reliable measures could be obtained with both techniques and that equivalent effects of repeated testing also occurred with both. Perhaps the most interesting result was seen when the subjects were divided into two groups depending on whether or not they touched the snake at the end of their trials. Subjects who ended up touching the snake approached it steadily (same speed of movement at each point of proximity) or pressed the button with similar latencies at each successive step. In contrast, for subjects who did not touch the snake, the closer they got to the snake, the longer was the latency for them to take the next step (traditional test) or to press the button again (PTA).

Craighead (1973a) used the PTA to study a group of college women who had failed to touch a 1.22-m-long kingsnake in a pretest, this being the usual criterion for defining ophiophobia. The mean heart rate was 85.0 beats/min and mean respiration rate was 20.3 breaths/min in the last minutes of a 10-minute acclimation period during which the women were seated in the PTA. They did not know about the snake's presence at this time, nor could they see the covered snake at the opposite end of the PTA. Also, the subjects did not know the specific purpose of the experiment; they were told only that the study was about the relation between physiological responses and emotion. Upon completion of the last PTA step, when subjects knew about the snake and could see it, the mean heart rate was 96.1 beats/min and mean respiration rate was 41 breaths/min. At this point, subjects also rated their subjective fear on a 10-point scale called the "fear thermometer" (Walk 1956) and the mean was 7.4. Heart rate increase was related both to the proximity of the snake and the subjective fear. Respiration rate increase was related only to proximity to the snake. In general, subjects who showed the greatest increase in physiological arousal reported the greatest fear and stopped the snake further from them than did subjects who showed smaller increases in arousal.

These findings led Craighead (1973a) to hypothesize that, although all subjects terminated the test without touching the snake, some subjects probably did so because of high levels of physiological arousal (e.g., fearful or phobic), whereas others did so because of external cues without experiencing high arousal (e.g., avoidant). This point is of theoretical importance because it implies that the PTA could distinguish between truly phobic individuals and individuals who were suggestible or who might exaggerate at the behavioral level their actual level of fear at the physiological level. The traditional walk-up test cannot make this distinction, mainly because it is incapable of continuous monitoring of arousal and fear. Today we could employ much more sophisticated measures.

The most common therapeutic approach to phobias is systematic desensitization, which involves muscular relaxation and the visualization of feared objects or events in an order of increasing intensity. Although the whole therapeutic package alleviates or reduces fear and avoidance behaviors (Rachman 1967; Paul 1969a, 1969b), theorists argued about the importance of the various components, particularly muscular relaxation (Lang 1969). The results of testing visualization with and without relaxation were ambiguous, with some studies producing positive results but most showing that visualization without relaxation was no less effective than visualization with relaxation, provided that the duration of visualization was held constant in the two conditions (Craighead 1973b).

Although Craighead (1973b) and others cited earlier left little doubt that systematic desensitization reduced ophiophobia, they did not reveal the psychological processes responsible for the change. Indeed, therapy outcome research and process research are quite different, and the positive results of desensitization experiments would seem to justify additional research on the psychological mechanisms underlying people's reactions to snakes. Theoretical candidates include Wolpe's (1958) counterconditioning hypothesis, in which therapy is seen as conditioning new (nonavoidance) responses to the initially feared stimulus. Another hypothesis is that fear and avoidance responses are extinguished as a consequence of elicitation through visualization without being followed by punishing (negative) reinforcement (Wolpin and Raines 1966). Closely related to this is the notion that visualization results in the habituation of phobic responses (Lader and Matthews 1968). Morris and Morris speculated that yet another approach might prove useful:

> We are still reacting towards the snake as if we were a bunch of medieval peasants....Familiarity with snakes should effectively reduce our instinctive dislike of them. It should also make it more difficult for us to make them into fiendish symbols. Knowledge of the details of snake biology should enable us to get them into perspective as forms of animal life that are worthy of our objective interest. (1965:215)

Although increased knowledge might prevent some people from being fearful of and indifferent to the fate of snakes, there is little evidence that instruction can help truly phobic individuals overcome their animal phobias. Because the problem is rooted in irrationality, it appears to be more or less immune to rational appeals.

Clinical psychological research on ophiophobia ended rather abruptly in the 1970s, not to be resumed until recently, due to a sea change in attitudes and values among clinical researchers. Ophiophobia had been a convenient phenomenon to study as a model of other pathologies that might respond to systematic desensitization and related therapies. This paradigm shift tended away from nonpatient populations and toward populations with more serious psychiatric conditions (e.g., Hopko et al. 2007). As researchers concentrated their efforts on patient populations, animal phobias were left relatively unfunded and unstudied.

The preparedness hypothesis of learning to fear snakes popularized by Wilson (1984, 1994) and evolutionary psychologists is still controversial. Initially, the preparedness hypothesis was supported using physiological measures of responses to pictures of snakes and other stimuli (Öhman et al. 1975); conversely, Merckelbach (1989) found no support for the preparedness theory.

Today, however, the seemingly rapid acquisition of snake fears as compared with more arbitrary stimuli has been coupled with resurgent work in evolutionary psychology to suggest that there are evolutionarily derived "fear modules" in the brains of people and other primates (Öhman and Mineka 2001) that are "prepared" to be activated through appropriate stimuli. Studies with adult humans presented with pictures of fear relevant (e.g., snake) and fear irrelevant (e.g., flower) stimuli in arrays on a computer screen showed that detection of a snake among flowers was faster than detecting a flower among snakes, supporting the preparedness approach (Öhman et al. 2001), which is a finding that has been replicated (LoBue and DeLoache 2008). More recently, children as young as 3 years old similarly were faster at detecting snakes than flowers, even those with virtually no experience with snakes or whose parents claimed to be unafraid of snakes (LoBue and DeLoache 2008). The results support the prepared recognition of evolutionarily relevant fear stimuli.

It should be noted that the "fear module" research is somewhat separate from the issue of fears and phobias. The PFD approach is still alive and well in this research. Even more recently, researchers found that adult captive-reared Japanese Monkeys (*Macaca fuscata*) raised indoors with no opportunity to ever experience a snake also show such recognition using very similar touch-screen computer methodology (Shibasaki and Kawai 2009).

A problem in these and other studies is that various species of snakes and animals with similar features, such as lizards, are not used as controls. Given the ability of nonhuman primates to discriminate among snakes posing different kinds of risks, more biological sophistication is needed in experiments on human snake fears and their ontogeny. In short, there is still little that can be concluded about the PFD, preparedness, innate recognition, and emotional fear and phobia hypotheses in responses to snakes by humans or even nonhuman primates. We suspect that all may be involved. This being said, studies are needed to develop effective treatments for individuals with extreme and incapacitating fears, such as those involving snakes (Hopko et al. 2007) as well as for populations with normal, rather than pathological, dislikes and fears (Olson and Kendrick 2008; Walther and Langer 2008). It is promising that a recent study conducted with both lay people and those who had experience with snakes showed that while the former group feared all snakes, the latter reacted more fearfully to dangerous than to harmless snakes. Furthermore, implicit measures of snake fear supported the explicit statements of the participants (Purkis and Lipp 2007). Thus, familiarity and knowledge may be useful, and it is to efforts toward this end that we now turn.

Visitor Behavior in Zoo Reptile Buildings

Zoological parks and related animal exhibits are readily available settings to assess public attitudes toward snakes and to test ideas about educating visitors and altering their perceptions of snakes (Dodd 1993b). Several studies attest to the promise and challenges of such attempts. At the London Zoo, visitors were asked to rank their favorite exhibits; two of the three most popular were the aquarium and the reptile house (Balmford et al. 1996; Balmford 2000). In another study, youngsters ranging from 4 to 14 years of age were asked to identify the animals they liked least and the ones they liked most; 50,000 submitted their choices (Morris and Morris 1965). From this group, a random sample of 2200 was used—100 boys and 100 girls from each age group. Snakes ranked first (28%), followed distantly by spiders (10%), as the most disliked animals. At the Zoologicka Zahrada Bratislava in the Czech Republic, 4123 children were asked the same questions—22% listed snakes as the most unpopular, followed by the rat (4%) (Šurinová 1971). The underlying message seems to be that many people dislike snakes but are compelled to look at them nonetheless—a form of morbid curiosity.

Visitors often come to the zoo in family groups (White and Marcellini 1986). It is therefore interesting to note what happens once they arrive at displays housing snakes. Hoff and Maple (1982) examined the characteristics of visitors to the reptile buildings at Zoo Atlanta, Georgia, and the Sacramento Zoo, California, and found that (1) female patrons refused to enter the buildings more often than males and spent less time in the exhibits than males and (2) teenagers entered the buildings more often than other age classes and spent more time viewing exhibits. In our experience, mothers often sat outside, saying that snakes were "too ugly" to even view. Children were often told to hurry through the exhibit. Young children would initially say that snakes were really beautiful and interesting. The parents immediately confronted them, warning them about how dangerous snakes were, how ugly, and so on. By the time the tour was finished, it was a rare child who still maintained his or her initial feelings. Most were completely cowed by their parents and notably silent at the end. Adolescents used snakes as psychological tools in many instances. Young males were protectors, holding their female companions tightly as they walked through the building. Young women screamed and swooned, allowing the males to display courage and calm them while confronting these "terrifying" creatures.

Once the zoo visitor enters the reptile house, how much learning occurs? The mean time that visitors to the herpetology exhibit at the Smithsonian National Zoological Park (NZP) spent in the building viewing exhibits was astonishing low—15 and 8 minutes, respectively (Marcellini and Jenssen 1988). The average time spent looking at an exhibit was 8.1 seconds, very little time to locate the animal or read the accompanying educational graphics. Crocodilians were the most popular, followed by snakes and turtles; larger animals were looked at longer than smaller ones. Phillpot (1996) reported similar findings for median visitor time spent in the exhibit space at the Jersey Wildlife Preservation Trust. The average viewing time at the Birmingham Zoo, Alabama, was 16.6 seconds, with large boas and pythons being the most popular. Not surprisingly, the holding power of an exhibit was directly related to animal activity (Bitgood et al. 1986). Meaningful education about snakes is impossible if the public remains in a reptile building for only brief periods of time.

To counteract this dilemma and provide a more stimulating and interactive environment for family learning, HERPlab was designed and administered by the NZP staff (White and Barry 1984; White and Marcellini 1986). The John Ball Zoo, Grand Rapids, Michigan, and the Philadelphia Zoo, Pennsylvania, joined the project for 3 months as field test sites. Eighty percent of the visitors were family groups and the median length of time spent in the reptile exhibit was 27.5 (NZP), 30 (John Ball Zoo), and 21 (Philadelphia Zoo) minutes, a considerable increase over previous amounts of time spent at reptile exhibits. The major challenge for the zoo herpetologist was to keep the viewer interested in an animal that was often immobile or difficult to see. Human interpreters successfully interacted with the public by using live snakes as demonstrations. Any lecturer who has used live snakes as props has certainly been impressed by their incredible drawing power. The Arizona Sonora Desert Museum, outside of Tucson, Arizona, recently launched an educational demonstration using live rattlesnakes indigenous to the area, which draws large crowds (Johnson 2007; C. Ivanyi, pers. comm.).

What other changes can turn reptile houses into discovery centers and counteract the fear of snakes by demystifying them? A Reptile Discovery Center—an informal science education initiative—was tried at the Dallas Zoo, Texas; Zoo Atlanta, and the NZP (Doering 1994; Marcellini and Murphy 1998). Among its purposes were to improve visitors' experiences,

expand their understanding of the collections, and counteract negative feelings toward snakes and other reptiles. The study assessed public responses
by conducting surveys and observations in the reptile buildings in 1991 before twelve learning activity modules were installed. Would there be changes
in 1992 after the installation of these interactive modules? Four separate
free-standing modules asked questions about snakes and humans:

1. Are you bothered by my long, thin shape?
2. Does my scaled skin disturb you?
3. Does my face look unfriendly to you?
4. Where did you get your attitudes toward reptiles?

Other modules covered thermoregulation, reproduction, anatomy, physiology, senses, and feeding behavior. Total visit time increased 20.5% from one
year to the next. In 1992, visitors spent a higher proportion of time engaged
in the exhibits, the average number of stops at exhibits increased, the average
stop duration was over one-quarter longer, and the interactive exhibits were
most attractive to children, although all visitors reported enjoying them.
Doering (1994) concluded that the modules had positive effects on visitors'
viewing behavior, attentiveness, affect, and knowledge of reptiles.

Because some humans do not realize that snakes are capable of feeling
pain and therefore deserving of moral consideration, it is sometimes necessary to shock the zoo visitor into comprehension. A graphic and disturbing
exhibit at the Dallas Zoo about the rattlesnake roundups so prevalent in
the southwestern United States was developed using photographs of snakes
being gassed from dens, dumped in pits, beheaded by children for a fee,
eviscerated while alive, and prepared as curios. The title was "Diamond-
backs Can't Scream," referring to the fact that snakes do not vocalize pain.
The herpetological staff was prepared for public outrage because the im-
ages were unsettling, but, surprisingly, the overwhelming response was that
people appeared to be appalled by the practice and felt that it was important
to treat snakes humanely (J. B. M., pers. obs.).

Nonetheless, rattlesnake roundups, still common in many states in the
United States, continue to be problematic. In the long run, they are not sustainable; they do not seem to assuage the public's fear of snakes but, instead,
augment the view that snakes are expendable and can be cruelly captured,
kept, and treated. Surprisingly, animal welfare groups, state wildlife authorities, and well-known popular authors have, historically, been uninterested
in opposing these roundups, although roundups are getting a bit more attention recently. Still, there is no emotional outrage over them comparable to
the use of animals in other "recreation," such as dog and cock fights.

Can zoos desensitize a patron's fear of snakes? Murphy and Chiszar
(1989) recommended the installation of a "Fear Room," which was to be
constructed with the consultation of psychiatrists and architects in order to

design a nonthreatening facility. They envisioned a circular structure with eight separate rooms, in each of which a myth or question about snakes would be addressed, such as:

- Can snakes "outrun" humans, and how fast can they move?
- Are snakes carriers of disease?
- Do snakes hypnotize prey?
- How dangerous are snakes to humans?
- Can snakes feel pain?
- Should snakes be protected?

Each of these topics would involve video presentations, interactive displays, computers, or current audiovisual technology. The visitor who had viewed all eight presentations could be given other options, such as meeting a staff member or volunteer holding a live snake, providing opportunities to ask questions and touch the reptile. This novel approach, renamed "Fear Zone," was incorporated in the renovation of the 1936 reptile building at the Staten Island Zoo in New York and opened to the public in 2007. It is hoped that research will be done using this facility.

Zoo outreach programs can include the Internet, television, traveling exhibits, temporary exhibits, libraries, schools, and unusual venues such as shopping malls and casinos. Several decades ago, the herpetological staff at the Dallas Zoo used living snakes, artwork and artifacts depicting snakes, interactive modules, and hourly lectures describing snake biology and conservation in a temporary exhibit in the Dallas Public Library. This resulted in the highest attendance of any exhibit ever held in that setting to date. Zoos are not the only sources of outreach programs. Nature centers, although smaller than most zoos, are numerous and growing in local impact across the United States. They are probably, as a group, more focused on education and local outreach as their main objectives than zoos and are typically less bureaucratic and more open to volunteer-initiated projects. Faculty members at universities can also play important roles, as has been done on behalf of rattlesnakes in New York (Greene 2003).

What Can Be Done?

An appreciation of snakes can be encouraged and reinforced in zoos and aquariums if attention is focused on providing a stimulating and enjoyable experience. Nonetheless, the long-term effects of such exhibits are undocumented. The most important question remains, can these approaches change the overall negative view of snakes held by humans quickly enough to make a difference in their protection and long-term survival? Sadly, we think the challenge is great and the prognosis discouraging. But all is not lost. For instance, the commonwealth of Massachusetts in late 2006 adopted the Eastern Gartersnake (*Thamnophis s. sirtalis*) as its state reptile.

Some endangered snakes are gaining public support and protection. The Aruba Island Rattlesnake (*Crotalus unicolor*) has become a flagship species for conservation in the Dutch Antilles—captive breeding in zoos of the last members of the species enabled their survival in the wild (Odum and Goode 1994). In Ohio, attempts to save endangered and isolated populations of the Plains Gartersnake (*T. radix*) and the Lake Erie Watersnake (*Nerodia sipedon insularum*) have been featured in a variety of exhibits and public media outlets, and this has led to effective community and government protection efforts (e.g., King et al. 2006a). In these cases, however, the areas involved are small. Protection of isolated populations of Butler's Gartersnake (*T. butleri*) in four counties in urbanized southeast Wisconsin has generated much protest from developers, civic boosters, and the media (G. M. B., pers. obs.). Only a political change through recent elections prevented politicians from, for the first time in U.S. history, delisting an endangered species without even a superficial reliance on scientific data. Science was considered irrelevant, especially for a mere gartersnake species. Fortunately, this led to vigorous opposition from conservation and environmental organizations, as well as from the Wisconsin Department of Natural Resources (WDNR; Burghardt et al. 2006). Unfortunately, the WDNR, under political pressure, is now in the process of "reinterpreting" the protection regulations to remove crucial habitat protection, so the future of the protection plan being prepared based on solid scientific data representing genetics, morphology, behavior, ecological needs, and population viability analysis is in doubt even as the scientific value of these populations is on the rise (Fitzpatrick et al. 2008). Similar protection endeavors, and attendant controversies, are going on in many parts of the world. But even if conservationists "win," are such victories sufficient?

Humans are far from the rational animals that Aristotle and Descartes envisioned. The tragedy of modern cognitive science is that, until recently, the emotional underpinnings of attitudes, learning, and intelligence were neglected. The writings of Panksepp (1998) are particularly valuable as premonitions of the impending radical reorientation in how we view the human mind. What does this means for changing human perceptions of snakes and encouraging their conservation? Taking seriously the wonder, mystery, and fascination with snakes, as well as our varying fears of them, could be the key to further progress. We should not leave it to the commercially motivated roadside zoos, reptile farms, serpentariums, and modern-day sideshow-style attractions to educate the public. In addition, many fictional movies, video games, and television shows sensationalize, demonize, harm, or belittle snakes. There is, in fact, much fine footage in shows produced by the BBC, National Geographic, and others that could be repackaged into more emotional appeals, using snakes as icons of our planet and all the mysteries it yet holds, for the need to take care of and save all the pieces of the planet.

Given the entrenched human apprehension of snakes (sometimes warranted and evolutionarily explicable) that hinders snake conservation, radical action may be the only hope. Saving snakes may require the reinstatement of a form of reverence toward them, such as still occurs in indigenous societies around the world (i.e., India, Africa, and Australia). Scientists are adamant about objectivity in their work and often view it as far removed from religious and spiritual values. This could be a serious mistake. Again, we emphasize that we cannot conserve snakes by relying solely on rational discourse and objective facts. There are too many people who will never be reached by the most enlightened educational efforts.

One approach is to incorporate a modern scientific understanding of nature, including snakes, into a spiritual message, an approach vigorously pursued by Wilson (2006), a giant of modern conservation. As mentioned earlier, snakes were once the most prominent animals involved in religious practices across the world. Africa is home to the oldest human populations, and serpent worship was widespread (Hambley 1931). Pythons were by far the most revered snakes, and in some tribes causing injury to a python, even inadvertently, could lead to ostracism or even the death of that person. Sheila Coulson, a Norwegian archaeologist, has found evidence supporting the existence of python worship in Africa 70,000 years ago—a cave with a huge pythonlike stone entrance (Vogt 2006). If true, this would be the first documented religious ritual site of any type!

Almost 70,000 years later, snakes were still revered worldwide, and were even at the center of a brutally suppressed early Christian Gnostic sect, the Ophites. Their worldview involved snakes in a reformulated Christianity in which the true Father God sent the serpent to show humans that they contain true divine light if only they would look for it. Thus, it was important that they ate from the tree of knowledge (George 1995).

It is necessary to respect the human populations who need, and find value and meaning in, stories that explain the big picture. What we want to achieve is that sufficient numbers of people value snakes and strive to conserve them and their habitats. Herpetologists should recall what initially fascinated and intrigued them about snakes, especially the kinds of emotional experiences with snakes that we do not feel comfortable talking about in our quest to be objective scientists (Bowers and Burghardt 1992). Similar intellectual and emotional commitments are needed to change the attitudes of those who are prone to fear and loathe snakes.

One of us (G. M. B.) lives in the southeastern United States, where some Christian churches have serpent-handling worship services (Brown and McDonald 2000). The snake-handling aspect of the service is an emotional event for the participants. Dozens of people handle Copperheads and Timber Rattlesnakes, and this occurs along with dancing and loud music. There are still mysteries to be solved concerning why this practice persists and why bites are rare given the ample opportunities for them to occur.

Although people might consider snake handling in Christian churches to be a primitive and archaic practice, it actually developed less than a century ago in Tennessee as part of the rise of Pentecostalism (Covington 1995). Why only venomous snakes are used is unclear; the motivation may be similar to that found in some herpetologists (e.g., Wiley and Dickinson, as cited in Murphy and Jacques 2006), some television personalities (e.g., Steve Irwin), and many zoos and serpentariums around the world—that is, giant and "killer" snakes are used to draw paying customers, to show off the handler's bravery and talents, or to demonstrate "God's forbearance." Perhaps it would be more useful to advocate snakes as integral aspects of spiritual attitudes toward nature and thus elevate the plane of debate on their value and conservation. Certainly, as scientists, we are skeptical proponents of reason and truth. All animals and plants contain many truths, however, and snakes perhaps more than most animals. Furthermore, time is short and, as conservationists, we need to balance the means of effective conservation against the ultimate results.

Other approaches also deserve exploration. There is a vast and growing literature on the ease of learning to be prejudiced against those who differ from one's group in race, ethnicity, language/dialect, religion, sexual orientation, and even diet and clothing (Kunda 1999). This literature on stereotypy and stigma could be mined for tools to do a bit of social engineering on attitudes toward animals. *Teaching Tolerance* is a magazine distributed to teachers throughout the country. Little in this magazine is devoted to teaching tolerance for different species, although stories for young children included in the magazine often use animal themes. The goal never is teaching tolerance toward snakes, however, or other widely feared animals such as bats and spiders. Perhaps we need to be more aggressive in our outreach. Ongoing questionnaire and interview studies (e.g., Christoffel 2007) may provide useful information. Surveys have been distributed to thousands of residents in Michigan and Minnesota to assess the potential of educational outreach programs.

Fear of snakes may actually contribute to their conservation in some cultures. In Korea, for example, fear of snakes causes rural people to avoid them altogether, and this mutual avoidance may be beneficial for both (M. H., pers. obs.; see also Tanaka et al. 1999). Of course, the challenges here are particularly complex. How can we increase human compassion and caring for animals (e.g., Goodall and Bekoff 2003) while at the same time developing an understanding of what it is going to take to conserve them in a human-dominated world? Ironically, this must be done at the same time that we must reduce populations of nonnative species. Current examples include pythons, boa constrictors, caimans, green and spiny-tailed iguanas, Nile monitor lizards, Cuban frogs, and brown anoles in Florida, to give but one locale. Compassion may be the first step to conservation, but compassion alone is limited as a conservation tool (Hutchins 2007a). We need to generate informed concern to supplement and direct emotional connections to animals and allow for necessary management efforts (Hutchins 2007b). A better understanding of the origins and nature of human attitudes toward and perceptions of snakes and other creatures might help us to do a better job in this regard (see Cooper 1999).

If widespread concern by humans sufficient to ensure the conservation of snakes is not attainable, then we suggest focusing more attention on conserving entire geographical landscapes that snakes inhabit. For example, the hotspot approach favored by some conservation organizations (Mittermeier et al. 2004) seeks to set aside areas with a high degree of endemism and high densities of species; many of these areas are inhabited by snakes. In addition, other more charismatic and wide-ranging flagship species such as tigers, rhinos, elephants, and great apes, are found in the same geographical areas inhabited by snakes, and they can be used to conserve snakes by proxy. Snakes, as representatives of Earth's incredible diversity and as important pieces of its ecological puzzle, certainly deserve to be conserved, regardless of human attitudes toward them.

Acknowledgments

We thank J. Block and M. Olson for reading and commenting on earlier versions of this chapter. S. and V. Donnelley and The Center for Humans and Nature deserve thanks for their inspiration and support. P. Lasker, the Smithsonian's National Zoological Park's librarian, assisted with the literature search. A. Mori provided observations on his work in Madagascar, the relatives of M. H. provided observations from Korea, and Z. Tadic provided and translated newspaper accounts of the Long-nosed Viper bite.

11

Snake Conservation, Present and Future

RICHARD A. SEIGEL AND STEPHEN J. MULLIN

In the preceding volumes of this series, the co-editors were careful to include at least one chapter that dealt specifically with the conservation of snakes (Dodd 1987, 1993b). The relatively low proportion of chapters devoted to this field did not mean that they felt that snake conservation was unimportant—quite the opposite was true. Instead, this reflected the relatively low number of studies devoted to the conservation and management of snakes before 1993. As noted in the Introduction to this volume, both of us felt that the time had come to devote an entire volume to this topic, for two reasons. First, like most field biologists, we have watched as populations of the animals to which we have devoted our professional lives have slowly (and in some cases, not so slowly) dwindled in numbers. Second, the number of studies that are directly or indirectly applicable to snake conservation and management has grown substantially over the last 15 years. Clearly, we have reached the time when an entire volume with a conservation theme now makes sense.

The preceding chapters in this volume have done an excellent job in both reviewing the literature and pointing out the directions in which specific fields need to move. In this chapter, we provide an overview of where we stand with snake conservation; where our strengths and weaknesses lie in studying snake populations; and where, in our opinion, we should be going to assure a future for these fascinating reptiles. Our comments reflect our personal experiences and should not be taken as the only viewpoints available; others may have very different ideas about what we have covered here. Our intent is to provoke debate and discussion. This chapter should be read in that light.

An Overriding Issue: Many People Do *Not* Like Snakes

In the summary to a past volume (Seigel 1993), we noted that snake ecologists sometimes operate under a feeling of inferiority compared with those who study the other group of squamate reptiles, the lizards. Thankfully, this feeling of inferiority (to the degree to which it existed) is now largely a thing of the past, and some authors have noted that snakes are the new model organisms (Shine and Bonnet 2000). Still, those who wish to contribute to the conservation and management of snakes operate under a very real limitation. Many people (both in the lay public and in the management community) simply do not like snakes, whereas their opinions of other reptiles and amphibians are less negative (see Burghardt et al., Chapter 10). This opinion is especially forceful when dealing with species viewed by the public as dangerous, such as rattlesnakes (*Crotalus* and *Sistrurus*), copperheads, and cottonmouths (*Agkistrodon*), as well as many nonvenomous snakes such as pythons, boas, and watersnakes (*Natrix* and *Nerodia*) that are perceived as being dangerous. At the very least, this antipathy toward snakes makes the researcher's life more complicated. In working with Massasaugas (*S. catenatus*), the members of one of our lab groups (R. A. S.) have had to justify our work to a skeptical public, both in one-on-one situations in the field (usually starting with the question "You work with *what*?" and ending with the question "You do kill them, correct?") and in group sessions with local school groups and natural history societies. Indeed, any field researcher working with snakes can verify the observation of Burghardt et al. (Chapter 10) that the public is at once fascinated and repelled by snakes, at least judging by the number of people who attend snake talks at local study sites.

Unfortunately, this dislike of snakes by the public can have negative and severe consequences both for research opportunities and for the kinds of management recommendations we are able to get adopted once studies are completed. An excellent example of this was a major research program on radiotelemetry of Timber Rattlesnakes (*C. horridus*) at a city park in Virginia that was terminated by pressure from local residents. One state delegate noted that capturing and then releasing radiotagged snakes in the park was "reckless endangerment" and likened the study to finding a loaded gun and not removing it (Virginian-Pilot 1994). Despite support from the state wildlife agency and the production of published data (Cross and Peterson 2001), the study was terminated much earlier than planned. With the possible exception of wolves, it is unlikely that any other organism would have generated the same level of public outcry. Other notable examples of negative public perceptions influencing conservation decisions include the controversy over the protection of Butler's Gartersnake (*Thamnophis butleri*) in Michigan (detailed by Burghardt et al., Chapter 10), the way that Brazos River Watersnakes (*Nerodia harteri*) were vilified by local residents when it was thought the species might stop a dam project (Mathews 1989), and the continued practice of rattlesnake roundups in the central and western United States (Fitch 1998).

The impact of negative public perceptions should not be seen as a rationale for not vigorously pursuing studies on snake conservation biology. First, many state and federal agencies are highly supportive of research on snake conservation. For example, the state of Maryland recently proposed a major restoration project for the Northern Pinesnake (*Pituophis m. melanoleucus*). In a public hearing, the only controversy was over whether the species was, in fact, native to Maryland not whether protecting snakes was a good use of state tax dollars (R. A. S., pers. obs.). Second, even rattlesnakes can receive a better reception by the public. The Toronto Zoo has been very successful in getting visitors to Canadian National Parks to accept the presence of Massasaugas (*S. catenatus*) via a combination of workshops, naturalist talks, and informative displays (Johnson 1999). Some visitors even "sponsor" individual snakes by donating funds to research programs. Thus, those working on snake conservation should not be daunted by the issue of a negative press, but should look for proactive ways to overcome these obstacles.

The Need for More Data: Academic Training versus Management Reality

The training of most academic biologists makes us cautious by nature. During editorial reviews of our research, we are frequently told to avoid overinterpretation of our data and to steer clear of statements that are overly broad or speculative. Conclusions are often tempered by hedges such as "more data are necessary to support this conclusion fully" or "the reader should keep in mind that the conclusions should not be interpreted beyond these data." Although laudable and commonplace in the academic world, such statements and attitudes sometimes result in difficulties when dealing with conservation and management issues. Because funds for conservation research are scarce and management decisions often have to be made within a narrow time frame, it is not unreasonable for resource managers, when they support a study on the status or biology of an imperiled or listed taxa, to expect the study to provide specific recommendations as to what actions should or should not be taken. When the results of the report essentially state "we need more data," there can be a culture clash, with resource managers wondering whether academic biologists really understand their needs and whether all academic biologists really want is more funding.

One way of avoiding this potential clash is to assess two questions: (1) What is the current state of our knowledge of snake biology (as it relates to conservation), and (2) when do we have enough data on a specific topic so that additional studies are unwarranted from a conservation perspective? Obviously, these are very broad questions and cannot be dealt with fully here. The preceding chapters (as well as those in the first two volumes of this series) provide at least a reasonable approximation of the state of the art of some fields, and, in some cases at least, these chapters can help focus our efforts on the aspects of snake biology that simultaneously have the most importance for conservation issues and are the least well known.

As an example of what we mean, consider the data requirements to complete a population viability analysis (PVA; Boyce 1992; Shaffer 1994; Beissinger and Westphal 1998; Dorcas and Willson, Chapter 1; Shine and Bonnet, Chapter 6), a technique that is fairly common for many species of imperiled taxa but that is rarely applied to snakes. This technique requires data on several key life-history variables, most notably age at sexual maturity, clutch size, frequency of reproduction, and, especially, age- or stage-specific survival rates. From one perspective, these data demand a very broad and long-term life-history study of each population of concern on the justifications that populations of the same species often show strong intraspecific variation in these traits and that, even within single populations, these traits vary strongly over time (see Dorcas and Willson, Chapter 1; Shine and Bonnet, Chapter 6). Thus, it might take many years before sufficient data can be collected to provide an adequately detailed PVA.

To a limited degree at least, we challenge this conclusion. Although we are ardent proponents of long-term field studies and would never dispute that such studies are invaluable to our understanding of snake ecology, the idea that every aspect of every population requires a long-term study to make valid conservation or management recommendations is one that we find questionable and that furthers the perception that academic biologists have to study something to death before making conclusions. For example, the basic reproductive biology of many species of snakes (especially in the United States, Canada, Europe, and Australia) is reasonably well known, especially in terms of clutch size and reproductive frequency. Although the exact clutch size may vary from year to year or among populations, the level of impact of such variation on a PVA tends to be relatively minor (e.g., Seigel and Sheil 1999), suggesting that long-term data, while useful, may not be essential when making conservation recommendations. If the species (or a close phylogenetic relative) has been studied in detail elsewhere, at least a preliminary model can be constructed without direct data on the population or species in question (e.g., Whitley et al. 2006). Thus, justifying yet another study of the basic reproductive biology of (to name three well-studied groups) gartersnakes (*Thamnophis*), watersnakes (*Nerodia*), or Massasaugas (*Sistrurus catenatus*) to construct a PVA is difficult. This approach might then free up time and resources for a detailed study of age- or stage-specific survival rates, data that both are limited for snakes (Parker and Plummer

1987; Rossman et al. 1996) and have a strong impact on the results of PVA models (Seigel and Sheil 1999).

Note that we are arguing not that an understanding of reproductive biology (or movement patterns or mating behavior) is unimportant to conservation, but rather, that snake biologists who justify or obtain funding for their work on the basis of its importance to conservation and management need to carefully show an explicit link between the data that are being collected and conservation and management needs. Using reproductive biology as an example again, a descriptive study of the reproductive patterns of rattlesnakes in a protected area might have only a limited link with conservation. But a study comparing the reproductive potential of one part of a rattlesnake population that is subjected to a land use change (e.g., controlled burns or land conversion) with that of another part of the rattlesnake population for which land use remains unchanged might have very strong management implications. Thus, we need to look not simply at the data being collected but at how those data will be used to further conservation goals (see also Dorcas and Willson, Chapter 1).

Prioritizing Research for Conservation

One corollary of the ideas just noted is the need for a set of research priorities as they relate to conservation. Naturally, any such attempt must be regarded as an initial step in an evolving process, but, as a first step in this direction, we have asked the authors of the chapters of this book to share with us the topics they view as their top research needs in snake conservation. What follows is a distillation of these suggestions, combined with our own views. Again, our intent is to provoke debate and discussion; these areas are listed in no particular order.

Landscape Ecology and Climate Change

Several chapters in this book (e.g., Jenkins et al., Chapter 4; Weatherhead and Madsen, Chapter 5; Shoemaker et al., Chapter 8; Beaupre and Douglas, Chapter 9) note the importance of examining snake conservation in a broader context (i.e., not as isolated populations at one point in space but, instead, as part of what is happening on a broader spatial scale). For example, the studies reviewed by Weatherhead and Madsen (Chapter 5) showing the relationship between the availability of critical nesting habitat for Black Ratsnakes and landscape changes caused by human settlement are an excellent example of this approach. Examining the short- and long-term effects of timber harvesting and other causes of habitat alteration on snake populations is another area that needs much more work (Gardner et al. 2007; Shoemaker et al., Chapter 8). Finally, the overriding impact of

climate change on everything from hibernation duration to the timing of reproduction requires much more attention than has been apparent (Weatherhead and Madsen, Chapter 6). Testing specific hypotheses regarding climate
change will no doubt be difficult, but imaginative studies (such as those by
Shine and co-workers on temperature and incubation effects) will be most
welcome (Shine and Bonnet, Chapter 6).

Impacts of Roads on Snake Populations

The impacts of roads on wild populations has received considerable recent
attention, both for snakes (Andrews and Gibbons 2005; Weatherhead and
Madsen, Chapter 5; Shine and Bonnet, Chapter 6) and for other taxa (Aresco
2005; Marsh et al. 2005; Ramp et al. 2006). Much of what we know about
snakes is reviewed in this volume (Weatherhead and Madsen, Chapter 5;
Shine and Bonnet, Chapter 6; Shoemaker et al., Chapter 8), but there are
some remaining issues that need attention. Chief among these is the direct
impact of road mortality on population viability. Although this has been addressed at least indirectly by some authors (e.g., Rosen and Lowe 1994; Sullivan 2000), the only study that we are aware of that directly addresses this
issue is Row et al. (2007). Consequently, we lack the kind of data needed to
determine whether road mortality is a critical issue for most snake populations. Despite the large numbers of snakes often seen dead on roads, it is
difficult to ask resource managers to take specific actions (e.g., road closures
and road barriers) that may be costly or politically difficult without more
concrete information.

Even as we begin to understand how snake behavior is impacted by roads
(e.g., Shine et al. 2004a; Andrews and Gibbons 2005), our understanding
of why snakes cross roads remains poor. Although this may seem trivial, it
can have important consequences. For example, imagine two very different
scenarios for why snakes cross a highway. In the first case, a population has
a hibernation site located some distance from a foraging site, with the road
in between the two areas. Individuals thus cross the road twice per year and
spend little time near the road except when moving to and from hibernation
sites (e.g., Western Rattlesnakes, *Crotalus viridis,* in Wyoming; Duvall et al.
1985). Road mortality in this case would be limited to the period of the migratory movements and might be predictable based on local climatic conditions. Thus, measures to reduce mortality would be fairly easy to implement
because it would be more acceptable to close a road for a few days per year
than for longer periods. In the second case, a population is located in an
area where the road bisects a foraging habitat, so that snakes move across
the road routinely. Snakes in this second situation would be subject to the
almost daily probability of being killed—and this would be especially severe
if drivers targeted snakes for collisions (Ashley et al. 2007). Thus, we might
predict that road mortality would have a more substantial impact in this

situation, perhaps leading to a source-sink scenario (Smith and Green 2005). It would also likely be more difficult to take management action in such a situation because mortality would occur over a much longer time interval.

Clearly, this is not an exhaustive list of possible scenarios, but the overall point remains the same—we need a better understanding of both how snake populations are impacted by roads and why snakes cross roads. Better mark-recapture data, combined with detailed radiotelemetry, seems to be an appropriate starting place for such studies.

Better Integration of Molecular Methods with Ecological and Conservation Studies

Although some field biologists may be reluctant to admit it, there is no doubt that information from molecular studies can be essential in understanding snake populations. Examples include understanding the effects of genetic drift and population fragmentation (Burbrink and Castoe, Chapter 2; King, Chapter 3), determining mating success and sexual selection (Gibbs and Weatherhead 2001), and documenting the impact of population augmentations (Madsen et al. 1999). Recent papers (albeit not on snakes) have even integrated molecular methods into such emerging fields as landscape ecology (e.g., Spear et al. 2005). We do not think it is pushing the envelope too much to suggest that many of the apparent problems faced by snake biologists may be overcome through more collaborative studies with our colleagues in the molecular fields.

Experimental Tests of Manipulative Conservation Measures

As reviewed by Shoemaker et al. (Chapter 8) and Seigel and Dodd (2000), resource managers often rely on manipulative methods to manage imperiled taxa. This includes translocations and reintroductions (see Kingsbury and Attum, Chapter 7), as well as habitat manipulations such as controlled burning and the creation of artificial hibernation sites (Shoemaker et al., Chapter 8). Unfortunately, data testing the effectiveness of such manipulative methods are extremely limited, and we echo the call made by Shoemaker et al. (Chapter 8) for increased experimental tests of such methods. We are especially eager to see the use of common, nonthreatened species as surrogates for working with endangered taxa (e.g., Dodd and Franz 1993). For example, King and Stanford (2006) recently used the relatively abundant Plains Gartersnake (*Thamnophis radix*) to test whether head-starting snakes to increase relocation success was a valid management tool. Additional, well-replicated experimental tests of manipulative conservation practices are needed badly for snakes (as for other species of reptiles and amphibians).

Better Data on Population Demography

We have already noted that the lack of detailed information on snake demography (especially survival rates) remains a serious impediment to understanding snake populations and their status. Although modern analytical tools such as PVA and the program MARK have been used widely for other groups, their use for snakes remains limited (Dorcas and Willson, Chapter 1). Despite gloomy discussions about how the "secretive" nature of snakes prevents solid work in this area, some very good population studies recently have been conducted (King and Stanford 2006). Thus, we renew earlier calls for more attention to this area of snake biology (Parker and Plummer 1987).

Other Topics

In addition to the areas already noted, here we list some of the other topics suggested to us by the authors of this book. Again, these are not listed in any particular order.

- We need research on a much broader array of snake taxa, both geographically and taxonomically. As one author noted, comparing "the relative research output on ecology of scolecophidians worldwide versus massasaugas in the northern USA [is]...scary." Note that this does not mean that work on well-studied species is not valuable, only that such research should tell us something new about snake biology or conservation.
- We need a better understanding of the cognitive capacities of snakes. This will go a long way toward designing better captive habitats as well as giving us a feel for the critical characteristics of appropriate natural habitats.
- We need a greater understanding of urban ecology and related areas. This will encourage the cohabitation of people and wildlife in managed settings.
- We still have a poor understanding of the ecology and behavior of small snakes (both small-bodied species and juveniles). Specific needs with regards to juveniles is to learn about dispersal behavior and elucidate patterns of mortality.
- We still have a very limited ability to monitor snake populations and how they change over time. Related (but not identical) to our lack of understanding of snake demography, the absence of a standardized way of monitoring population status remains a major obstacle in the management of snake populations (see Dorcas and Willson, Chapter 1).
- We need to develop better ways of using genetic markers to assess population size and compare those estimates with what we learn from mark-recapture studies. Additional genetic data are also needed for assessing landscape-level impacts of roads and other habitat disturbances.
- We need an evaluation of the effectiveness of different types of education programs for reptiles and amphibians, especially related to snakes.

Conserving Snakes for the Future

When asked about their profession, many herpetologists explain that what they do provides them the opportunity to avoid growing up. Although the likelihood that studying snakes will become the elixir of youth is, at best, tenuous, those of us who regularly work with these animals feel fortunate to be doing so. These animals fascinated us at an age when we were only beginning to understand the world's biodiversity, and they continue to do so now that we appreciate the scope of that diversity. We recognize that snakes have contributed to the evolution of their respective ecosystems (Beaupre and Douglas, Chapter 9) and that the activities associated with an ever-increasing human population threaten the animals we care most about and, by association, the ecological dynamics within a wide variety of habitats. Regardless of a person's feelings about snakes, we are certain that allowing these threats to further impact snake populations is neither ethically nor scientifically responsible.

Having just celebrated the three hundredth anniversary of his birth, we think it is unfortunate that the contributions Carolus Linnaeus made to science should be tempered by his view that snakes are "foul and loathsome" (1758). Most humans seem unable to rid themselves of this mind-set, and wariness of snakes might well have shaped the evolution of the human brain (Isbell 2006). As such, the task before anyone wishing to conserve snakes is a daunting one, but one we are convinced is worth undertaking.

There are several tactics that can be employed here.

1. Convince the public that the common perception of snakes is ill-deserved. Compared with the reptiles roaming the planet in the Mesozoic, snakes are neither large nor fierce predators. Indeed, a species fitting such a description is a rare thing among today's fauna (Colinvaux 1978). An animal as secretive as the typical snake cannot, at the same time, be perceived as ferocious. Rather, snakes are rarely aggressive, and such behavior is often displayed as a defensive strategy of last resort (i.e., if they cannot flee from their attacker).

2. Promote the awareness that a failure to conserve snake populations has significant evolutionary implications. As was recently described for amphibians (Blaustein and Bancroft 2007), snakes have several features inherent in their natural history (e.g., requirements for successful reproduction; see Shine and Bonnet, Chapter 6) that can experience only limited adaptation to changes in the surrounding environment. Given the time frame required for most evolutionary changes to occur (Mayr 1963; but, see Phillips and Shine 2006), snakes will simply fail to keep pace with the continued direct or indirect persecution of their populations. Their absence from their respective ecosystems will set into motion a cascade of ecological failure within those habitats (Schmitz et al. 2000; Terborgh et al. 2001), which ultimately will impact the quality of human life.

3. Publicize the fact that the applied benefits of snakes are too numerous to ignore. Although the alarm has already been raised (e.g., Gibbons et al. 2000), the loss of a snake species has unforeseen negative consequences. Providing cultural (Whitaker 1989), economic (Shine et al. 1999), or medical (Albuquerque et al. 1979; Bonta et al. 1979; Christensen 1979) benefits, snakes have many current positive impacts on human livelihood. If snake species are allowed to slip toward extinction, current and future benefits from human interactions with snakes will also be lost.

4. Include snakes in campaigns promoting environmental stewardship. Among those people unwilling to accept the notion that snakes offer a number of direct and indirect benefits to humans, there are probably those who identify with the plight of several taxa that have been the proverbial poster children for biodiversity conservation (e.g., pandas, sea turtles, and hyacinth macaws). If someone is willing to embrace an environmental ethic that includes all biota (Leopold 1949; Wilson 1984), then protecting snakes from being killed with a hoe should not be different from preventing baby seals from being killed with a club.

One of the conclusions we wish to impart here is that there are no reasons why snake ecologists cannot take some of the enthusiasm they apply to their studies of these animals and allot it to promoting snake conservation. Perhaps this is a theme that some readers of this book are tired of hearing, but getting the word out about conservation is never a task in which we should lose interest. As an example (and one in which conservation arguably receives more attention than in most other countries), federal and state spending for habitat conservation in the United States is woefully inadequate (Lerner et al. 2007), even when the voting public advocates such spending through open-space ballot measures (Szabo 2007).

To end on a more positive note, we are pleased by the advances in snake ecology—both methodologically and conceptually—since the second volume of this series was published. We are encouraged to see much of that work applied toward the conservation of snakes. Even more promising is the fact that the field continues to attract so many bright and enthusiastic researchers to the field. The burden of living with "lizard envy" (Seigel 1993) has been alleviated, and some of the most critically endangered snake species appear to be hanging on in spite of predictions about their extinction (Dodd 1993b). Equally clear to us is that there are so many more opportunities available to study snakes. The questions yet to be addressed in species distributed outside North America, Europe, and Australia are especially tantalizing. If you are among the ecologists fortunate enough to contribute to this body of knowledge, remember to use a fraction of the youthful vigor that you feel when searching for snakes to educate the general public about all the fascinating facets of snake biology.

References

Abbatius, B.A. 1589. *De admirabili viiperae natura, et de mirificis eiusdem facultatibus liber.* Urbino. B. Ragusio.

Adolph, S.C., and W.P. Porter. 1993. Temperature, activity, and lizard life-histories. Am. Nat. 142:273–295.

Aebischer, N.J., and P.A. Robertson. 1992. Practical aspects of compositional analysis as applied to pheasant habitat utilisation. *In* I.G. Pried and S.M. Swift, eds., Wildlife telemetry: Remote monitoring and tracking of animals, pp. 285–293. Chilchester, U.K.: Ellis Horwood Ltd.

Aebischer, N.J., P.A. Robertson, and R.E. Kenwood. 1993. Compositional analysis of habitat use from animal radio-tracking data. Ecology 74:1313–1325.

Affre, A. 2003. Trade studies in live reptiles. Traffic Bull. 19(3):12.

Aitchison, J. 1986. The statistical analysis of compositional data. London: Chapman and Hall.

Albuquerque, E.X., A.T. Eldefrawi, and M.E. Eldefrawi. 1979. The use of snake toxins for the study of the acetylcholine receptor and its ion-conductance modulator. *In* C.-Y. Lee, ed., Handbook of experimental pharmacology, vol. 52: Snake venoms, pp. 377–402. St. Louis: Sigma Chemical Co.

Aldridge, R.D. 1979. Female reproductive cycles of the snakes *Arizona elegans* and *Crotalus viridis*. Herpetologica 35:256–261.

Aldridge, R.D., and W.S. Brown. 1995. Male reproductive cycle, age at maturity, and cost of reproduction in the Timber Rattlesnake (*Crotalus horridus*). J. Herpetol. 29:399–407.

Alfaro, M.E., D.R. Karns, H.K. Voris, E. Abernathy, and S.L. Sellins. 2004. Phylogeny of *Cerberus* (Serpentes: Homalopsinae) and phylogeography of *Cerberus rynchops:* Diversification of a coastal marine snake in Southeast Asia. J. Biogeogr. 31:1277–1292.

Ali, B.A., T.H. Huang, D.N. Qin, and X.M. Wang. 2004. A review of random amplified polymorphic DNA (RAPD) markers in fish research. Rev. Fish Biol. Fish. 14:443–453.

Allen, C.R., K.G. Rice, D.P. Wojcik, and H.F. Percival. 1997. Effect of red imported fire ant envenomization on neonatal American alligators. J. Herpetol. 31:318–321.

Allen, C.R., E.A. Forys, K.G. Rice, and D.P. Wojcik. 2001. Effects of fire ants (Hymenoptera: Formicidae) on hatchling turtles and prevalence of fire ants on sea turtle nesting beaches in Florida. Fla. Entomol. 84:250–253.

Allen, E.A., and K.E. Omland. 2003. Novel intron phylogeny supports plumage convergence in orioles (*Icterus*). Auk 120:961–969.

Allendorf, F.W., and G. Luikart. 2007. Conservation and the genetics of populations. Malden, MA: Blackwell.

Almli, L.M., and G.M. Burghardt. 2006. Environmental enrichment alters the behavioral profile of ratsnakes (*Elaphe*). J. Appl. Anim. Welfare Sci. 9:85–109.

Altwegg, R., S. Dummermuth, B.R. Anholt, and T. Flatt. 2005. Winter weather affects Asp viper *Vipera aspis* population dynamics through susceptible juveniles. Oikos 110:55–66.

Anderka F.W., and P.J. Weatherhead. 1983. A radiotransmitter and implantation technique for snakes. Proc. Int. Conf. Wildl. Biotelem. 4:47–56.

Anderson, C.D. 2006. Utility of a set of microsatellite primers developed for the Massasauga Rattlesnake (*Sistrurus catenatus*) for population genetic studies of the Timber Rattlesnake (*Crotalus horridus*). Mol. Ecol. Notes 6:514–517.

Anderson, N.L., R.F. Wack, L. Calloway, T.E. Hetherington, and J.B. Williams. 1999. Cardiopulmonary effects and efficacy of propofol as an anethetic agent in brown tree snakes, *Boiga irrgularis*. Bull. Assoc. Reptil. Amphib. Vet. 9:9–15.

Andre, M. 2003. Genetic population structure by microsatellite DNA analysis of the Eastern Massasauga Rattlesnake (*Sistrurus catenatus catenatus*) at Carlyle Lake. MS thesis, Northern Illinois University, DeKalb.

Andrewartha, H.G., and L.C. Birch. 1954. The distribution and abundance of animals. Chicago: University of Chicago Press.

Andrews, K.M., and J.W. Gibbons. 2005. How do highways influence snake movement? Behavioral responses to roads and vehicles. Copeia 2005:772–782.

Arbogast, B.S., and G.J. Kenagy. 2001. Comparative phylogeography as an integrative approach to historical biogeography. J. Biogeogr. 28:819–825.

Archibald, J.K., M.E. Mort, and D.J. Crawford. 2003. Bayesian inference of phylogeny: A non-technical primer. Taxon 52:187–191.

Aresco, M.J. 2005. Mitigation measures to reduce highway mortality of turtles and other herpetofauna at a north Florida lake. J. Wildl. Manage. 69:549–560.

Aris-Brosou, S., and L. Excoffier. 1996. The impact of population expansion and mutation rate heterogeneity on DNA sequence polymorphism. Mol. Biol. Evol. 13:494–504.

Arnold, S.J. 1981. Behavioral variation in natural populations. II. The inheritance of a feeding response in crosses between geographic races of the garter snake *Thamnophis elegans*. Evolution 35:510–515.

Arnold, S.J. 1993. Foraging theory and prey-size—predator-size relations in snakes. *In* R.A. Seigel and J.T. Collins, eds., Snakes: Ecology and behavior, pp. 87–115. New York: McGraw-Hill.

Arnold, S.J., and C.R. Peterson. 2002. A model for optimal reaction norms: The case of the pregnant garter snake and her temperature-sensitive embryos. Am. Nat. 160:306–316.

Arnold, S.J., and P.C. Phillips. 1999. Hierarchical comparison of genetic variance-covariance matrices. II. Coastal-inland divergence in the garter snake, *Thamnophis elegans*. Evolution 53:1516–1527.

Ashley, E.P., and J.T. Robinson. 1996. Road mortality of amphibians, reptiles and other wildlife on the Long Point causeway, Lake Erie, Ontario. Can. Field-Nat. 110:403–412.

Ashley, E.P., A. Kosloski, and S.A. Petrie. 2007. Incidence of intentional vehicle-reptile collisions. Hum. Dimens. Wildl. 12:137–143.

Ashley, M.V., M.F. Willson, O.R W. Rergams, D.J. O'Dowd, S.M. Gende, and J.S. Brown. 2003. Evolutionarily enlightened management. Biol. Conserv. 111:115–123.

Ashton, K.G., and R.L. Burke. 2007. Long-term retention of a relocated population of gopher tortoises. J. Wildl. Manage. 71:783–787.

Ashton, K.G., and A. de Queiroz. 2001. Molecular systematics of the Western Rattlesnake, *Crotalus viridis* (Viperidae), with comments on the utility of the D-Loop in phylogenetic studies of snakes. Mol. Phylogenet. Evol. 21:176–189.

Aubret, F., X. Bonnet, R. Shine, and S. Maumelat. 2003. Clutch size manipulation, hatching success and offspring phenotype in the ball python (*Python regius*, Pythonidae). Biol. J. Linn. Soc. 78:263–272.

Aubret, F., R. Shine, and X. Bonnet. 2004. Adaptive developmental plasticity in snakes. Nature 431:261–262.

Avery, R.A. 1982. Field studies of body temperatures. *In* C. Gans and F.H. Pough, eds., Biology of the Reptilia, Vol. 12, pp. 93–166. New York: Academic Press.

Avise, J.C. 1998. The history and purview of phylogeography: A personal reflection. Mol. Ecol. 7:371–379.

Avise, J.C. 2000. Phylogeography: The history and formation of species. Cambridge, Mass.: Harvard University Press.

Avise, J.C. 2004. Molecular markers, natural history, and evolution, 2nd ed. Sunderland, Mass.: Sinauer.

Avise, J.C., and K. Wollenberg. 1997. Phylogenetics and the origin of species. Proc. Natl. Acad. Sci. U.S.A. 94:7748–7755.

Avise, J.C., J. Arnold, R.M. Ball, E. Bermingham, T. Lamb, J.E. Neigel, C.A. Reeb, and N.C. Saunders. 1987. Intraspecific phylogeography: The mitochondrial DNA bridge between population genetics and systematics. Annu. Rev. Ecol. Syst. 18:489–522.

Aymar, B., ed. 1956. Treasury of snake lore. New York: Greenberg.

Ayres, F.A., and S.J. Arnold. 1983. Behavioral variation in natural populations. IV. Mendelian models and heritability of a feeding response in the garter snake *Thamnophis elegans*. Heredity 51:405–414.

Bahlo, M., and R.C. Griffiths. 2000. Inference from gene trees in a subdivided population. Theor. Popul. Biol. 57:79–95.

Bailey, L.L., T.R. Simons, and K.H. Pollock. 2004a. Comparing population size estimators for plethodontid salamanders. J. Herpetol. 38:370–380.

Bailey, L.L., T.R. Simons, and K.H. Pollock. 2004b. Estimating detection probability parameters for plethodon salamanders using the robust capture-recapture design. J. Wildl. Manage. 68:1–13.

Bailey, L.L., T.R. Simons, and K.H. Pollock. 2004c. Estimating site occupancy and species detection probability parameters for terrestrial salamanders. Ecol. Appl. 14:692–702.

Bailey, L.L., T.R. Simons, and K.H. Pollock. 2004d. Spatial and temporal variation in detection probability of plethodon salamanders using the robust capture-recapture design. J. Wildl. Manage. 68:14–24.

Bailey, R.M. 1949. Temperature toleration of gartersnakes in hibernation. Ecology 30:238–242.

Bakken, G.S., and D.M. Gates. 1975. Heat transfer analysis of animals: Some implications for field ecology, physiology, and evolution. *In* D.M. Gates and R.B. Schmerl, eds., Perspectives of biophysical ecology, pp. 255–290. New York: Springer-Verlag.

Balmford, A. 2000. Separating fact from artifact in analyzes of zoo visitor preferences. Conserv. Biol. 14:1193–1195.

Balmford, A., G.M. Mace, and N. Leader-Williams. 1996. Designing the ark: Setting priorities for captive breeding. Conserv. Biol. 10:719–727.

Barker, D.G. 1992. Variation, infraspecific relationships and biogeography of the Ridgenose Rattlesnake, *Crotalus willardi*. *In* J.A Campbell and E.D. Brodie Jr., eds., Biology of the pitvipers, pp. 89–105. Tyler, Tex.: Selva.

Barlow, C.E. 1999. Habitat use and spatial ecology of Blanding's turtles (*Emydoidea blandingii*) and spotted turtles (*Clemmys guttata*) in northeastern Indiana. MS thesis, Purdue University, Fort Wayne.

Baskett, M.L., S.A. Levin, S.D. Gaines, and J. Dushoff. 2005. Marine reserve design and the evolution of size at maturation in harvested fish. Ecol. Appl. 15:882–901.

Bauerle, B., D.L. Spencer, and W. Wheeler. 1975. The use of snakes as a pollution indicator species. Copeia 1975:366–368.

Bearhop, S., C.E. Adams, S. Waldron, R.A. Fuller, and H. Macleod. 2004. Determining trophic niche width: A novel approach using stable isotope analysis. J. Anim. Ecol. 73:1007–1012.

Beaumont, M.A. 1999. Detecting population expansion and decline using microsatellites. Genetics 153:2013–2029.

Beaupre, S.J. 1995a. Comparative ecology of the Mottled Rock Rattlesnake, *Crotalus lepidus*, in Big Bend National Park. Herpetologica 51:45–56.

Beaupre, S.J. 1995b. Effects of geographically variable thermal environment on bioenergetics of Mottled Rock Rattlesnakes. Ecology 76:1655–1665.

Beaupre, S.J. 1996. Field metabolic rate, water flux, and energy budgets of Mottled Rock Rattlesnakes, *Crotalus lepidus*, from two populations. Copeia 1996:319–329.

Beaupre, S.J. 2002. Modeling time-energy allocation in vipers: Individual responses to environmental variation and implications for populations. In G.W. Schuett, M. Höggren, M.E. Douglas, and H.W. Greene, eds., Biology of the vipers, pp. 463–481. Eagle Mountain, Utah: Eagle Mountain Publishing.

Beaupre, S.J. 2008. Annual variation in time-energy allocation by Timber Rattlesnakes (*Crotalus horridus*) in relation to food acquisition. In W.K. Hayes, K.R. Beaman, M.D. Cardwell, and S.P. Bush, eds., The biology of rattlesnakes, pp. 111–122. Loma Linda, Calif.: Loma Linda University Press.

Beaupre, S.J., and R.W. Beaupre. 1994. An inexpensive data-collection system for temperature telemetry. Herpetologica 50:509–516.

Beaupre, S.J., and D.J. Duvall. 1998a. Integrative biology of rattlesnakes: Contributions to biology and evolution. BioScience 48:531–538.

Beaupre, S.J., and D.J. Duvall. 1998b. Variation in oxygen consumption of the Western Diamondback Rattlesnake (*Crotalus atrox*): Implications for sexual size dimorphism. J. Comp. Physiol. B Biochem. Syst. Environ. Physiol. 168:497–506.

Beaupre, S.J., and F.I. Zaidan. 2001. Scaling of CO_2 production in the Timber Rattlesnake (*Crotalus horridus*), with comments on cost of growth in neonates and comparative patterns. Physiol. Biochem. Zool. 74:757–768.

Beçek, M.L., N.N. Rabello-Gay, W. Beçek, M. Soma, R.F. Batistic, and I. Trajtengertz. 1990. The W chromosome during the evolution and in sex abnormalities of snakes, DNA content, C-banding. In E. Olmo, ed., Cytogenetics of amphibians and reptiles, pp. 221–240. Basel: Birkhäuser Verlag.

Bechtel, H.B. 1995. Reptile and amphibian variants: Colors, patterns, and scales. Malabar, Fla.: Krieger.

Beck, B.B., L.G. Rapaport, M.R. Stanley Price, and A.C. Wilson. 1994. Reintroduction of captive born animals. In P.J.S. Olney, G.M., Mace, and A.T.C. Fiestner, eds., Creative conservation: Interactive management of wild and captive animals, pp. 265–286. London: Chapman & Hall.

Beebee, T.J.C. 1995. Amphibian breeding and climate. Nature 374:219–220.

Beerli, P. 1998. Estimation of migration and population sizes in geographically structured populations. In G.E. Carvalho, ed., Advances in molecular ecology, pp. 39–53. Amsterdam: IOS Press.

Beerli, P. 2006. Comparison of Bayesian and maximum likelihood inference of population genetic parameters. Bioinformatics 22:341–345.

Beerli, P., and J. Felsenstein. 1999. Maximum likelihood estimation of migration rates and effective population numbers in two populations using a coalescent approach. Genetics 152:763–773.

Beerli, P., and J. Felsenstein. 2001. Maximum likelihood estimation of a migration matrix and effective population sizes in n subpopulations by using a coalescent approach. Proc. Natl. Acad. Sci. U.S.A. 98:4563–4568.

Beissinger, S.R., and M.I. Westphal. 1998. On the use of demographic models of population viability analysis in endangered species management. J. Wildl. Manage. 62:821–841.

Bellemin, J., G. Adest, G.C. Gorman, and M. Aleksiuk. 1978. Genetic uniformity in northern populations of *Thamnophis sirtalis* (Serpentes: Colubridae). Copeia 1978:150–151.

Bensch, S., and M. Åkesson. 2005. Ten years of AFLP in ecology and evolution: Why so few animals? Mol. Ecol. 14:2899–2914.

Berglind, S.A. 2000. Demography and management of relict sand lizard *Lacerta agilis* populations on the edge of extinction. Ecol. Bull. 48:123–142.

Berkhoudt, K. 2003. Hot trade in cool creatures: Live reptile trade in the European Union. TRAFFIC dispatches, February 2003, no. 20.

Bermingham, E., and C. Moritz. 1998. Comparative phylogeography: Concepts and applications. Mol. Ecol. 7:367–369.

Bernardino, F.S., Jr., and G.H. Dalrymple. 1992. Seasonal activity and road mortality of the snakes of the Pa-hay-okee wetlands of Everglades National Park, USA. Biol. Conserv. 62:71–75.

Berry, O., M.D. Tocher, and S.D. Sarre. 2004. Can assignment tests measure dispersal? Mol. Ecol. 13:551–561.

Berthier, P., M.A. Beaumont, J-M. Cornuet, and G. Luikart. 2002. Likelihood-based estimation of the effective population size using temporal changes in allele frequencies: A genealogical approach. Genetics 160:741–751.

Bertorelle, G., and M. Slatkin. 1995. The number of segregating sites in expanding human populations, with implications for estimates of demographic parameters. Mol. Biol. Evol. 12:887–892.

Bierlein, J.F. 1994. Parallel myths. New York: Ballantine Wellspring.

Bieser, N.D. 2008. Spatial ecology and survival of resident juvenile and headstarted Eastern Massasauga (*Sistrurus catenatus catenatus*)in northern Michigan. MS thesis, Purdue University, Fort Wayne, Indiana.

Biggins, D.E., J.L. Godbey, L.R. Hanebury, B. Luce, P.E. Marinari, M.R. Matchett, and A. Vargas. 1998. The effect of rearing methods on survival of reintroduced black-footed ferrets. J. Wildl. Manage. 62:643–653.

Birky, C.W., Jr. 1991. Evolution and population genetics of organelle genes: Mechanisms and models. *In* R.K. Selander, A.G. Clark, and T.S. Whittam, eds., Evolution at the molecular level, pp. 112–134. Sunderland, Mass.: Sinauer.

Bitgood, S., D. Patterson, and A. Benefield. 1986. Understanding your visitors: Ten factors that influence visitor behavior. AAZPA Annu. Conf. Proc. 1986:726–743.

Bittner, T.D. 2000. The evolutionary significance of melanism in the Common Garter Snake, *Thamnophis sirtalis*. PhD dissertation, Northern Illinois University, DeKalb.

Bittner, T.D. 2003. Polymorphic clay models of *Thamnophis sirtalis* suggest patterns of avian predation. Ohio J. Sci. 103:62–66.

Bittner, T.D., and R.B. King. 2003. Gene flow and melanism in garter snakes revisited: A comparison of molecular markers and island vs. coalescent models. Biol. J. Linn. Soc. 79:389–399.

Bittner, T.D., R.B. King, and J. Kerfin. 2002. Effects of body size and melanism on the thermal biology of garter snake (*Thamnophis sirtalis*). Copeia 2002:477–482.

Blanchard, B.M., and R.R. Knight. 1995. Biological consequences of relocating grizzly bears in the Yellowstone ecosystem. J. Wildl. Manage. 59:560–565.

Blanchard, F., and F. Blanchard. 1940. The inheritance of melanism in the garter snake *Thamnophis sirtalis sirtalis* (Linnaeus) and some further evidence of effective autumn mating. Mich. Acad. Sci. Arts Lett. 26:177–193.

Blaustein, A.R., and B.A. Bancroft. 2007. Amphibian population declines: Evolutionary considerations. BioScience 57:437–444.

Blaustein, A.R., L.K. Belden, D.H. Olson, D.M. Green, T.L. Root, and J.M. Kiesecker. 2001. Amphibian breeding and climate change. Conserv. Biol. 15:1804–1809.

Block, W.M., L.A. Brennan, and R.J. Gutierrez. 1986. The use of guilds and guild-indicator species for assessing habitat suitability. *In* J. Verner, M.L. Morrison, and C.J. Ralph, eds., Wildlife 2000: Modeling habitat relationships of terrestrial vertebrates, pp. 109–114. Madison: University of Wisconsin Press.

Blomquist, S.M., J.D. Zydlewski, and M.L. Hunter Jr. 2008. Efficacy of PIT tags of tracking the terrestrial anurans *Rana pipiens* and *Rana sylvatica*. Herpetol. Rev. 39:174–179.

Blouin, M.S. 2003. DNA-based methods for pedigree reconstruction and kinship analysis in natural populations. Trends Ecol. Evol. 18:503–511.

Blouin-Demers, G., and H.L. Gibbs. 2003. Isolation and characterization of microsatellite loci in the Black Rat Snake (*Elaphe obsoleta*). Mol. Ecol. Notes 3:98–99.

Blouin-Demers, G., and P.J. Weatherhead. 2000. A novel association between a beetle and a snake: Parasitism of *Elaphe obsoleta* by *Nicrophorus pustulatus*. Ecoscience 7:395–397.

Blouin-Demers, G., and P.J. Weatherhead. 2001a. An experimental test of the link between foraging, habitat selection and thermoregulation in Black Rat Snakes *Elaphe obsoleta obsoleta*. J. Anim. Ecol. 70:1006–1013.

Blouin-Demers, G., and P.J. Weatherhead. 2001b. Habitat use by Black Rat Snakes (*Elaphe obsoleta obsoleta*) in fragmented forests. Ecology 82:2882–2896.

Blouin-Demers, G., and P.J. Weatherhead. 2002a. Habitat-specific behavioural thermoregulation by Black Rat Snakes (*Elaphe obsoleta obsoleta*). Oikos 97:59–68.

Blouin-Demers, G., and P.J. Weatherhead. 2002b. Implications of movement patterns for gene flow in Black Rat Snakes (*Elaphe obsoleta*). Can. J. Zool. 80:1162–1172.

Blouin-Demers, G., K.J. Kissner, and P.J. Weatherhead. 2000. Plasticity in preferred body temperature of young snakes in response to temperature during development. Copeia 2000:841–845.

Blouin-Demers, G., P.J. Weatherhead, and J. R. Row. 2004. Phenotypic consequences of nest-site selection in Black Rat Snakes (*Elaphe obsoleta*). Can. J. Zool. 82:449–456.

Blouin-Demers, G., H.L. Gibbs, and P.J. Weatherhead. 2005. Genetic evidence for sexual selection in Black Ratsnakes, *Elaphe obsoleta*. Anim. Behav. 69:225–234.

Bloxam, Q.M.C., and S.J. Tonge. 1995. Amphibians: Suitable candidates for breeding release programs. Biodivers. Conserv. 4:636–644.

Boarman, W.I., M.L. Beigel, G.C. Goodlett, and M. Sazaki. 1998. A passive integrated transponder system for tracking animal movements. Wildl. Soc. Bull. 26:886–891.

Bodkin, J.L., B.E. Ballachey, M.A. Cronin, and K.T. Scribner. 1999. Population demographics and genetic diversity in remnant and translocated populations of sea otters. Conserv. Biol. 13:1378–1385.

Bollback, J.P. 2002. Bayesian model adequacy and choice in phylogenetics. Mol. Biol. Evol. 19:1171–1180.

Bond, J.M., R. Porteous, S. Husghe, R.G. Mogg, M.G. Gardner, and J.C. Reading. 2005. Polymorphic microsatellite markers, isolated using a simple enrichment procedure, in the threatened Smooth Snake (*Coronella austriaca*) Mol. Ecol. Notes 5:42–44.

Bonnet, X., and G. Naulleau. 1996. Catchability in snakes: Consequences for estimates of breeding frequency. Can. J. Zool. 74:233–239.

Bonnet, X., D. Bradshaw, R. Shine, and D. Pearson. 1999a. Why do snakes have eyes? The (non-) effect of blindness in island Tigersnakes. Behav. Ecol. Sociobiol. 46:267–272.

Bonnet, X., G. Naulleau, and R. Shine. 1999b. The dangers of leaving home: Dispersal and mortality in snakes. Biol. Conserv. 89:39–50.

Bonnet, X., G. Naulleau, R. Shine, and O. Lourdais. 1999c. What is the appropriate timescale for measuring costs of reproduction in a "capital breeder" such as the aspic viper? Evol. Ecol. 13:485–497.

Bonnet, X., G. Naulleau, R. Shine, and O. Lourdais. 2001. Short-term versus long-term effects of food intake on reproductive output in a viviparous snake, *Vipera aspis*. Oikos 92:297–308.

Bonnet, X., O. Lourdais, R. Shine, and G. Naulleau. 2002a. Reproduction in a typical capital breeder: Costs, currencies, and complications in the aspic viper. Ecology 83:2124–2135.

Bonnet, X., D. Pearson, M. Ladyman, O. Lourdais, and D. Bradshaw. 2002b. "Heaven" for serpents? A mark-recapture study of Tiger Snakes (*Notechis scutatus*) on Carnac Island, Western Australia. Austral Ecol. 27:442–450.

Bonnet, X., R. Shine, O. Lourdais, and G. Naulleau. 2003. Measures of reproductive allometry are sensitive to sampling bias. Funct. Ecol. 17:39–49.

Bonta, I.L., B.B. Vargaftig, and G.M. Böhm. 1979. Snake venoms as an experimental tool to induce and study models of microvessel damage. In C.-Y. Lee, ed., Handbook of experimental pharmacology, vol. 52: Snake venoms, pp. 629–683. St. Louis: Sigma Chemical Co.

Boonstra, R., and T.R. Redhead. 1994. Population dynamics of an outbreak population of house mice (Mus domesticus) in the irrigated rice-growing area of Australia. Wildl. Res. 21:583–598.

Borkovec, T.D., and W.E. Craighead. 1971. The comparison of two methods of assessing fear and avoidance behavior. Behav. Res. Ther. 9:285–291.

Bouskila, A. 1995. Interactions between predation risk and competition: A field study of kangaroo rats and snakes. Ecology 76:165–178.

Bowers, B.B., and G.M. Burghardt. 1992. The scientist and the snake: Relationships with reptiles. In H. Davis and D. Balfour, eds., The inevitable bond: Examining scientist-animal interactions, pp. 250–263. Cambridge, U.K.: Cambridge University Press.

Bowers, C.F., H.G. Hanlin, D.C. Guynn Jr., J.P. McLendon, and J.R. Davis. 2000. Herpetofaunal and vegetational characterization of a thermally-impacted stream at the beginning of restoration. Ecol. Eng. 15:S101–S114.

Boyce, M.S. 1992. Population viability analysis. Annu. Rev. Ecol. Syst. 23:481–506.

Boyd, I.L., and A.W.A. Murray. 2001. Monitoring a marine ecosystem using responses of upper trophic level predators. J. Anim. Ecol. 70:747–760.

Bradley, N.L., A.C. Leopold, J. Ross, and W. Huffaker. 1999. Phenological changes reflect climate change in Wisconsin. Proc. Nat. Acad. Sci. U.S.A. 96:9701–9704.

Branch, W.R. 1998. A field guide to the snakes and other reptiles of southern Africa. Cape Town: Struik.

Brandley, M.C., A. Schmitz, and T.W. Reeder. 2005. Partitioned Bayesian analyses, partition choice, and the phylogenetic relationships of scincid lizards. Syst. Biol. 54:373–390.

Braverman, J.M., R.R. Hudson, N.L. Kaplan, C.H. Langley, and W. Stephan. 1995. The hitchhiking effect on the site frequency spectrum of DNA polymorphisms. Genetics 140:783–796.

Brawn, J.D., S.K. Robinson, and F.R. Thompson III. 2001. The role of disturbance in the ecology and conservation of birds. Annu. Rev. Ecol. Syst. 32:251–276.

Breiman, L., J.H. Friedman, R.A. Olshen, and C.J. Stone. 1984. Classification and regression trees. Boca Raton, Fla.: Chapman and Hall.

Bright, P.W., and P.A. Morris. 1994. Animal translocation for conservation: Performance of dormice in relation to release methods, origin, and season. J. Appl. Ecol. 31:699–708.

Brito, J. 2004. Feeding ecology of Vipera latastei in northern Portugal: Ontogenetic shifts, prey size and seasonal variation. Herpetol. J. 14:13–19.

Broderip, W.J. 1835. Observations on the habits of a male chimpanzee. Proc. Zool. Soc. Lond. 1835:160–168.

Brodie, E.D., III, and T. Garland Jr. 1993. Quantitative genetics of snake populations. In R.A. Seigel and J.T. Collins, eds., Snakes: Ecology and behavior, pp. 315–362. New York: McGraw-Hill.

Brodie, E.D., Jr., B.J. Ridenhour, and E.D. Brodie III. 2002. The evolutionary response of predators to dangerous prey: Hotspots and coldspots in the geographic mosaic of coevolution between garter snakes and newts. Evolution 56:2067–2082.

Bronikowski, A.M. 2000. Experimental evidence for the adaptive evolution of growth rate in the garter snake Thamnophis elegans. Evolution 54:1760–1767.

Bronikowski, A.M., and S.J. Arnold. 1999. The evolutionary ecology of life history variation in the garter snake Thamnophis elegans. Ecology 80:2314–2325.

Bronikowski, A.M., and S.J. Arnold. 2001. Cytochrome b phylogeny does not match subspecific classification in the western Terrestrial Garter Snake, Thamnophis elegans. Copeia 2001:508–513.

Brook, B.W., J.J. O'Grady, A.P. Chapman, M.A. Burgman, H.R. Akcakaya, and R. Frankham. 2000. Predictive accuracy of population viability analysis in conservation biology. Nature 404:385–387.

Brook, B.W., M.A. Burgman, H.R. Akcakaya, J.J. O'Grady, and R. Frankham. 2002. Critiques of PVA ask the wrong questions: Throwing the heuristic baby out with the numerical bath water. Conserv. Biol. 16:262–263.

Brooks, R.J., G.P. Brown, and D.A. Galbraith. 1991. Effects of a sudden increase in natural mortality of adults on a population of the common snapping turtle (*Chelydra serpentina*). Can. J. Zool. 69:1314–1320.

Broughton, R.E., and R.G. Harrison. 2003. Nuclear gene genealogies reveal historical, demographic and selective factors associated with speciation in field crickets. Genetics 163:1389–1401.

Brown, C., and R.L. Day. 2002. The future of stock enhancements: Lessons for hatchery practice from conservation biology. Fish Fish. (Oxford) 3:79–94.

Brown, F., and J. McDonald. 2000. The serpent handlers: Three families and their faith. Winston-Salem, N.C.: John F. Blair.

Brown, G.P., and R. Shine. 2004. Maternal nest-site choice and offspring fitness in a tropical snake (*Tropidonophis mairii*, Colubridae). Ecology 85:1627–1634.

Brown, G.P., and R. Shine. 2005a. Female phenotype, life history, and reproductive success in free-ranging snakes (*Tropidonophis mairii*). Ecology 86:2763–2770.

Brown, G.P., and R. Shine. 2005b. Nesting snakes (*Tropidonophis mairii*, Colubridae) selectively oviposit in sites that provide evidence of previous successful hatching. Can. J. Zool. 83:1134–1137.

Brown, G.P., and R. Shine. 2007. Like mother, like daughter: Inheritance of nest site location in snakes. Biol. Lett. 3:131–133.

Brown, G.P., and P.J. Weatherhead. 1999. Demography and sexual size dimorphism in Northern Water Snakes, *Nerodia sipedon*. Can. J. Zool. 77:1358–1366.

Brown, G.P., and P.J. Weatherhead. 2000. Thermal ecology and sexual size dimorphism in Northern Water Snakes, *Nerodia sipedon*. Ecol. Monogr. 70:311–330.

Brown, W.S. 1991. Female reproductive ecology in a northern population of the Timber Rattlesnake, *Crotalus horridus*. Herpetologica 47:101–115.

Brown, W.S. 1993. Biology, status and management of the Timber Rattlesnake (*Crotalus horridus*): A guide for conservation. Herpetol. Circ. no. 22:1–78.

Brown, W.S., and W.S. Parker. 1976. A ventral scale clipping system for permanently marking snakes (Reptilia, Serpentes). J. Herpetol. 10:247–249.

Brown, W.S., D.W. Pyle, K.R. Greene, and J.B. Friedlander. 1982. Movements and temperature relationships of Timber Rattlesnakes (*Crotalus horridus*) in northeastern New York. J. Herpetol. 16:151–161.

Browning, D.M., S.J. Beaupre, and L. Duncan. 2005. Using partitioned Mahalanobis $D^2(K)$ to formulate a GIS-based model of Timber Rattlesnake hibernacula. J. Wildl. Manage. 69:33–44.

Brumfield, R.T., P. Beerli, D.A. Nickerson, and S.V. Edwards. 2003. The utility of single nucleotide polymorphisms in inferences of population history. Trends Ecol. Evol. 18:249–256.

Buckland, S.T., D.R. Anderson, K.P. Burnham, J.L. Laake, D.L. Borchers, and L. Thomas. 2001. Introduction to distance sampling: Estimating abundance of biological populations. New York: Oxford University Press.

Buckland, S.T., D.R. Anderson, K.P. Burnham, J.L. Laake, D.L. Borchers, and L. Thomas. 2004. Advanced distance sampling: Estimating abundance of biological populations. Oxford: Oxford University Press.

Buhlmann, K.A., and G. Coffman. 2001. Fire ant predation of turtle nestlings and implications for the strategy of delayed emergence. J. Elisha Mitchell Sci. Soc. 117:94–100.

Burbrink, F.T. 2001. Systematics of the Eastern Ratsnake complex (*Elaphe obsoleta*). Herpetol. Monogr. 15:1–53.

Burbrink, F.T. 2002. Phylogeographic analysis of the Cornsnake (*Elaphe guttata*) complex as inferred from maximum likelihood and Bayesian analyses. Mol. Phylogenet. Evol. 25:465–476.

Burbrink, F.T., and R. Lawson. 2007. How and when did Old World ratsnakes disperse into the New World? Mol. Phylogenet. Evol. 43:173–189.

Burbrink, F.T., R. Lawson, and J.B. Slowinski. 2000. Mitochondrial DNA phylogeography of the polytypic North American Rat Snake (*Elaphe obsoleta*): A critique of the subspecies concept. Evolution 54:2107–2218.

Burbrink, F.T., F. Fontanella, R.A. Pyron, T.J. Guiher, and C. Jimenez. 2008. Phylogeography across a continent: The evolutionary and demographic history of the North American racer (Serpentes: Colubridae: *Coluber constrictor*). Mol. Phylogenet. Evol. 46:484–502.

Burgdorf, S.J., D.C. Rudolph, R.N. Conner, D. Saenz, and R.R. Schaefer. 2005. A successful trap design for capturing large terrestrial snakes. Herpetol. Rev. 36:421–424.

Burger, J. 1998a. Antipredator behaviour of hatchling snakes: Effects of incubation temperature and simulated predators. Anim. Behav. 56:547–553.

Burger, J. 1998b. Effects of incubation temperature on hatchling Pine Snakes: Implications for survival. Behav. Ecol. Sociobiol. 43:11–18.

Burger, J. 2001. The behavioral response of basking Northern Water (*Nerodia sipedon*) and Eastern Garter (*Thamnophis sirtalis*) Snakes to pedestrians in a New Jersey park. Urban Ecosyst. 5:119–129.

Burger, J., and R.T. Zappalorti. 1986. Nest site selection by Pine Snakes, *Pituophis melanoleucus*, in the New Jersey pine barrens. Copeia 1986:116–121.

Burger, J., and R.T. Zappalorti. 1988a. Effects of incubation temperature on sex ratios in Pine Snakes: Differential vulnerability of males and females. Am. Nat. 132:492–505.

Burger, J., and R.T. Zappalorti. 1988b. Habitat use in free-ranging Pine Snakes, *Pituophis melanoleucus*, in New Jersey pine barrens. Herpetologica 44:48–55.

Burger, J., and R.T. Zappalorti. 1991. Nesting behavior of Pine Snakes (*Pituophis m. melanoleucus*) in the New Jersey pine barrens. J. Herpetol. 25:152–160.

Burger, J., R.T. Zappalorti, J. Dowdell, T. Georgiadis, J. Hill, and M. Gochfeld. 1992. Subterranean predation on Pine Snakes. J. Herpetol. 26:259–263.

Burghardt, G.M., and J.M. Schwartz. 1999. Geographic variations on methodological themes in comparative ethology: A snake perspective. *In* S.A. Foster and J.A. Endler, eds., Geographic variation in behavior: Perspectives on evolutionary mechanisms, pp. 69–94. New York: Oxford University Press.

Burghardt, G.M., J.S. Placyk, G.S. Casper, R.L. Small, and K. Taylor. 2006. Genetic structure of Great Lakes region *Thamnophis butleri* and *Thamnophis radix* based on mtDNA sequence data: Conservation implications for Wisconsin Butler's Gartersnake. Report to the Wisconsin Department of Natural Resources, Madison.

Burke, R.L. 1991. Relocations, repatriations, and translocations of amphibians and reptiles: Taking a broader view. Herpetologica 47:350–357.

Burnham, K.P., and D.R. Anderson. 2002. Model selection and multimodel inference: A practical information-theoretic approach, 2nd ed. New York: Springer-Verlag.

Burns, E.L., and B.A. Houlden. 1999. Isolation and characterization of microsatellite markers in the Broad-headed Snake *Hoplocephalus bungaroides*. Mol. Ecol. 8:520–521.

Busack, S.D. 1986. Biogeographic analysis of the herpetofauna separated by the formation of the Strait of Gibraltar. Nat. Geogr. Res. 2:17–36.

Busch, J.D., P.M. Waser, and J.A. DeWoody. 2007. Recent demographic bottlenecks are not accompanied by a genetic signature in banner-tailed kangaroo rats (*Dipodomys spectabilis*). Mol. Ecol. 16:2450–2463.

Bushar, L.M., H.K. Reinert, and L. Gelbert. 1998. Genetic variation and gene flow within and between local populations of the Timber Rattlesnake, *Crotalus horridus*. Copeia 1998:411–422.

Bushar, L.M., M. Maliga, and H.K. Reinert. 2001. Cross-species amplification of *Crotalus horridus* microsatellites and their application in phylogenetic analysis. J. Herpetol. 35:532–537.

Butler, H., B. Malone, and N. Clemann. 2005. Activity patterns and habitat preferences of translocated and resident Tiger Snakes (*Notechis scutatus*) in a suburban landscape. Wildl. Res. 32:157–163.

Callicott, J.B., and K. Mumford. 1997. Ecological sustainability as a conservation concept. Conserv. Biol. 11:32–40.

Callicott, J.B., L.B. Crowder, and K. Mumford. 1999. Current normative concepts in conservation. Conserv. Biol. 13:22–35.

Callicott, J.B., L.B. Crowder, and K. Mumford. 2000. Normative concepts in conservation biology: Reply to Willers and Hunter. Conserv. Biol. 14:575–578.

Calow, P. 2000. Critics of ecosystem health misrepresented. Ecosyst. Health 6:3–4.

Camin, J.H., and P.R. Ehrlich. 1958. Natural selection in water snakes (*Natrix sipedon* L.) on islands in Lake Erie. Evolution 12:504–511.

Camin, J.H., C.A. Triplehorn, and H.J. Walter. 1954. Some indications of survival value in the type "A" pattern of the island water snakes of Lake Erie. Chicago Acad. Sci. Nat. Hist. Misc. 131:1–3.

Campbell, E.W., III. 1999. Barriers to movements of the Brown Tree Snakes (*Boiga irregularis*). *In* G.H. Rodda, Y. Sawai, D. Chizar, and H. Tanaka, eds., Problem snake management: The habu and the Brown Tree Snake, pp. 306–312. Ithaca: Cornell University Press.

Campbell, K.R., T.S. Campbell, and J. Burger. 2005. Heavy metal concentrations in Northern Water Snakes (*Nerodia sipedon*) from East Fork Poplar Creek and the Little River, East Tennessee, USA. Arch. Environ. Contam. Toxicol. 49:239–248.

Camper, J.D., and J.R. Dixon. 1988. Evaluation of a microchip system for amphibians and reptiles. Texas Parks and Wildlife Dept., Research Publication 7100-159, pp. 1–22.

Cardozo, G., P.C. Rivera, M. Lamfri, M. Scavuzzo, C. Gardenal, and M. Chiaraviglio. 2007. Effects of habitat loss on the genetic structure of populations of the Argentine boa constrictor, *Boa constrictor constrictor. In* R.W. Henderson and R. Powell, eds., Biology of boas and pythons, pp. 329–338. Eagle Mountain, Utah: Eagle Mountain Publishing.

Carfagno, G.L.F., and P.J. Weatherhead. 2006. Intraspecific and interspecific variation in use of forest-edge habitat by snakes. Can. J. Zool. 84:1440–1452.

Carfagno, G.L.F., E.J. Heske, and P.J. Weatherhead. 2006. Does mammalian prey abundance explain forest-edge use by snakes? Ecoscience 13:293–297.

Carlsson, M., and H. Tegelström. 2002. Phylogeography of Adders (*Vipera berus*) from Fennoscandia. *In* G.W. Schuett, M. Höggren, M.E. Douglas, and H.W. Greene, eds., Biology of the vipers, pp. 1–10. Eagle Mountain, Utah: Eagle Mountain Publishing.

Carlsson, M., M. Isaksson, M. Höggren, and H. Tegelström. 2003. Characterization of polymorphic microsatellite markers in the Adder, *Vipera berus*. Mol. Ecol. Notes 3:73–75.

Carlsson, M., L. Soderber, and H. Tegelström. 2004. The genetic structure of Adders (*Vipera berus*) in Fennoscandia: Congruence between different kinds of genetic markers. Mol. Ecol. 13:3147–3152.

Caro, T. 1998. The significance of behavioral ecology for conservation biology. *In* T. Caro, ed., Behavioral ecology and conservation biology, pp. 3–26, New York: Oxford University Press.

Carpenter, R.A. 1996. Ecology should apply to ecosystem management: A comment. Ecol. Appl. 6:1373–1377.

Carranza, S., E.N. Arnold, E. Wadec, and S. Fahdd. 2004. Phylogeography of the false smooth snakes, *Macroprotodon* (Serpentes, Colubridae): Mitochondrial DNA sequences show European populations arrived recently from Northwest Africa. Mol. Phylogenet. Evol. 33:523–532.

Carranza, S., E.N. Arnold, and J.M. Pleguezuelos. 2006. Phylogeny, biogeography, and evolution of two Mediterranean snakes, *Malpolon monspessulanus* and *Hemorrhois hippocrepis* (Squamata, Colubridae), using mtDNA sequences. Mol. Phylogenet. Evol. 40:532–546.

Carstens, B.C., and L.L. Knowles. 2007. Estimating species phylogeny from gene-tree probabilities despite incomplete lineage sorting: An example from *Melanoplus* grasshoppers. Syst. Biol. 56:400–411.

Castelloe, J., and A.R. Templeton. 1994. Root probabilities for intraspecific gene trees under neutral coalescent theory. Mol. Phylogenet. Evol. 2:102–113.

Castoe, T.A., and C.L. Parkinson. 2006. Bayesian mixed models and the phylogeny of pitvipers (Serpentes: Viperidae). Mol. Phylogenet. Evol. 39:91–110.

Castoe, T.A., P.T. Chippindale, J.A. Campbell, L.A. Ammerman, and C.L. Parkinson. 2003. The evolution and phylogeography of the Middle American jumping pitvipers, genus *Atropoides*, based on mtDNA sequences. Herpetologica 59:421–432.

Castoe, T.A., T.M. Doan, and C.L. Parkinson. 2004. Data partitions and complex models in Bayesian analysis: The phylogeny of gymnophthalmid lizards. Syst. Biol. 53:448–469.

Castoe, T.A., M. Sasa, and C.L. Parkinson. 2005. Modeling nucleotide evolution at the mesoscale: The phylogeny of the neotropical pitvipers of the *Porthidium* group (Viperidae: Crotalinae). Mol. Phylogenet. Evol. 37:881–898.

Castoe, T.A., E.N. Smith, R.M. Brown, and C.L. Parkinson. 2007a. Higher-level phylogeny of Asian and American Coralsnakes, their placement within the Elapidae (Squamata), and the systematic affinities of the enigmatic Asian Coralsnake *Hemibungarus calligaster*. Zool. J. Linn. Soc. 151:809–831.

Castoe, T.A., C.L. Spencer, and C.L. Parkinson. 2007b. Phylogeographic structure and historical demography of the Western Diamondback Rattlesnake (*Crotalus atrox*): A perspective on North American desert biogeography. Mol. Phylogenet. Evol. 42:193–212.

Castoe, T.A., J.M. Daza, E.N. Smith, M. Sasa, U. Kuch, P.T. Chippindale, J.A. Campbell, and C.L. Parkinson. 2008. Comparative phylogeography of pitvipers suggests a consensus of ancient Middle American highland biogeography. J. Biogeogr. 35:22–47.

Cattaneo, A. 1975. *Presenzza di* Elaphe longissima longissima *(Laurenti, 1768) melanica a Castelfusano (Roma)*. Atti della Societa Italiana di Scienze Naturali e del Museo Civico di Storia Naturale di Milano 116:251–262.

Caughley, G. 1977. Analysis of vertebrate populations. New York: John Wiley & Sons.

Caughley, G. 1994. Directions in conservation biology. J. Anim. Ecol. 63:215–244.

Cavalli-Sforza, L.L., and A.W.F. Edwards. 1964. Analysis of human evolution. Genetics Today 3:923–933.

Cavitt, J.F. 2000. Fire and a tallgrass prairie reptile community: Effects on relative abundance and seasonal activity. J. Herpetol. 34:12–20.

Charas, M. 1677. New experiments on vipers. London: Mark Pardoe.

Charland, M.B. 1995. Thermal consequences of reptilian viviparity—thermoregulation in gravid and nongravid garter snakes (*Thamnophis*). J. Herpetol. 29:383–390.

Charland, M.B., and P.T. Gregory. 1989. Feeding rate and weight gain in postpartum rattlesnakes: Do animals that eat more always grow more? Copeia 1989:211–214.

Charland, M.B., and P.T. Gregory. 1990. The influence of female reproductive status on thermoregulation in a viviparous snake, *Crotalus viridis*. Copeia 1990:1089–1098.

Chase, M.K., W.B. Kristan III, A.J. Lynam, M.V. Price, and J.T. Rotenberry. 2000. Single species as indicators of species richness and composition in California coastal sage scrub birds and small mammals. Conserv. Biol. 14:474–487.

Chen, C., E. Durand, F. Forbes, and O. Francois. 2007. Bayesian clustering algorithms ascertaining spatial population structure: A new computer program and a comparison study. Mol. Ecol. Notes 7:747–756.

Cheney, D.L., and R.M. Seyfarth. 1990. How monkeys see the world. Chicago: University of Chicago Press.

Chenna, R.H.S., T. Koike, R. Lopez, T.J. Gibson, D.G. Higgins, and J.D. Thompson. 2003. Multiple sequence alignment with the Clustal series of programs. Nucleic Acids Res. 31:3497–3500.

Chijiwa, T., M. Deshimaru, I. Nobuhisa, M. Nakai, T. Ogawa, N. Oda, K. Nakashima, Y. Fukumaki, Y. Shimohigashi, S. Hattori, and M. Ohno. 2000. Regional evolution of venom-gland phospholipase A₂ isoenzymes of *Trimeresurus flavoviridis* snakes in the southwestern islands of Japan. Biochem. J. 347:491–499.

Chisholm, I.M., and W.A. Hubert. 1985. Expulsion of dummy transmitters by rainbow trout. Trans. Am. Fish. Soc. 114:766–767.

Chiszar, D., H.M. Smith, and C.W. Radcliffe. 1993. Zoo and laboratory experiments on the behavior of snakes: Assessment of competence in captive-raised animals. Am. Zool. 33:109–116.

Christensen, N.L., A.M. Bartuska, J.H. Brown, S. Carpenter, C. D'Antonio, R. Francis, J.F. Franklin, J.A. MacMahon, R.F. Noss, D.J. Parsons, C.H. Peterson, M.G. Turner, and R.G. Woodmansee. 1996. The report of the Ecological Society of America committee on the scientific basis for ecosystem management. Ecol. Appl. 6:665–691.

Christensen, P.A. 1979. Production and standardization of antivenin. *In* C.-Y. Lee, ed., Handbook of experimental pharmacology, vol. 52: Snake venoms, pp. 825–846. St. Louis: Sigma Chemical Co.

Christoffel, R.A. 2007. Using human dimensions insights to improve conservation efforts for the Eastern Massasauga Rattlesnake (*Sistrurus catenatus catenatus*) in Michigan and the Timber Rattlesnake (*Crotalus horridus horridus*) in Minnesota. PhD dissertation, Michigan State University, East Lansing.

Ciofi C., and G. Chelazzi 1991. Radiotracking of *Coluber viridiflavus* using external transmitters. J. Herpetol. 25:37–40.

Clark, A.M., P.E. Moler, E.E. Possardt, A.H. Savitzky, W.S. Brown, and B.W. Bowen. 2003. Phylogeography of the Timber Rattlesnake (*Crotalus horridus*) based on mtDNA sequences. J. Herpetol. 37:145–154.

Clark, D.R., Jr. 1970. Ecological study of the Worm Snake *Carphophis vermis* (Kennicott). Museum of Natural History, University of Kansas Publishing, no. 19:85–194.

Clark, J.A., and R.M. May. 2002. Taxonomic bias in conservation research. Science 297:191–192.

Clark, J.R. 1999. Diversity: The ecosystem approach from a practical point of view. Conserv. Biol. 13:679–681.

Clark, R.W. 2004. Timber Rattlesnakes (*Crotalus horridus*) use chemical cues to select ambush sites. J. Chem. Ecol. 30:607–617.

Clark, R.W., 2006. Fixed vidography to study predation behavior of an ambush foraging snake, *Crotalus horridus*. Copeia 2006:181–187.

Clark, R.W., W.S. Brown, R. Stechert, and K.R. Zamudio. 2007. Integrating individual behaviour and landscape genetics: The population structure of Timber Rattlesnake hibernacula. Mol. Ecol. 17:719–730.

Clement, M., D. Posada, and K.A. Crandall. 2000. TCS: A computer program to estimate gene genealogies. Mol. Ecol. 9:1657–1659.

Cobb, V.A. 1994. The ecology of pregnancy in free-ranging Great Basin Rattlesnakes (*Crotalus viridis lutosus*). PhD dissertation, Idaho State University, Pocatello.

Colinvaux, P. 1978. Why big fierce animals are rare: An ecologist's perspective. Princeton, N.J.: Princeton University Press.

Coltman, D.W., P. O'Donoghue, J.T. Jorgenson, J.T. Hogg, C. Strobeck, and M. Festa-Bianchet. 2003. Undesirable evolutionary consequences of trophy hunting. Nature 426:655–658.

Conant, R. 1948. Regeneration of clipped subcaudal scales in a Pilot Black Snake. Nat. Hist. Misc. 13:1–2.

Congalton, R.G., and K. Green. 1999. Assessing the accuracy of remotely sensed data: Principles and practices. Boca Raton, Fla: Lewis Publications.

Congdon, J.D., A.E. Dunham, and D.W. Tinkle. 1982. Energy budgets and the life histories of reptiles. *In* C. Gans and F.H. Pough, eds., Biology of the Reptilia, pp. 233–271. London: Academic Press.

Congdon, J.D., A.E. Dunham, and R.C. van Loben Sels. 1994. Demographics of common snapping turtles (*Chelydra serpentina*)—implications for conservation and management of long-lived organisms. Am. Zool. 34:397–408.

Conner, L.M., and B.W. Plowman. 2001. Using Euclidean distance to assess nonrandom habitat use. *In* J.J. Millspaugh and J.M. Marzluff, eds., Radio tracking and animal populations, pp. 275–290. San Diego, Calif.: Academic Press.

Conners, J.S. 1998a. Natural history notes: *Chelydra serpentina* (predation). Herpetol. Rev. 29:235.

Conners, J.S. 1998b. Natural history notes: *Opheodrys aestivus* (egg predation). Herpetol. Rev. 29:243.

Cook, M., and S. Mineka. 1990. Selective associations in the observational conditioning of fear in rhesus monkeys. J. Exp. Psychol. Anim. Behav. Process 16:372–389.

Cook, S.A. 1911. Serpent worship. *In* H. Chisholm, ed., Encyclopedia britannica, 11th ed., Vol. 24, pp. 676–682. New York: Horace Edward Hooper.

Cooper, D.E. 1999. Human sentiment and the future of wildlife. *In* F.L. Dollins, ed., Attitudes to animals: Views in animal welfare, pp. 231–243. Cambridge, U.K.: Cambridge University Press.

Cooper-Doering, S. 2005. Modeling rattlesnake hibernacula on the Idaho National Laboratory, Idaho. MS thesis, Idaho State University, Pocatello.

Coppin, P., I. Jonckheere, K. Nackaerts, and B. Muys. 2004. Digital change detection methods in ecosystem monitoring: A review. Int. J. Remote Sens. 25:1565–1596.

Coppola, C.J. 1999. Spatial ecology of southern populations of the Copperbelly Water Snake, *Nerodia erythrogaster neglecta*. MS thesis, Purdue University, Fort Wayne.

Corander, J., P. Waldmann, P. Marttinen, and M.J. Sillanpaa. 2004. BAPS 2: Enhanced possibilities for the analysis of genetic population structure. Bioinformatics 20:2363–2369.

Cornuet, J.M., and G. Luikart. 1996. Description and power analysis of two tests for detecting recent population bottlenecks from allele frequency data. Genetics 144:2001–2014.

Costanzo, J.P. 1989. Effects of humidity, temperature, and submergence behavior on survivorship and energy use in hibernating garter snakes, *Thamnophis sirtalis*. Can. J. Zool. 67:2486–2492.

Costanza, R., and M. Mageau. 1999. What is a healthy ecosystem? Aquat. Ecol. 33:105–115.

Courchamp, F., E. Angulo, P. Rivalan, R.J. Hall, L. Signoret, L. Bull, and Y. Meinard. 2006. Rarity value and species extinction: The anthropogenic Allee effect. PLoS Biol. 4:2405–2410.

Covacevich, J., and C. Limpus. 1972. Observations on community egg-laying by the Yellow-Faced Whip Snake, *Demansia psammophis* (Schlegel) 1837 (Squamata: Elapidae). Herpetologica 28:208–210.

Covington, D. 1995. Salvation on Sand Mountain: Snake handling and redemption in southern Appalachia. Reading, Mass.: Addison-Wesley.

Craighead, W.E. 1973a. The assessment of avoidance responses on the Levis phobic test apparatus. Behav. Ther. 4:235–240.

Craighead, W.E. 1973b. The role of muscular relaxation in systematic desensitization. *In* R. Rubin, J.P. Brady, and J.D. Henderson, eds., Advances in behavior therapy, Vol. 5, pp. 177–197. New York: Academic Press.

Crandall, K.A., D. Posada, and D. Vasco. 1999. Effective population size: Missing measures and missing concepts. Anim. Conserv. 2:317–319.

Crandall, K.A., O.R.P. Bininda-Edmonds, G.M. Mace, and R.K. Wayne. 2000. Considering evolutionary processes in conservation biology. Trends Ecol. Evol. 16:290–295.

Creer S., A. Malhotra, R.S. Thorpe, and W.H. Chou. 2001. Multiple causation of phylogeographical pattern as revealed by nested clade analysis of the bamboo viper (*Trimeresurus stejnegeri*) within Taiwan. Mol. Ecol. 10:1967–1981.

Crisci, J.V., L. Katinas, and P. Posadas. 2003. Historical biogeography: An introduction. Cambridge, Mass.: Harvard University Press.

Crnokrak, P., and D.A. Roff. 1999. Inbreeding depression in the wild. Heredity 83:260–270.

Croes, B.M., W.F. Laurance, S.A. Lahm, L. Tchignoumba, A. Alonso, M.E. Lee, P. Campbell, and R. Buij. 2007. The influence of hunting on antipredator behavior in central African monkeys and duikers. Biotropica 39:257–263.

Cromer, R.B., J.D. Lanham, and H.G. Hanlin. 2002. Herpetofaunal response to gap and skidder-rut wetland creation in a southern bottomland hardwood forest. For. Sci. 48:407–413.

Cross, C.L., and C.E. Peterson. 2001. Modeling snake microhabitat from radiotelemetry studies using polytomous logistics regression. J. Herpetol. 35: 590–597.

Crosswhite, D.L., S.F. Fox, and R.E. Thill. 2004. Herpetological habitat relations in the Ouachita Mountains, Arkansas. In J.M. Guldin, ed., Ouachita and Ozark Mountains Symposium: Ecosystem management research, October 26–28, 1999, Hot Springs, Arkansas. General Technical Report SRS-74, pp. 273–282. Asheville, N.C.: U.S. Forest Service, Southern Research Station.

Crother, B.I., ed. 2008. Scientific and standard English names of amphibians and reptiles of North America north of México, pp. 1–84. SSAR Herpetological Circular 37.

Crother, B.I., J. Boundy, J.A. Campbell, K. de Queiroz, D.R. Frost, R.H. Highton, J.B. Iverson, P.A. Meylan, T.W. Reeder, M.E. Seidel, J.W. Sites Jr., T.W. Taggart, S.G. Tilley, and D.B. Wake. 2000. Scientific and standard English names of amphibians and reptiles of North America north of Mexico, with comments regarding confidence in our understanding. Herpetol. Circ. no. 29:iv, 82.

Crother, B.I., J. Boundy, J.A. Campbell, K. de Quieroz, D.R. Frost, D.M. Green, R. Highton, J.B. Iverson, R.W. McDiarmid, P.A. Meylan, T.W. Reeder, M.E. Seidel, J.W. Sites Jr., S.G. Tilley, and D.B. Wake. 2003. Scientific and standard English names of amphibians and reptiles of North America North of Mexico: Update. Herpetol. Rev. no. 34:196–203.

Cundall, D., and S.J. Beaupre. 2001. Field records of predatory strike kinematics in Timber Rattlesnakes, Crotalus horridus. Amphib-Reptilia 22:492–498.

Currie, D.J. 2001. Projected effects of climate change on patterns of vertebrate and tree species richness in the coterminous United States. Ecosystems 4:216–225.

Cushman, S.A., K.S. McKelvey, J. Hayden, and M.K. Schwartz. 2006. Gene flow in complex landscapes: Testing multiple hypotheses with causal modeling. Am. Nat. 168:486–499.

Dalrymple, G.H. 1984. Management of an endangered species of snake in Ohio, USA. Biol. Conserv. 30:195–200.

Dalrymple, G.H. 1994. Non-indigenous amphibians and reptiles. In D.C. Schmitz and T.C. Brown (project directors), An assessment of invasive non-indigenous species in Florida's public lands. Technical Report no. TSS-94-100. Tallahassee, Fla: Florida Department of Environmental Protection.

Daltry, J.C., Q. Bloxam, G. Cooper, M.L. Day, J. Hartley, M. Henry, K. Lindsay, and B.E. Smith. 2001. Five years of conserving the "world's rarest snake," the Antiguan racer Alsophis antiguae. Oryx 35:119–127.

Daltry, J.C., W. Wüster, and R.S. Thorpe. 1996. Diet and snake venom evolution. Nature 379:537–540.

Darwin, C. 1872. The expression of the emotions in man and animals. London: John Murray.

Davidson, C., H.B. Shaffer, and M.R. Jennings. 2002. Spatial tests of the pesticide drift, habitat destruction, UV-B, and climate-change hypotheses for California amphibian declines. Conserv. Biol. 16:1588–1601.

Davis, J.M., and J.A. Stamps. 2004. The effect of natal experience on habitat preferences. Trends Ecol. Evol. 19:411–416.

DeAngelis, D.L., L. Godbout, and B.J. Shuter. 1991. An individual-based approach to predicting density-dependent dynamics in smallmouth bass populations. Ecol. Model. 57:91–115.

deBry, R.W., and S. Seshadri. 2001. Nuclear intron sequences for phylogenetics of closely related mammals: An example using the phylogeny of *MUS*. J. Mammal. 82:280–288.

Deeming, D.C. 2004. Post-hatching phenotypic effects of incubation on reptiles. *In* D.C. Deeming, ed., Reptilian incubation: Environment, evolution and behaviour, pp. 229–251. Nottingham, U.K.: Nottingham University Press.

DeGregorio, B.A. 2008. Response of the Eastern Massasauga (*Sistrurus catenatus catenatus*) to clear-cutting. MS thesis, Purdue University, Fort Wayne, Indiana.

De Leo, G.A., and S.A. Levin. 1997. The multifaceted aspects of ecosystem integrity. Conserv. Ecol. 1. Available at: http://www.consecol.org/vol1/iss1/art3/.

Demers, M.N. 2002. Fundamentals of geographic information systems. New York: John Wiley & Sons.

de Queiroz, A., R. Lawson, and J.A. Lemos-Espinal. 2002. Phylogenetic relationships of North American garter snakes (*Thamnophis*) based on four mitochondrial genes: How much DNA sequence is enough? Mol. Phylogenet. Evol. 22:315–329.

de Queiroz, K. 1998. The general lineage concept of species, species criteria, and the process of speciation: A conceptual unification and terminological recommendations. *In* D.J. Howard and S.H. Berlocher, eds., Endless forms: Species and speciation, pp. 57–75. Oxford, U.K.: Oxford University Press.

Dessauer, H.C., C.J. Cole, and M.S. Hafner. 1996. Collection and storage of tissues. *In* D.M. Hillis, C. Moritz, and B.K. Mable, eds., Molecular systematics, pp. 29–47. Sunderland, Mass.: Sinauer.

DeVault, T.L., and O.E.J. Rhodes. 2002. Identification of vertebrate scavengers of small mammal carcasses in forested landscape. Acta Theriol. 42:185–192.

Devitt, T.J. 2006. Phylogeography of the Western Lyresnake (*Trimorphodon biscutatus*): Testing aridland biogeographical hypotheses across the nearctic–neotropical transition. Mol. Ecol. 15:4387–4407.

Diamond, J. 1989. Overview of recent extinctions. *In* D. Western and M. Pearl, eds., Conservation for the twenty-first century, pp. 37–41. New York: Oxford University Press.

Diller, L.V., and R.L. Wallace. 2002. Growth, reproduction, and survival in a population of *Crotalus viridis oreganus* in north central Idaho. Herpetol. Monogr. 16:26–45.

Dodd, C.K., Jr. 1987. Status, conservation, and management. *In* R.A. Seigel, J.T. Collins, and S.S. Novak, eds., Snakes: Ecology and evolutionary biology, pp. 478–513. New York: McGraw-Hill.

Dodd, C.K., Jr. 1993a. Population structure, body mass, activity, and orientation of an aquatic snake (*Seminatrix pygaea*) during a drought. Can. J. Zool. 71:1281–1288.

Dodd, C.K., Jr. 1993b. Strategies for snake conservation. *In* R.A. Seigel and J.T. Collins, eds., Snakes: Ecology and behavior, pp. 363–393. New York: McGraw-Hill.

Dodd, C.K., Jr., and R. Franz. 1993. The need for status information on common herpetofaunal species. Herpetol. Rev. 24:47–49.

Dodd, C.K., Jr., and R.A. Seigel. 1991. Relocation, repatriation, and translocation of amphibians and reptiles: Are they conservation strategies that work? Herpetologica 47:336–350.

Dodd, C.K., Jr., K.M. Enge, and J.N. Stuart. 1989. Reptiles on highways in north-central Alabama, USA. J. Herpetol. 23:197–200.

Dodd, C.K., Jr., W.J. Barichivich, and L.L. Smith. 2004. Effectiveness of a barrier wall and culverts in reducing wildlife mortality on a heavily traveled highway in Florida. Biol. Conserv. 118:619–631.

Doering, Z. 1994. From reptile houses to reptile discovery centers. Institutional Studies Smithsonian Institution Report 94-4, Washington D.C.

Dong, S., and Y. Kumazawa. 2005. Complete mitochondrial DNA sequences of six snakes: Phylogenetic relationships and molecular evolution of genomic features. J. Mol. Evol. 61:12–22.

Donnelly, P., and S. Tavaré. 1997. Progress in population genetics and human evolution. New York: Springer-Verlag.

Dorcas, M.E., and C.R. Peterson. 1997. Head-body temperature differences in free-ranging rubber boas. J. Herpetol. 31:87–93.

Dorcas, M.E., and C.R. Peterson. 1998. Daily body temperature variation in free-ranging rubber boas. Herpetologica 54:88–103.

Dorcas, M.E., C.R. Peterson, and M.E.T. Flint. 1997. The thermal biology of digestion in rubber boas (Charina bottae): Behavior, physiology, and environmental constraints. Physiol. Zool. 70:292–300.

Dorcas, M.E., W.A. Hopkins, and J.H. Roe. 2004. Effects of body mass and temperature on standard metabolic rate in the Eastern Diamondback Rattlesnake (Crotalus adamanteus). Copeia 2004:145–151.

Douglas, M.E., M.R. Douglas, G.W. Schuett, L.W. Porras, and A.T. Holycross. 2002. Phylogeography of the Western Rattlesnake (Crotalus viridis) complex, with emphasis on the Colorado Plateau. In G.W. Schuett, M. Höggren, M.E. Douglas, and H.W. Greene, eds., Biology of the vipers, pp. 11–50. Eagle Mountain, Utah: Eagle Mountain Publishing.

Douglas, M.E., M.R. Douglas, G.W. Schuett, and L.W. Porras. 2006. Evolution of rattlesnakes (Viperidae; Crotalus) in the warm deserts of western North America shaped by neogene vicariance and Quaternary climate change. Mol. Ecol. 15:3353–3374.

Driscoll, D.A. 2004. Extinction and outbreaks accompany fragmentation of a reptile community. Ecol. Appl. 14:220–240.

Driscoll, D.A., and C.M. Hardy. 2005. Dispersal and phylogeography of the agamid lizard Amphibolurus nobbi in fragmented and continuous habitat. Mol. Ecol. 14:1613–1629.

Drummond, A.J., and A. Rambaut. 2006. BEAST v 1.4. Available at: http://beast.bio.ed.ac.uk/.

Drummond A.J., S.Y.W. Ho, M.J. Phillips, and A. Rambaut. 2006. Relaxed phylogenetics and dating with confidence. PLoS Biology 4:5.

Drummond, A.J., G.K. Nicholls, A.G. Rodrigo, and W. Solomon. 2002. Estimating mutation parameters, population history and genealogy simultaneously from temporally spaced sequence data. Genetics 161:1307–1320.

Drummond, A.J., A. Rambaut, B. Shapiro, and O.G. Pybus. 2005. Bayesian coalescent inference of past population dynamics from molecular sequences. Mol. Biol. Evol. 22:1185–1192.

Duellman, W.E. 1999. Perils of permits: Procedures and pitfalls. Herpetol. Rev. 30:12–16.

Dufrene, M., and P. Legendre. 1997. Species assemblages and indicator species: The need for a flexible asymmetrical approach. Ecol. Monogr. 67:345–366.

Duncan, L., and J.E. Dunn. 2001. Partitioned Mahalanobis D^2 to improve GIS classification. SAS User's Group Int. Conf. Proc., Paper no. 198-26. Cary, N.C.: SAS Institute.

Dunham, A.E. 1993. Population responses to environmental change: Operative environments, physiologically-structured models, and population dynamics. In P.M. Kareiva, J.G. Kingsolver, and R.B. Huey, eds., Biotic interactions and global change, pp. 95–119. Sunderland, Mass.: Sinauer.

Dunham, A.E., and S.J. Beaupre. 1998. Ecological experiments: Scale, phenomenology, mechanism, and the illusion of generality. In W.J.J. Resetarits and J. Bernardo, eds., Experimental ecology: Issues and perspectives, pp. 27–49. New York: Oxford University Press.

Dunham, A.E., and K.L. Overall. 1994. Population responses to environmental change: Life history variation, individual-based models, and the population dynamics of short-lived organisms. Am. Zool. 34:382–396.

Dunham, A.E., B.W. Grant, and K.L. Overall. 1989. Interfaces between biophysical and physiological ecology and the population ecology of terrestrial vertebrate ectotherms. Physiol. Zool. 62:335–355.

Dunham, K.D. 1997. Population growth of mountain gazelles Gazella gazella reintroduced to central Arabia. Biol. Conserv. 81:205–214.

Dunham, K.D. 2001. Status of the reintroduced population of mountain gazelles *Gazella gazella* in central Arabia: Management lessons from an arid land reintroduction. Oryx 35:111–118.

Dunn, P.O., and D.W. Winkler. 1999. Climate change has affected breeding date of tree swallows throughout North America. Proc. R. Soc. Lond. B Biol. Sci. 266:2487–2490.

Durner, G.M., and J.E. Gates. 1993. Spatial ecology of Black Rat Snakes on Remington Farms, Maryland. J. Wildl. Manage. 57:812–826.

Duvall, D., M.B. King, and K.J. Gutzwiller. 1985. Behavioral ecology and ethology of the Prairie Rattlesnake. Nat. Geog. Res. 1:80–111.

Duvall, D., M.J. Goode, W.K. Hayes, J.K. Leonhardt, and D.G. Brown. 1990. Prairie Rattlesnake vernal migration: Field experimental analyses and survival value. Nat. Geogr. Res. 6:457–469.

Duvall, D., G.W. Schuett, and S.J. Arnold. 1993. Ecology and evolution of snake mating systems. *In* R.A. Seigel and J.T. Collins, eds., Snakes: Ecology and behavior, pp. 165–200. New York: McGraw-Hill.

Eastridge, R., and J.D. Clark. 2001. Evaluation of 2 soft-release techniques to reintroduction black bears. Wildl. Soc. Bull. 29:1163–1174.

Edwards, S.E., L. Liang, and D.K. Pearl. 2007. High-resolution species trees without concatenation. Proc. Natl. Acad. Sci. U.S.A. 104:5936–5941.

Ehleringer, J.R., and P.W. Rundel. 1989. Stable isotopes: History, units, and instrumentation. *In* P.W. Rundel, J.R. Ehleringer, and K.A. Nagy, eds., Stable isotopes in ecological research, pp. 1–16. New York: Springer-Verlag.

Ehleringer, J.R., P.W. Rundel, and K.A. Nagy. 1986. Stable isotopes in physiological ecology and food web research. Trends Ecol. Evol. 1:42–45.

Ehrlich, P.R., and J.H. Camin. 1960. Natural selection in Middle Island Water Snakes (*Natrix sipedon* L.). Evolution 14:136.

Eller, E., and H.C. Harpending. 1996. Simulations show that neither population expansion nor population stationarity in a west African population can be rejected. Mol. Biol. Evol. 13:1155–1157.

Elsey, R.M., T. Joanen, L. McNease, and N. Kinler. 1992. Growth rates and body condition factors of *Alligator mississippiensis* in coastal Louisiana wetlands: A comparison of wild and farm-released juveniles. Comp. Biochem. Physiol. 103A:667–672.

Elton, C.S. 1958. The ecology of invasions by animals and plants. London: Methuen.

Emerson, B.C., E. Paradis, and C. Thébaud. 2001. Revealing the demographic histories of species using DNA sequences. Trends Ecol. Evol. 16:707–716.

Enge, K.M. 2001. The pitfalls of pitfall traps. J. Herpetol. 35:467–478.

Enge, K.M., and W.R. Marion. 1986. Effects of clearcutting and site preparation on herpetofauna of a north Florida flatwoods. For. Ecol. Manage. 14:177–192.

Engeman, R.M., and M.A. Linnell. 2004. The effect of trap spacing on the capture of Brown Tree Snakes on Guam. Int. Biodeterior. Biodegrad. 54:265–267.

England, P.R., J-M. Cornuet, P. Berthier, D.A. Tallmon, and G. Luikart. 2006. Estimating effective population size from linkage disequilibrium: Severe bias in small samples. Conserv. Genet. 7:303–308.

Erwin, W.J., and R.H. Stasiak. 1977. Vertebrate mortality during the burning of a reestablished prairie in Nebraska. Am. Midl. Nat. 101:247–249.

Evanno, G., S. Regnaut, and J. Goudet. 2005. Detecting the number of clusters of individuals using the software Structure: A simulation study. Mol. Ecol. 14:2611–2620.

Excoffier, L., and S. Schneider. 1999. Why hunter-gatherer populations do not show signs of Pleistocene demographic expansions. Proc. Natl. Acad. Sci. U.S.A. 96:10597–10602.

Excoffier, L., G. Laval, and S. Schneider. 2005. Arlequin (ver. 3.0): An integrated software package for population genetics data analysis. Evol. Bioinf. Online 2005:47–50.

Falush, D., M. Stephens, and J.K. Pritchard. 2003. Inference of population structure using mulitlocus genotype data: Linked loci and correlated allele frequencies. Genetics 164:1567–1587.

Fazey, I., J.G. Salisbury, D.B. Lindenmayer, J. Maindonald, and R. Douglas. 2004. Can methods applied in medicine be used to summarize and disseminate conservation research? Environ. Conserv. 31:190–198.

Feldman, C.R., and G.S. Spicer. 2002. Mitochondrial variation in Sharp-tailed Snakes (*Contia tenuis*): Evidence of a cryptic species. J. Herpetol. 36:648–655.

Feldman, C.R., and G.S. Spicer. 2006. Comparative phylogeography of woodland reptiles in California: Repeated patterns of cladogenesis and population expansion. Mol. Ecol. 15:2201–2222.

Felsenstein, J. 1973. Maximum likelihood and minimum-steps methods for estimating evolutionary trees from data on discrete characters. Syst. Zool. 22:240–249.

Felsenstein, J. 1978. Cases in which parsimony or compatibility methods will be positively misleading. Syst. Zool. 27:401–410.

Felsenstein, J. 1985. Confidence limits on phylogenies: An approach using the bootstrap. Evolution 39:783–791.

Felsenstein, J. 1992. Estimating effective population size from samples of sequences: Inefficiency of pairwise and segregating sites as compared to phylogenetic estimates. Genet. Res. 59:139–147.

Felsenstein, J. 2004. Inferring phylogenies. Sunderland, Mass.: Sinauer.

Ferrer, M., and J.A. Donzar. 1996. Density-dependent fecundity by habitat heterogeneity in an increasing population of Spanish Imperial Eagles. Ecology 77:69–74.

Ferriere, R., F. Sarrazin, S. Legendre, and J.P. Baron. 1996. Matrix population models applied to viability analysis and conservation: Theory and practice using the ULM software. Acta Oecol. 17:629–656.

Ficetola, G.F., and F.D. Bernardi. 2005. Supplementation or in situ conservation? Evidence of local adaptation in the Italian agile frog *Rana latastei* and consequences for the management of populations. Anim. Conserv. 8:33–40.

Filippi, E., and L. Luiselli. 2000. Status of the Italian snake fauna and assessment of conservation threats. Biol. Conserv. 93:219–225.

Fischer, J., and D.B. Lindenmayer. 2000. An assessment of the published results of animal relocations. Biol. Conserv. 96:1–11.

Fischer, J., D.B. Lindenmayer, and A. Cowling. 2004. The challenge of managing multiple species at multiple scales: Reptiles in an Australian grazing landscape. J. Appl. Ecol. 41:32–44.

Fisher, R.A. 1930. The genetical theory of natural selection. Oxford, U.K.: Claredon Press.

Fitch, H.S. 1951. A simplified type of funnel trap for reptiles. Herpetologica 7:77–80.

Fitch, H.S. 1982. Reproductive cycles in tropical reptiles. Occas. Pap. Mus. Nat. Hist. Univ. Kansas 96:1–53.

Fitch, H.S. 1987a. Collecting and life-history techniques. *In* R.A. Seigel, J.T. Collins, and S.S. Novak, eds., Snakes: Ecology and evolutionary biology, pp. 143–164. New York: McGraw-Hill.

Fitch, H.S. 1987b. Resources of a snake community in prairie-woodland habitat of northeastern Kansas. *In* N.J. Scott Jr., ed., Herpetological communities: A symposium of the Society for the Study of Amphibians and Reptiles and the Herpetologists' League, August 1977, pp. 83–98. Washington, D.C.: U.S. Fish & Wildlife Service.

Fitch, H.S. 1992. Methods for sampling snake populations and their relative success. Herpetol. Rev. 23:17–19.

Fitch, H.S. 1998. The Sharon Springs roundup and Prairie Rattlesnake demography. Trans. Kansas Acad. Sci. 101:101–113.

Fitch, H.S. 1999. A Kansas snake community: Composition and change over 50 years. Malabar, Fla.: Krieger.

Fitch, H.S. 2006. Ecological succession on a natural area in northeastern Kansas from 1948 to 2006. Herpetol. Conserv. Biol. 1:1–5.

Fitch, H.S., and H.W. Shirer. 1971. A radiotelemetric study of spatial relationships in some common snakes. Copeia 1971:118–128.

Fitzgerald, L.A., and C.W. Painter. 2000. Rattlesnake commercialization: Long-term trends, issues, and implications for conservation. Wildl. Soc. Bull. 28:235–253.

Fitzgerald, M., and R. Shine. 2004. Life history attributes of the threatened Australian snake (Stephen's Banded Snake, *Hoplocephalus stephensii*, Elapidae). Biol. Conserv. 119:121–128.

Fitzpatrick, B.M., J.S. Placyk, M.L. Neimiller, G.S. Casper, and G.M. Burghardt. 2008. Distinctiveness in the face of gene flow: Hybridization between generalist and specialist gartersnakes. Mol. Ecol. 17:4107–4117.

Flannery, T.F. 1994. The future eaters. Sydney: Reed Books.

Fleet, R.R., and F.W. Plapp Jr. 1978. DDT residues in snakes decline since DDT ban. Bull. Environ. Contam. Toxicol. 19:383–388.

Fleet, R.R., D.R. Clark Jr., and F.W. Plapp Jr. 1972. Residues of DDT and dieldrin in snakes from two Texas agro-systems. BioScience 22:664–665.

Fleishman, E., J.R. Thomson, R. Mac Nally, D.D. Murphy, and J.P. Fay. 2005. Using indicator species to predict species richness of multiple taxonomic groups. Conserv. Biol. 19:1125–1137.

Floyd, T.M., K.R. Russell, C.E. Moorman, D.H. Van Lear, D.C. Guynn Jr., and J.D. Lanham. 2002. Effects of prescribed fire on herpetofauna within hardwood forests of the upper piedmont of South Carolina: A preliminary analysis. *In* Proc. Eleventh Biennial Southern Silvicultural Res. Conf. General Technical Report SRS-48, pp. 123–127. Asheville, N.C.: U.S. Forest Service, Southern Research Station.

Fontanella, F.F., C.R. Feldman, M.E. Siddall, and F.T. Burbrink. 2008. Phylogeography of *Diadophis punctatus*: Extensive lineage diversity and repeated patterns of historical demography in a trans-continental snake. Mol. Phylogenet. Evol. 46:1049–1070.

Ford, N.B., and R.A. Seigel. 1989. Phenotypic plasticity in reproductive traits: Evidence from a viviparous snake. Ecology 70:1768–1774.

Ford, N.B., and R.A. Seigel. 1994. An experimental study of the trade-offs between age and size at maturity: Effects of energy availability. Funct. Ecol. 8:91–96.

Ford, W.M., A.M. Menzel, D.W. McGill, J. Laerm, and T.S. McCay. 1999. Effects of a community restoration fire on small mammals and herpetofauna in the southern Appalachians. For. Ecol. Manage. 114:233–243.

Forman, R.T.T., and L.E. Alexander. 1998. Roads and their major ecological effects. Annu. Rev. Ecol. Syst. 29:207–231.

Forsman, A. 1993. Growth rate in different color morphs of the Adder, *Vipera berus*, in relation to yearly weather variation. Oikos 66:279–285.

Forsman, A., and L.E. Lindell. 1997. Responses of a predator to variation in prey abundance: Survival and emigration of Adders in relation to vole density. Can. J. Zool. 75:1099–1108.

Fraker, M.A. 1970. Home range and homing in the watersnake, *Natrix sipedon sipedon*. Copeia 1970:665–673.

Frankham, R. 1995. Effective population size/adult population size ratios in wildlife: A review. Genet. Res. 66:95–107.

Frankham, R. 2008. Genetic adaptation to captivity in species conservation programs. Mol. Ecol. 17:325–333.

Frankham, R., and J. Kingsolver. 2004. Responses to environmental change: Adaptation or extinction. *In* R. Ferrière, U. Dieckmann, and D. Couvet, eds., Evolutionary conservation biology, pp. 85–100. New York: Cambridge University Press.

Frankham, R., J.D. Ballou, and D.A. Briscoe. 2002. Introduction to conservation genetics. New York: Cambridge University Press.

Fredericksen, N.J., and T.S. Fredericksen. 2002. Terrestrial wildlife responses to logging and fire in a Bolivian tropical humid forest. Biodiv. Conserv. 11:27–38.

Freeland, W.J. 1985. The rate of expansion by *Bufo marinus* in northern Australia. Aust. Wildl. Res. 12:555–559.

Freeland, W.J. 1986. Populations of cane toads *Bufo marinus* in relation to time since colonization. Aust. Wildl. Res. 13:321–330.

Frier, J.A., and R.T. Zappalorti. 1983. Reptile and amphibian management techniques. Trans. Wildl. Soc. 40:142–148.

Fritts, S.H., W.J. Paul, and D.L. Mech. 1984. Movements of translocated wolves in Minnesota. J. Wildl. Manage. 48:709–721.

Fritts, T.H., N.J. Scott, and J.A. Savidge. 1987. Activity of the arboreal Brown Tree Snake (*Boiga irregularis*) on Guam as determined by electrical power outages. Snake 19:51–58.

Fu, Y.X. 1997. Statistical tests of neutrality of mutations against population growth, hitchhiking and background selection. Genetics 147:915–925.

Fu, Y.X., and W.H. Li. 1993. Statistical tests of neutrality of mutations. Genetics 133:693–709.

Fu, Y.X., and W.H. Li. 1999. Coalescing into the 21st century: An overview and prospects of coalescent theory. Theor. Popul. Biol. 56:1–10.

Futuyma, D. 1998. Evolutionary biology, 3rd ed. Sunderland, Mass.: Sinauer.

Galligan, J.H., and W.A. Dunson. 1979. Biology and status of Timber Rattlesnake (*Crotalus horridus*) populations in Pennsylvania. Biol. Conserv. 15:13–58.

Gander, H., and P. Ingold. 1997. Reactions of alpine chamois *Rupicapra r. rupicapra* to hikers, joggers and mountainbikers. Biol. Conserv. 79:107–109.

Gannes, L.Z., D.M. O'Brien, and C. Martinez del Rio. 1997. Stable isotopes in animal ecology: Assumptions, caveats, and a call for more laboratory experiments. Ecology 78:1271–1276.

Gannes, L.Z., C. Martinez del Rio, and P. Koch. 1998. Natural abundance variations in stable isotopes and their potential uses in animal physiological ecology. Comp. Biochem. Physiol. 119:725–737.

Gans, C. 1986. Locomotion of limbless vertebrates: Pattern and evolution. Herpetologica 42:33–46.

Gardner, T.A., J. Barlow, and C.A. Peres. 2007. Paradox, presumption and pitfalls in conservation biology: The importance of habitat change for amphibians and reptiles. Biol. Conserv. 138:166–179.

Garner, T.W.J. 1998. A molecular investigation of population structure and paternity in the Common Garter Snake, *Thamnophis sirtalis*. MS thesis, University of Victoria, Victoria, Canada.

Garner, T.W.J., P.T. Gregory, G.F. McCracken, G.M. Burghardt, B.F. Koop, S E. McLain, and R.J. Nelson. 2002. Geographic variation of multiple paternity in the Common Garter Snake (*Thamnophis sirtalis*). Copeia 2002:15–23.

Garner, T.W.J., P.B. Pearman, P.T. Gregory, G. Tomio, S.G. Wishniowski, and D.J. Hosken. 2004. Microsatellite markers developed from *Thamnophis elegans* and *Thamnophis sirtalis* and their utility in three species of garter snakes. Mol. Ecol. Notes 4:3–69.

Gartside, D.F., J.S. Rogers, and H.C. Dessauer. 1977. Speciation with little genic and morphological differentiation in the ribbon snakes *Thamnophis proximus* and *T. sauritus* (Colubridae). Copeia 1977:697–705.

Garza, J.C., and E.G. Williamson. 2001. Detection of reduction in population size using data from microsatellite loci. Mol. Ecol. 10:305–318.

Gaston, K.J. 2003. The structure and dynamics of geographic ranges. Oxford, U.K.: Oxford University Press.

Gates, S. 2002. Review of methodology of quantitative reviews using meta-analysis in ecology. J. Anim. Ecol. 71:547–557.

Gautschi, B., A. Widmer, and J. Koella. 2000. Isolation and characterization of microsatellite loci in the Dice Snake (*Natrix tessellata*). Mol. Ecol. 9:2191–2193.

Gautschi, B., J. Joshi, A. Widmer, and J.C. Koella. 2002. Increased frequency of scale anomalies and loss of genetic variation in serially bottlenecked populations of the Dice Snake, *Natrix tessellata*. Conserv. Genet. 3:235–245.

Geer, J.H. 1965. The development of a scale to measure fear. Behav. Res. Ther. 3:45–53.

Geffeney, S.L., E.D. Brodie Jr., P.C. Ruben, and E.D. Brodie III. 2002. Mechanisms of adaptation in a predator-prey arms race: TTX-resistant sodium channels. Science 297:1336–1339.

Geffeney, S.L., E. Fujimoto, E.D. Brodie III, E.D. Brodie Jr., and P.C. Ruben. 2005. Evolutionary diversification of TTX-resistant sodium channels in a predator-prey interaction. Nature 434:759–763.

George, L. 1995. The encyclopedia of heresies and heretics. London: Robson Books.

Gerber, L.R. 2006. Including behavioral data in demographic models improves estimates of population viability. Front. Ecol. Environ. 4:419–427.

Germano, O.E., and D.F. Williams. 1993. Field evaluations of using passive integrated transponders (PIT) tags to permanently mark lizards. Herpetol. Rev. 24:54–56.

Giannasi, N., R.S. Thorpe, and A. Malhotra. 2001. The use of amplified fragment length polymorphism in determining species trees at fine taxonomic levels: Analysis of a medically important snake, *Trimeresurus albolabris*. Mol. Ecol. 10:419–426.

Gibbons, J.W., and K.M. Andrews. 2004. PIT tagging: Simple technology at its best. BioScience 54:447–454.

Gibbons, J.W., and R.D. Semlitsch. 1982. Terrestrial drift fences with pitfall traps: An effective technique for quantitative sampling of animal populations. Brimleyana 1982:1–16.

Gibbons, J.W., D.E. Scott, T.J. Ryan, K.A. Buhlmann, T.D. Tuberville, B.S. Metts, J.L. Greene, T. Mills, Y. Leiden, S. Poppy, and C.T. Winne. 2000. The global decline of reptiles, déjà vu amphibians. BioScience 50:653–666.

Gibbons, J.W., C.T. Winne, D.E. Scott, J.D. Willson, X. Glaudas, K.M. Andrews, B.D. Todd, L.A. Fedewa, L. Wilkinson, R.N. Tsaliagos, S.J. Harper, J.L. Greene, T.D. Tuberville, B.S. Metts, M.E. Dorcas, J.P. Nestor, C.A. Young, T. Akre, R N. Reed, K.A. Buhlmann, J. Norman, D.A. Croshaw, C. Hagen, and B.B. Rothermel. 2006. Remarkable amphibian biomass and abundance in an isolated wetland: Implications for wetland conservation. Conserv. Biol. 20:1457–1465.

Gibbons, W., and M. Dorcas. 2005. Snakes of the southeast. Athens: University of Georgia Press.

Gibbs, H.L., and P.J. Weatherhead. 2001. Insights into population ecology and sexual selection in snakes through the application of DNA-based genetic markers. J. Hered. 92:173–179.

Gibbs, H.L., K.A. Prior, and P.J. Weatherhead. 1994. Genetic analysis of populations of threatened snake species using RAPD markers. Mol. Ecol. 3:329–227.

Gibbs, H.L., K.A. Prior, P.J. Weatherhead, and G. Johnson. 1997. Genetic structure of populations of the threatened Eastern Massasauga Rattlesnake, *Sistrurus c. catenatus:* Evidence from microsatellite DNA markers. Mol. Ecol. 6:1123–1132.

Gibbs, H.L., K.A. Prior, and C. Parent. 1998. Characterization of DNA microsatellite loci from a threatened snake: The Eastern Massasauga Rattlesnake (*Sistrurus c. catenatus*) and their use in population studies. J. Hered. 89:169–173.

Gibbs, H.L., S.J. Corey, G. Blouin-Demers, K.A. Prior, and P.J. Weatherhead. 2006. Hybridization between mtDNA-defined phylogeographic lineages of Black Ratsnakes (*Pantherophis* sp.). Mol. Ecol. 15:3755–3767.

Gibbs, J.P., and A.R. Breisch. 2001. Climate warming and calling phenology of frogs near Ithaca, New York, 1900–1999. Conserv. Biol. 15:1175–1178.

Gibbs, J.P., H.L. Snell, and C.E. Causton. 1999. Effective monitoring for adaptive wildlife management: Lessons from the Galápagos Islands. J. Wildl. Manage. 63:1055–1065.

Gibson, A.R., and J.B. Falls. 1979. Thermal biology of the Common Garter Snake *Thamnophis sirtalis* (L.). II. The effects of melanism. Oecologia 43:99–109.

Gibson, A.R., and J.B. Falls. 1988. Melanism in the Common Garter Snake: A Lake Erie phenomenon. *In* J.F. Downhower, ed., The biogeography of the island region of western Lake Erie, pp. 233–245. Columbus: Ohio State University Press.

Giese, M. 1996. Effects of human activity on Adelie penguin *Pygoscelis adeliae* beeding success. Biol. Conserv. 75:157–164.

Gilks, W.R., S. Richardson, and D.J. Speigelhalter. 1996. Markov chain Monte Carlo in practice. London: Chapman and Hall.

Gillingham, J.C. 1987. Social behavior. *In* R.A. Seigel, J.T. Collins, and S.S. Novak, eds., Snakes: Ecology and evolutionary biology, pp. 184–209. New York: McGraw-Hill.

Gillingham, J.C., and C.C. Carpenter. 1978. Snake hibernation: Construction of and observations on a man-made hibernaculum (Reptilia, Serpentes). J. Herpetol. 12:495–498.

Gilpin, M.E., and M.E. Soulé. 1986. Minimum viable populations: Processes of species extinction. In M.E. Soulé, ed., Conservation biology: The science of scarcity and diversity, pp. 19–34. Sunderland, Mass.: Sinauer.

Godley, J.S. 1980. Foraging ecology of the Striped Swamp Snake, *Regina alleni*, in southern Florida. Ecol. Monogr. 50:411–436.

Goldberg, C.S., T. Edwards, M.E. Kaplan, and M. Goode. 2003. PCR primers for microsatellite loci in the Tiger Rattlesnake (*Crotalus tigris*, Viperidae). Mol. Ecol. Notes 3:539–541.

Goldfarb, G., E. Rogier, C. Gebauer, D. Lassen, D. Bernau, P. Jolis, and G. Feldmann. 1989. Comparative effects of halothane and isoflurane anesthesia on the ultrastructure of human hepatic cells. Anesth. Analg. 69:491–495.

Goldingay, R.L., G. Daly, and F. Lemckert. 1996. Assessing the impacts of logging on reptiles and frogs in the montane forests of southern New South Wales. Wildl. Res. 23:495–510.

Goldstein, D.B., A. Ruiz Linares, L.L. Cavalli-Sforza, and M.W. Feldman. 1995. Genetic absolute dating based on microsatellites and the origin of modern humans. Proc. Nat. Acad. Sci. U.S.A. 92:6723–6727.

Gooch, M.M., A.M. Heupel, S.J. Price, and M.E. Dorcas. 2006. The effects of survey protocol on detection probabilities and site occupancy estimates of summer breeding anurans. Appl. Herpetol. 3:129–142.

Goodall, J., and M. Bekoff. 2003. The ten trusts: What we must do to care for the animals we love. New York: Harper Collins.

Goris, R.C. 1971. The hibernation of captive snakes. Snake 3:65–69.

Goudet, J., N. Perrin, and P. Waser. 2002. Tests for sex-biased dispersal using bi-parentally inherited genetic markers. Mol. Ecol. 11:1103–1114.

Grafen, A. 1988. On the use of data on lifetime reproductive success. *In* T.H. Clutton-Brock, ed., Reproductive success, pp. 454–471. Chicago: University of Chicago Press.

Grant, B.W., A.D. Tucker, J.E. Lovich, A.M. Mills, P.M. Dixon, and J.W. Gibbons. 1992. The use of coverboards in estimating patterns of reptile and amphibian biodiversity. *In* D.R. McCullough and R.H. Barrett, eds., Wildlife 2001: Populations, pp. 379–403. London: Elsevier.

Graves, B.M. 1989. Defensive behavior of female Prairie Rattlesnakes (*Crotalus viridis*) changes after parturition. Copeia 1989:791–794.

Graves, B.M., and D. Duvall. 1993. Reproduction, rookery use, and thermoregulation in free-ranging *Crotalus v. viridis*. J. Herpetol. 27:33–41.

Graves, B.M., and D. Duvall. 1995. Aggregation of squamate reptiles associated with gestation, oviposition, and parturition. Herpetol. Monogr. 9:102–119.

Graves, J.E., S.D. Ferris, and A.E. Dizon. 1984. Close genetic similarity of Atlantic and Pacific skipjack tuna (*Katsuwonus pelamis*) demonstrated with restriction endonuclease analysis of mitochondrial DNA. Mar. Biol. 79:315–319.

Grayson, K.L., and M.E. Dorcas. 2004. Seasonal temperature variation in the painted turtle (*Chrysemys picta*). Herpetologica 60:325–336.

Grazziotin, F.G., M. Monzel, S. Echeverrigary, and S.L. Bonatto. 2006. Phylogeography of the *Bothrops jararaca* complex (Serpentes: Viperidae): Past fragmentation and island colonization in the Brazilian Atlantic forest. Mol. Ecol. 15:3969–3982.

Green, D.M., and J.M. Swets. 1966. Signal detection theory and psychophysics. New York: John Wiley & Sons.

Green, J. 2005. Thermal biology of the Eastern Racer (*Coluber constrictor*) in middle Tennessee. MS thesis, Middle Tennessee State University, Murfreesboro.

Greenberg, C.H. 2000. Fire, habitat structure and herpetofauna in the southeast. *In* W.M. Ford, K.R. Russell, and C.E. Moorman, eds., The role of fire in nongame wildlife management and community restoration: Traditional uses and new directions. Proceedings of a special workshop, September 15, 2000, Nashville, Tenn., pp. 91–99. Newtown Square, Penn.: USDA Forest Service, Northeastern Research Station.

Greenberg, C.H. 2001. Response of reptile and amphibian communities to canopy gaps created by wind disturbance in the southern Appalachians. For. Ecol. Manage. 148:135–144.

Greenberg, C.H., D.G. Neary, and L.D. Harris. 1994. Effect of high-intensity wildfire and silvicultural treatments on reptile communities in sand-pine scrub. Conserv. Biol. 8:1047–1057.

Greene, H.W. 1986. Natural history and evolutionary biology. *In* M.E. Feder and G.V. Lauder, eds., Predator-prey relationships: Perspectives and approaches from the study of lower vertebrates, pp. 99–108. Chicago: University of Chicago Press.

Greene, H.W. 1989. Biological, evolutionary, and conservation implications of feeding biology in old world cat snakes, genus *Boiga* (Colubridae). Calif. Acad. Sci. 46:193–207.

Greene, H.W. 1994. Systematics and natural history, foundations for understanding and conserving biodiversity. Am. Zool. 34:48–56.

Greene, H.W. 1997. Snakes: The evolution of mystery in nature. Berkeley: University of California Press.

Greene, H.W. 2003. Appreciating rattlesnakes. Wild Earth 13(2–3):28–32.

Greene, H.W. 2005. Organisms in nature as a central focus for biology. Trends Ecol. Evol. 20:23–27.

Greene, M.J., and R.T. Mason. 1998. Chemically mediated sexual behavior of the Brown Tree Snake, *Boiga irregularis*. Ecoscience 5:405–409.

Greenwood, P.J. 1980. Mating systems, philopatry and dispersal in birds and mammals. Anim. Behav. 28:1140–1162.

Greer, A.E. 1997. The biology and evolution of Australian snakes. Sydney: Surrey Beatty & Sons.

Gregory, P.T. 1974. Patterns of spring emergence of the Red-sided Garter Snake (*Thamnophis sirtalis parietalis*) in the interlake region of Manitoba. Can. J. Zool. 52:1063–1069.

Gregory, P.T. 1982. Reptilian hibernation. *In* C. Gans and F.H. Pough, eds., Biology of the Reptilia, Vol. 13, pp. 53–154. New York: Academic Press.

Gregory, P.T. 2007. Biology and conservation of a cold-climate snake fauna. *In* C. Seburn and C. Bishop, eds., Ecology, conservation and status of reptiles in Canada, pp. 41–56. Salt Lake City, Utah: Society for the Study of Amphibians and Reptiles.

Gregory, P.T., and L.A. Isaac. 2004. Food habits of the Grass Snake in southeastern England: Is *Natrix natrix* a generalist predator? J. Herpetol. 38:88–95.

Gregory, P.T., J.M. Macartney, and K.W. Larsen. 1987. Spatial patterns and movements. *In* R.A. Seigel, J.T. Collins, and S.S. Novak, eds., Snakes: Ecology and evolutionary biology, pp. 366–395. New York: McGraw-Hill.

Griffin, A.S., D.T. Blumstein, and C.S. Evans. 2000. Training captive-bred or translocated animals to avoid predators. Conserv. Biol. 14:1317–1326.

Griffith, B., J.M. Scott, J.W. Carpenter, and C. Reed. 1989. Translocation as a species conservation tool: Status and strategy. Science 245:477–480.

Griffith, R.C., and S. Tavaré. 1994. Sampling theory for neutral alleles in a varying environment. Philos. Trans. R. Soc. Lond. B 344:403–410.

Groombridge, B., and R. Luxmoore. 1991. Pythons in southeast Asia: A review of distributions, status and trade in three selected species. Report to CITES Secretariat, Lausanne, Switzerland.

Groot, T.B.M., E. Bruins, and J.A.J. Breeuwer. 2003. Molecular genetic evidence for parthenogenesis in the Burmese python, *Python molurus bivittatus*. Heredity 90:130–135.

Grudzien, T.A., and P.J. Owens. 1991. Genic similarity in the gray and brown color morphs of the snake *Storeria occipitomaculata*. J. Herpetol. 25:90–92.

Grumbine, R.E. 1994. What is ecosystem management? Conserv. Biol. 8:27–38.

Guicking, D., U. Joger, and M. Wink. 2002. Molecular phylogeography of the Viperine Snake (*Natrix maura*) and the Dice Snake (*Natrix tessellata*): First results. Biota 3:47–57.

Guicking, D., A. Herzberg, and M. Wink. 2004. Population genetics of the Dice Snake (*Natrix tessellata*) in Germany: Implications for conservation. Salamandra 40:217–234.

Guicking, D., R.A. Griffiths, R.D. Moore, U. Joger, and M. Wink. 2006. Introduced alien or persecuted native? Resolving the origin of the Viperine Snake (*Natrix maura*) on Mallorca. Biodiv. Conserv. 15:3045–3054.

Guiher, T.J., and F.T. Burbrink. 2008. Demographic and phylogeographic histories of two venomous North American snakes of the Genus *Agkistrodon*. Mol. Phylogenet. Evol. 48:543–553.

Gutzke, W.H.N., and G.C. Packard. 1987. Influence of the hydric and thermal environments on eggs and hatchlings of Bull Snakes, *Pituophis melanoleucus*. Physiol. Zool. 60:9–17.

Guyer, C., C.T. Meadows, S.C. Townsend, and L.G. Wilson. 1997. A camera device for recording vertebrate activity. Herpetol. Rev. 28:138–140.

Hager, H.A. 1998. Area-sensitivity of reptiles and amphibians: Are there indicator species for habitat fragmentation? Ecoscience 5:139–147.

Hahn M.W., M.D. Rausher, and C.W. Cunningham. 2002. Distinguishing between selection and population expansion in an experimental lineage of bacteriophage T7. Genetics 161:11–20.

Haila, Y. 2002. A conceptual genealogy of fragmentation research: From island biogeography to landscape ecology. Ecol. Appl. 12:321–334.

Hambler, C. 1994. Giant tortoise *Geochelone gigantea* translocation to Curieuse Island (Seychelles): Success or failure? Biol. Conserv. 69:293–299.

Hambly, W.D. 1931. Serpent worship in Africa. Field Mus. Nat. Hist. Pub. no. 289, Anthropol. Ser. 21:1–85.

Harden, L.A., N.A. DiLuzio, J.W. Gibbons, and M.E. Dorcas. 2007. The spatial and thermal biology of diamondback terrapins (*Malaclemys terrapin*) in a South Carolina salt marsh. J. North Carolina Acad. Sci. 123:154–162.

Hardy, D.L., Sr., and H.W. Greene. 2000. Inhalation anesthesia of rattlesnakes in the field for processing and transmitter implantation. Sonoran Herpetol. 13:109–113.

Hare, M.P. 2001. Prospects for nuclear gene phylogeography. Trends Ecol. Evol. 16:700–706.

Harpending, H.C. 1994. Signature of ancient population growth in a low resolution mitochondrial mismatch distribution. Hum. Biol. 66:131–137.

Harpending, H.C., S.T. Sherry, A.R. Rogers, and M. Stoneking. 1993. The genetic structure of ancient human populations. Curr. Anthropol. 34:483–496.

Harper, K.A, S.E. MacDonald, P.J. Burton, J. Chen, K.G. Brosofske, S.C. Saunders, E.S. Euskirchen, D. Roberts, M.S. Jaiteh, and P.-A. Esseen. 2005. Edge influence on forest structure and composition in fragmented landscapes. Conserv. Biol. 19:768–782.

Harper, S.J., and G.O. Batzli. 1996. Monitoring use of runways by voles with passive integrated transponders. J. Mammal. 77:364–369.

Harris, L.D. 1984. The fragmented forest: Island biogeography theory and the preservation of biotic diversity. Chicago: University of Chicago Press.

Harris, S.A. 1999. RAPDs in systematics—a useful methodology? *In* P.M. Hollingsworth, R.M. Bateman, and R.J. Gornall, eds., Molecular systematics and plant evolution, pp. 211–228. London: Taylor and Francis.

Hartl, D.L., and A.G. Clark. 2006. Principles of population genetics, 4th ed. Sunderland, Mass.: Sinauer.

Harvey, D.S., and P.J. Weatherhead. 2006. A test of the hierarchical model of habitat selection using Eastern Massasauga Rattlesnakes (*Sistrurus c. catenatus*). Biol. Conserv. 130:206–216.

Hasegawa, M., H. Kishino, and T. Yano. 1987. Man's place in Hominoidea as inferred from molecular clocks of DNA. J. Mol. Evol. 26:132–147.

Haskell, A., T.E. Graham, C.R. Griffin, and J.B. Hestbeck. 1996. Size related survival of headstarted redbelly turtles (*Pseudemys rubriventris*) in Massachusetts. J. Herpetol. 30:524–527.

Haslerud, G.M. 1938. The effect of movement of stimulus objects upon avoidance reactions in chimpanzees. J. Comp. Psychol. 25:507–528.

Hastings, W.K. 1970. Monte Carlo sampling methods using Markov chains and their applications. Biometrika 57:97–109.

Heath D.D., J.W. Heath, C.A. Bryden, R.M. Johnson, and C.W. Fox. 2003. Rapid evolution of egg size in captive salmon. Science 299:1738–1740.

Heath, J.E. 1964. Reptilian thermoregulation: Evaluation of field studies. Science 145:784–785.

Heckman, K.L., C.L. Mariana, R. Rasoloarison, and A.D. Yoder. 2007. Multiple nuclear loci reveal patterns of incomplete lineage sorting and complex species history within western mouse lemurs (*Microcebus*). Mol. Phylogenet. Evol. 43:353–367.

Hedges, S.B. 1992. The number of replications needed for accurate estimation of the bootstrap p value in phylogenetic studies. Mol. Biol. Evol. 9:366–369.

Hedrick, P.W. 1999. Perspective: Highly variable loci and their interpretation in evolution and conservation. Evolution 53:313–318.

Hedrick, P.W. 2000. Genetics of populations, 2nd ed. Sudbury, Mass.: Jones & Bartlett.

Hedrick, P.W. 2004. Recent developments in conservation genetics. For. Ecol. Manage. 197:3–19.

Hedrick, P.W. 2005a. "Genetic restoration": A more comprehensive perspective than "genetic rescue." Trends Ecol. Evol. 20:109.

Hedrick, P.W. 2005b. A standardized genetic differentiation measure. Evolution 59:1633–1638.

Heissenbuttel, A.E. 1996. Ecosystem management—principles for practical application. Ecol. Appl. 6:730–732.

Henderson, R.W. 2004. Lesser Antilliean snake faunas: Distribution, ecology and conservation concerns. Oryx 38:311–320.

Henderson, R.W., and R.A. Winstel. 1995. Aspects of habitat selection by an arboreal boa (*Corallus enydris*) in an area of mixed agriculture on Grenada. J. Herpetol. 29:272–275.

Henderson, R.W., M.H. Binder, R.A. Sajdak, and J.A. Buday. 1980. Aggregating behavior and exploitation of subterranean habitat by gravid Eastern Milk Snakes (*Lampropeltis t. triangulum*). Milw. Publ. Mus. Contrib. Biol. Geol. 32:1–9.

Henery, R.J. 1994. Classification. *In* D. Michie, D.J. Spiegelhalter, and C.C. Taylor, eds., Machine learning, neural and statistical classification, pp. 6–16. New York: Ellis Horwood.

Heppell, S.S. 1998. Application of life-history theory and population model analysis to turtle conservation. Copeia 1998:367–375.

Heppell, S.S., L.B. Crowder, and D.T. Crouse. 1996. Models to evaluate headstarting as a management tool for long lived turtles. Ecol. Appl. 6:556–565.

Herrington, R.E. 1988. Talus use by amphibians and reptiles in the Pacific Northwest. *In* R.C. Szaro, K.E. Severson, and D.R. Patton, eds., Management of amphibians, reptiles, and small mammals in North America. Proceedings of the symposium, July 19–21, 1988, Flagstaff, Arizona, pp. 216–221. Fort Collins, Colo.: U.S. Forest Service, Rocky Mountain Forest and Range Experiment Station.

Herzberg, A., and A.D. Schmidt. 2001. *Bericht zum Stand des Erprobungs- und Entwicklungsvorhabens "Würfelnatter" der DGHT*, Pt. 2: *Erprobungsstandort Lahn.* Elaphe 9:73–80.

Hey, J., and R. Nielsen. 2004. Multilocus methods for estimating population sizes, migration rates and divergence time, with applications to the divergenece of *Drosophila pseudoobscura* and *D. persimilis*. Genetics 167:747–760.

Hey, J., Y.-J. Won, A. Sivasundar, R. Nielsen, and J.A. Markert. 2004. Using nuclear haplotypes with microsatellites to study gene flow between recently separated cichlid species. Mol. Ecol. 13:909–919.

Hibbard, C.W. 1964. A brooding colony of the Blind Snake, *Leptotyphlops dulcis dissecta* Cope. Copeia 1964:222.

Hickerson, M.J., E. Stahl, and H.A. Lessios. 2006. Test for simultaneous divergence using approximate Bayesian computation. Evolution 60:2435–2453.

Hickerson, M.J., E. Stahl, and N. Takebayashi. 2007. msBayes: A flexible pipeline for comparative phylogeographic inference using approximate Bayesian computation (ABC). BMC Bioinf. 8:268.

Hill, W.G. 1981. Estimation of effective population size from data on linkage disequilibrium. Genet. Res. 38:209–216.

Hille, A. 1997. Biochemical variation between populations of the western and eastern Grass Snake (*Natrix natrix*) from the transition zone in Nordheim-Westfalen, Germany. *In* W. Bohme, W. Bischoff, and T. Siegler, eds., Herpetologia Bonnensis, pp. 177–184. Bonn: Societas Europaea Herpetologica.

Hille, A., I.A. Janssen, S.B. Menken, M. Schlegel, and R.S. Thorpe. 2002. Heterologous amplification of microsatellite markers from colubroid snakes in European natricines (Serpentes: Natricinae). J. Hered. 93:63–66.

Hillis, D.M., and J.J. Bull. 1993. An empirical test of bootstrapping as a method for assessing confidence in phylogenetic analysis. Syst. Biol. 42:182–192.

Hirsch, S.M., and R.C. Bolles. 1980. On the ability of prey to recognize predators. Z. Tierpsychol. 54:71–84.

Hoff, M.P., and T.L. Maple. 1982. Sex and age differences in the avoidance of reptile exhibits by zoo visitors. Zoo Biol. 1:263–269.

Holder, M., and P.O. Lewis. 2003. Phylogeny estimation: Traditional and Bayesian approaches. Nat. Rev. Genet. 4:275–284.

Holland, B.S., and R.H. Cowie. 2007. A geographic mosaic of passive dispersal: Population structure in the endemic Hawaiian amber snail *Succinea caduca* (Mighels, 1845). Mol. Ecol. 16:2422–2435.

Holway, D.A., L. Lach, A.V. Suarez, N.D. Tsutsui, and T.J. Case. 2002. The causes and consequences of ant invasions. Annu. Rev. Ecol. Syst. 33:181–233.

Holycross, A.T. 2002. Conservation biology of two rattlesnakes: *Crotalus willardi obscurus* and *Sistrurus catenatus edwardsi*. PhD dissertation, Arizona State University, Tempe.

Holycross, A.T., and M.E. Douglas. 2007. Geographic isolation, genetic divergence, and ecological non-exchangeability define ESUs in a threatened sky-island rattlesnake. Biol. Conserv. 134:142–154.

Holycross, A.T., and S.R. Goldberg. 2001. Reproduction in northern populations of the Ridgenose Rattlesnake, *Crotalus willardi* (Serpentes: Viperidae). Copeia 2001:473–481.

Holycross, A.T., M.E. Douglas, J.R. Higbee, and R.H. Bogden. 2002. Isolation and characterization of microsatellite loci from a threatened rattlesnake (New Mexico Ridgenosed Rattlesnake, *Crotalus willardi obscurus*). Mol. Ecol. Notes 2:537–539.

Homyack, J.D., and W.M. Giuliano. 2002. Effect of streambank fencing on herpetofauna in pasture stream zones. Wildl. Soc. Bull. 30:361–369.

Hooge, P.N., and B. Eichenlaub. 2000. Animal movement extension to Arcview (ver. 2.0). Alaska Science Center—Biological Science Office. Anchorage, AK: U.S. Geological Survey.

Hopkins, W.A., C.L. Rowe, and J.D. Congdon. 1999. Elevated trace element concentrations and standard metabolic rate in Banded Water Snakes (*Nerodia fasciata*) exposed to coal combustion wastes. Environ. Toxicol. Chem. 18:1258–1263.

Hopkins, W.A., J.H. Roe, J.W. Snodgrass, B.P. Jackson, D.E. Kling, C.L. Rowe, and J.D. Congdon. 2001. Nondestructive indices of trace element exposure in squamate reptiles. Environ. Pollut. 115:1–7.

Hopko, D.R., S.M.C. Robertson, L. Widman, and C.W. Lejuez. 2008. Specific phobias. *In* M. Hersen and J. Rosqvist, eds., Handbook of psychological assessment, case conceptualization, and treatment, vol. 1, pp. 139–170. New York: J. Wiley.

Hopley, C.G. 1882. Snakes: Curiosities and wonders of serpent life. London: Griffith & Farran.

Hosmer, D.W., and S. Lemeshow. 1989. Applied logistics regression. New York: John Wiley & Sons.

Houlihan, J.E., C.S. Findlay, B.R. Schmidt, A.H. Meyers, and S.L. Kuzmin. 2000. Quantitative evidence for global amphibian declines. Nature 404:752–755.

Houston, D.L., and R. Shine. 1993. Sexual dimorphism and niche divergence: Feeding habits of the Arafura Filesnake. J. Anim. Ecol. 62:737–749.

Howe, H.F. 1994. Managing species diversity in tallgrass prairie: Assumptions and implications. Conserv. Biol. 8:691–704.

Howes, B.J., B. Lindsay, and S.C. Lougheed. 2006. Range-wide phylogeography of a temperate lizard, the five-lined skink (*Eumeces fasciatus*). Mol. Phylogenet. Evol. 40:183–194.

Huang, S., S. He, Z. Peng, and E. Zhao. 2007. Molecular phylogeography of endangered sharp-snouted pitviper (*Deinagkistrodon acutus;* Reptilia, Viperidae) in mainland China. Mol. Phylogenet. Evol. 44:942–952.

Huang, T.F., and C. Ouyang. 1984. Action mechanism of the potent platelet aggregation inhibitor from *Trimeresurus gramineus* snake venom. Thromb. Res. 33:125–138.

Hudson, R.R. 1990. Gene genealogies and the coalescent process. Oxford Surv. Evol. Biol. 7:1–44.

Hudson, R.R., M. Slatkin, and W.P. Maddison. 1992. Estimation of levels of gene flow from DNA sequence data. Genetics 132:583–589.

Huelsenbeck, J.P.H., and D.M. Hillis. 1993. Success of phylogenetic methods in the four-taxon case. Syst. Biol. 42:247–264.

Huelsenbeck, J.P.H., B. Larget, R.E. Miller, and F. Ronquist. 2002. Potential applications and pitfalls of Bayesian inference of phylogeny. Syst. Biol. 51:673–688.

Huey, R.B. 1982. Temperature, physiology, and the ecology of reptiles. *In* C. Gans and F.H. Pough, eds., Biology of the Reptilia, Vol. 12, pp. 25–91. New York: Academic Press.

Huey, R.B., and J.G. Kingsolver. 1989. Evolution of thermal sensitivity of ectotherm performance. Trends Ecol. Evol. 4:131–135.

Huey, R.B., C.R. Peterson, S.J. Arnold, and W.P. Porter. 1989. Hot rocks and not-so-hot rocks: Retreat-site selection by garter snakes and its thermal consequences. Ecology 70:931–944.

Hughes, L. 2000. Biological consequences of global warming: Is the signal already apparent? Trends Ecol. Evol. 15:56–61.

Hughes, L. 2003. Climate change in Australia: Trends, projections and impacts. Aust. Ecol. 28:423–443.

Huhndorf, M.H., J.C. Kerbis Peterhans, and S.S. Loew. 2007. Comparative phylogeography of three endemic rodents from the Albertine Rift, east central Africa. Mol. Ecol. 16:663–674.

Humphries, M.M., D.W. Thomas, and J.R. Speakman. 2002. Climate-mediated energetic constraints on the distribution of hibernating mammals. Nature 418:313–316.

Hunter, M.L., Jr. 2000. Refining normative concepts in conservation. Conserv. Biol. 14:573–574.

Hurlbert, S.H. 1984. Pseudoreplication and the design of ecological field experiments. Ecol. Monogr. 54:187–211.

Hutchins, M. 2007a. The animal rights-conservation debate: Can zoos play a role? *In* A. Zimmermann, M. Hatchwell, L. Dickie, and C. West, eds., Zoos in the 21st century: Catalysts for conservation?, pp. 92–109. Cambridge, U.K.: Cambridge University Press.

Hutchins, M. 2007b. The limits of compassion. Wildl. Professional 1:42–44.

Hutchison, V.H., H.G. Dowling, and A. Vinegar. 1966. Thermoregulation in a brooding female Indian python, *Python molurus bivittatus*. Science 151:694–696.

Hyslop, N.R. 2001. Spatial ecology and habitat use of the Copperbelly Water Snake (*Nerodia erythrogaster neglecta*) in a fragmented landscape. MS thesis, Purdue University, Fort Wayne.

Ilmonen, J., and L. Paasiverta. 2005. Benthic macrocrustacean and insect assemblages in relation to spring habitat characteristics: Patterns in abundance and diversity. Hydrobiologia 533:99–113.

Ineich, I., X. Bonnet, R. Shine, T. Shine, F. Brischoux, M. Lebreton, and L. Chirio. 2006. What, if anything, is a "typical" viper? Biological attributes of basal viperid snakes (genus *Causus* Wagler, 1830). Biol. J. Linn. Soc. 89:575–588.

Irwin, K.J., T.E. Lewis, J.D. Kirk, S.L. Collins, and J.T. Collins. 2003. Status of the Eastern Indigo Snake (*Drymarchon couperi*) on St. Vincent National Wildlife Refuge, Franklin County, Florida. J. Kansas Herpetol. 7:13–20.

Isaac, L.A., and P.T. Gerogory 2004. Thermoregulatory behaviour of gravid and non-gravid female Grass Snakes (*Natrix natrix*) in a thermally limiting high-latitude environment. J. Zool. 264:403–409.

Isaaks, E.H., and R.M. Srivastava. 1989. An introduction to applied geostatistics. New York: Oxford University Press.

Isbell, L.A. 2006. Snakes as agents of evolutionary change in primate brains. J. Hum. Evol. 51:1–35.

Ishwar, N.M., R. Chellam, and A. Kumar. 2001. Distribution of forest floor reptiles in the rainforest of Kalakad-Mundanthurai Tiger Reserve, South India. Curr. Sci. 80:413–418.

[ITIS] Integrated Taxonomic Information System. 2006. Online at http://www.itis.usda.gov/ (accessed 14 Dec. 2006).

[IUCN] International Union for Conservation of Nature. 1998. Guidelines for re-introductions. Prepared by the IUCN/SSC re-introduction specialist group. Cambridge, U.K.: IUCN.

Jacobson, E. 1994. Veterinary procedures for the acquisition and release of captive-bred herpetofauna. *In* J.B. Murphy, K. Adler, and J.T. Collins, eds., Captive management and conservation of amphibians and reptiles, pp. 109–118. Ithaca: Society for the Study of Amphibians and Reptiles.

Jäggi, C., and B. Baur. 1999. Overgrowing forest as a possible cause for the local extinction of *Vipera aspis* in the northern Swiss Jura mountains. Amphib-Reptilia 20:25–34.

Jäggi, C., T. Wirth, and B. Baur. 2000. Genetic variability in subpopulations of the asp viper (*Vipera aspis*) in the Swiss Jura mountains: Implications for a conservation strategy. Biol. Conserv. 94:69–77.

James, C.D. 2003. Response of vertebrates to fenceline contrasts in grazing intensity in semi-arid woodlands of eastern Australia. Aust. Ecol. 28:137–151.

Jansen, K.P. 2001. Ecological genetics of the Salt Marsh Snake *Nerodia clarkii*. PhD dissertation, University of South Florida, Tampa.

Janzen, F.J., J.G. Frenz, T.S. Haselkorn, E.D. Brodie Jr., and E.D. Brodie III. 2002. Molecular phylogeography of Common Garter Snakes (*Thamnophis sirtalis*) in western North America: Implications for regional historical forces. Mol. Ecol. 11:1739–1751.

Jayne, B.C., and A.F. Bennett. 1990. Selection on locomotor performance capacity in a natural population of garter snakes. Evolution 44:1204–1229.

Jemison, S.C., L.A. Bishop, P.G. May, and T.M. Farrell. 1995. The impact of pit-tags on growth and movement of the rattlesnake, *Sistrurus miliarius*. J. Herpetol. 29:129–132.

Jenkins, C.L. 2007. Ecology and conservation of rattlesnakes in sagebrush steppe ecosystems: Landscape disturbance, small mammal communities, and Great Basin Rattlesnake reproduction. PhD dissertation, Idaho State University, Pocatello.

Jenkins, M., and S. Broad. 1994. International trade in reptile skins: A review and analysis of the main consumer markets, 1983–91. Cambridge, U.K.: TRAFFIC International.

Ji, X., and W.G. Du. 2001a. The effects of thermal and hydric environments on hatching success, embryonic use of energy and hatchling traits in a colubrid snake, *Elaphe carinata*. Comp. Biochem. Physiol. A Mol. Integr. Physiol. 129:461–471.

Ji, X., and W.G. Du. 2001b. Effects of thermal and hydric environments on incubating eggs and hatchling traits in the cobra, *Naja naja atra*. J. Herpetol. 35:186–194.

Ji, X., P.Y. Sun, S.Y. Fu, and H.S. Zhang. 1997. Utilization of energy and nutrients in incubating eggs and post-hatching yolk in a colubrid snake, *Elaphe carinata*. Herpetol. J. 7:7–12.

Jia, X.P., and J.A. Richards. 1999. Segmented principal components transformation for efficient hyperspectral remote-sensing image display and classification. IEEE Trans. Geosci. Remote Sens. 37:538–542.

Jiang, Z.J., T.A. Castoe, C.C. Austin, F.T. Burbrink, M.D. Herron, J.A. McGuire, C.L. Parkinson, and D.D. Pollock. 2007. Comparative mitochondrial genomics of snakes: Extraordinary substitution rate dynamics and functionality of the duplicate control region. BMC Evol. Biol. 7:123.

Jin, L., and M. Nei. 1990. Limitations of the evolutionary parsimony method of phylogenetic analysis. Mol. Biol. Evol. 7:82–102.

Johnson, B. 1999. Managing attitudes: Education strategies for the recovery of the Eastern Massasauga Rattlesnake. *In* B. Johnson and M. Wright, eds., Second international symposium and workshop on the conservation of the Eastern Massasauga Rattlesnake, *Sistrurus catenatus catenatus:* Population and habitat management issues in urban, bog, prairie, and forested ecosystems, pp. 33–41. Toronto: Toronto Zoo.

Johnson, D.H. 1980. The comparison of usage and availability measurements for evaluating resource preference. Ecology 61:65–71.

Johnson, G. 1995. Spatial ecology, habitat preferences, and habitat management of the eastern massasauga, *Sistrurus catenatus catenatus*, in a New York transition peatland. PhD dissertation, State University of New York, Syracuse.

Johnson, G., and A.R. Breisch. 1993. The Eastern Massasauga Rattlesnake in New York: Occurrence and habitat management. *In* B. Johnson and V. McKenzies, eds., Proceedings of the international symposium and workshop on the conservation of the Eastern Massasauga Rattlesnake, *Sistrurus catenatus catenatus*, pp. 48–54. West Hill, Canada: Toronto Zoo.

Johnson, G., and A.R. Breisch. 1999. Preliminary evaluation of a habitat management plan for the eastern massasauga in a New York peatland. *In* B. Johnson and M. Wright, eds., Second international symposium and workshop on the conservation of the Eastern Massasauga Rattlesnake, *Sistrurus catenatus catenatus:* Population and habitat management issues in urban, bog, prairie and forested ecosystems, pp. 155–159. Toronto: Toronto Zoo.

Johnson, G., and D.J. Leopold. 1998. Habitat management for the eastern massasauga in a central New York peatland. J. Wildl. Manage. 62:84–97.

Johnson, G., B. Kingsbury, R. King, C. Parent, R.A. Seigel, and J. Szymanski. 2000. The Eastern Massasauga Rattlesnake: A handbook for land managers. Fort Snelling, Minn.: U.S. Fish and Wildlife Service.

Johnson, J.P. 2007. Herpetology at the Arizona-Sonora Desert Museum. Herpetol. Rev. 38:7–12.

Johnson, K.P., and D.H. Clayton. 2000. Nuclear and mitochondrial genes contain similar phylogenetic signal for pigeons and doves (Aves: Columbiformes). Mol. Phylogenet. Evol. 14:141–151.

Johnston, J.J., P.J. Savarie, T.M. Primus, J.D. Eiseman, J.C. Hurley, and D.J. Kohler. 2002. Risk assessment of an acetaminophen baiting program for chemical control of the Brown Tree Snake on Guam: Evaluation of baits, snake residues, and potential primary and secondary hazards. Environ. Sci. Technol. 36:3827–3833.

Jones, B., S.F. Fox, D.M. Leslie Jr., D.M. Engle, and R.L. Lochmiller. 2000. Herpetofaunal responses to brush management with herbicide and fire. J. Range Manage. 53:154–158.

Jones, H.E., and M.C. Jones. 1928. Maturation and emotion: Fear of snakes. Child. Educ. 5:136–143.

Jones, J.M., and J.H. Witham. 1990. Post-translocation survival and movements of metropolitan white-tailed deer. Wildl. Soc. Bull. 18:434–441.

Jones, K.W., and L. Singh. 1985. Snakes and the evolution of sex chromosomes. Trends Genet. 1:55–61.

Jordan, P.W., A.E. Goodman, and S. Donnellan. 2002. Microsatellite primers for Australian and New Guinean pythons isolated with an efficient marker development method for related species. Mol. Ecol. Notes 2:78–82.

Joslin, J., H. Fletcher, and J. Emlen. 1964. A comparison of the responses to snakes of lab- and wild-reared rhesus monkeys. Anim. Behav. 12:348–352.

Jourdan, H., R.A. Sadlier, and A.M. Bauer. 2001. Little fire ant invasion (*Wasmannia auropuntata*) as a threat to New Caledonian lizards: Evidences from a Sclerophyll forest (Hymentoptera: Formicidae). Sociobiology 38:283–301.

Jukes, T.H., and C.R. Cantor. 1969. Evolution of protein molecules. *In* H.M. Munro, ed., Mammalian protein metabolism, pp. 21–132. New York: Academic Press.

Kavanagh, R., and M.A. Stanton. 2005. Vertebrate species assemblages and species sensitivity to logging in the forests of north-eastern New South Wales. For. Ecol. Manage. 209:309–341.

Kawai, N., and M. Shibasaki. 2009. Rapid detection of snakes by Japanese monkeys (*Macaca fuscata*): An evolutionarily predisposed visual system. J. Comp. Psychol. 123: in press.

Kearney, M. 2002. Hot rocks and much-too-hot rocks: Seasonal patterns of retreat-site selection by a nocturnal ectotherm. J. Therm. Biol. 27:205–218.

Keck, M.B. 1994a. A new technique for sampling semi-aquatic snake populations. Herpetol. Nat. Hist. 2:101–103.

Keck, M.B. 1994b. Test for detrimental effects of PIT tags in neonatal snakes. Copeia 1994:226–228.

Keenlyne, K.D., and J.R. Beer. 1973. Food habits of *Sistrurus catenatus catenatus*. J. Herpetol. 7:382–384.

Keller, L.F., and D.M. Waller. 2002. Inbreeding effects in wild populations. Trends Ecol. Evol. 17:230241.

Kelly, J.M., and M.R. Hodge. 1996. The role of corporations in ensuring biodiversity. Environ. Manage. 20:947–954.

Kendall, W.L., J.D. Nichols, and J.E. Hines. 1997. Estimating temporary emigration using capture-recapture data with Pollock's robust design. Ecology 78:2248–2248.

Keogh, J.S., D.G. Barker, and R. Shine. 2001. Heavily exploited but poorly known: Systematics and biogeography of commercially harvested pythons (*Python curtus* group) in Southeast Asia. Biol. J. Linn. Soc. 73:113–129.

Keogh, J.S., I.A.W. Scott, M. Fitzgerald, and R. Shine. 2003. Molecular phylogeny of the Australian venomous snake genus *Hoplocephalus* (Serpentes, Elapidae) and conservation genetics of the threatened *H. stephensii*. Conserv. Genet. 4:57–65.

Keogh, J.S., I.A.W. Scott, and C. Hayes. 2005. Rapid and repeated origin of insular gigantism and dwarfism in Australian Tiger Snakes. Evolution 59:226–233.

Keogh, J.S., J.K. Webb, and R. Shine. 2007. Spatial genetic analysis and long-term mark-recapture data demonstrate male-biased dispersal in a snake. Biol. Lett. 3:33–35.

Kephart, D.G., and S.J. Arnold. 1982. Microgeographic variation in the diets of garter snakes. Oecologia 52:287–291.

Kerfoot, W.C., and A. Sih. 1987. Predation: Direct and indirect impacts on aquatic communities. Hanover, N.H.: University Press of New England.

Kery, M. 2002. Inferring the absence of a species: A case study of snakes. J. Wildl. Manage. 66:330–338.

Kilpatrick, E.S., D.B. Kubacz, D.C. Guynn Jr., J.D. Lanham, and T.A. Waldrop. 2004. The effects of prescribed burning and thinning on herpetofauna and small mammals in the upper piedmont of South Carolina: Preliminary results of the National Fire and Fire

Surrogate Study. *In* K.F. Connor, ed., Proceedings of the 12th biennial southern silvicultural research conference, February 24–28, 2003, Biloxi, Mississippi. General Technical Report SRS-71, pp. 18–22. Asheville, N.C.: U.S. Forest Service, Southern Research Station.

King, M.B., and D. Duvall. 1990. Prairie Rattlesnake seasonal migrations: Episodes of movement, vernal foraging and sex differences. Anim. Behav. 39:924–935.

King, R., C. Berg, and B. Hay. 2004. A repatriation study of the eastern massasauga (*Sistrurus catenatus catenatus*) in Wisconsin. Herpetologica 60:429–437.

King, R.B. 1986. Population ecology of the Lake Erie Water Snake, *Nerodia sipedon insularum*. Copeia 1986:757–772.

King, R.B. 1987. Color pattern polymorphism in the Lake Erie Water Snake, *Nerodia sipedon insularum*. Evolution 41:241–255.

King, R.B. 1988. Polymorphic populations of the garter snake *Thamnophis sirtalis* near Lake Erie. Herpetologica 44:451–458.

King, R.B. 1992. Lake Erie Water Snakes revisited: Morph and age specific variation in relative crypsis. Evol. Ecol. 6:115–124.

King, R.B. 1993a. Color pattern variation in Lake Erie Water Snakes: Inheritance. Can. J. Zool. 71:1985–1990.

King, R.B. 1993b. Color pattern variation in Lake Erie Water Snakes: Prediction and measurement of natural selection. Evolution 47:1819–1833.

King, R.B. 1993c. Determinants of offspring number and size in the Brown Snake, *Storeria dekayi*. J. Herpetol. 27:175–185.

King, R.B. 2003. Mendelian inheritance of melanism in the garter snake *Thamnophis sirtalis*. Herpetologica 59:486–491.

King, R.B., and R. Lawson. 1995. Color pattern variation in Lake Erie Water Snakes: The role of gene flow. Evolution 49:885–896.

King, R.B., and R. Lawson. 1996. Sex-linked inheritance of fumarate hydratase alleles in natricine snakes. J. Hered. 87:81–83.

King, R.B., and R. Lawson. 2001. Patterns of population subdivision and gene flow in three sympatric natricine snakes. Copeia 2001:602–614.

King, R.B., and K.M. Stanford 2006. Headstarting as a management tool: A case study of the Plains Gartersnake. Herpetologica 62:282–292.

King, R.B., W.B. Milstead, H.L. Gibbs, M.R. Prosser, G.M. Burghardt, and G.F. McCracken. 2001. Application of microsatellite DNA markers to discriminate between maternal and genetic effects on scalation and behavior in multiply-sired garter snake litters. Can. J. Zool. 79:121–128.

King, R.B., A. Queral-Regil, and K.M. Stanford. 2006a. Population size and recovery criteria of the threatened Lake Erie Watersnake: Integrating multiple methods of population estimation. Herpetol. Monogr. 20:83–104.

King, R.B., J.M. Ray, and K.M. Stanford. 2006b. Gorging on gobies: Beneficial effects of alien prey on a threatened vertebrate. Can. J. Zool. 84:108–115.

Kingman, J.F.C. 1982. On the genealogy of large populations. J. Appl. Probab. 19A:27–43.

Kingsbury, B.A. 2002. Conservation approach for eastern massasauga (*Sistrurus c. catenatus*). Milwaukee, Wis.: USDA Forest Service, Eastern Region.

Kingston, T. 1995. Valuable modeling tool: RAMAS GIS. Conserv. Biol. 9:966–968.

Kissner, K.J., and P.J. Weatherhead. 2005. Phenotypic effects on survival of neonatal Northern Watersnakes *Nerodia sipedon*. J. Anim. Ecol. 74:259–265.

Kissner, K.J., M.R. Forbes, and D.M. Secoy. 1998a. Sexual dimorphism in size of cloacal glands of the garter snake, *Thamnophis radix haydeni*. J. Herpetol. 32:268–270.

Kissner, K.J., D.M. Secoy, and M.R. Forbes. 1998b. Rattling behavior of Prairie Rattlesnakes (*Crotalus viridis viridis*, Viperidae) in relation to sex, reproductive status, body size and temperature. Ethology 103:1042–1050.

Kissner, K.J., G. Blouin-Demers, and P.J. Weatherhead. 2000. Sexual dimorphism in malodorousness of musk secretions of snakes. J. Herpetol. 34:491–493.

Kjoss, V.A., and J.A. Litvaitis. 2001. Community structure of snakes in a human-dominated environment. Biol. Conserv. 98:285–292.

Kleiman, D.G. 1989. Reintroduction of captive animals for conservation. Bioscience 39:152–164.

Kling, G.W., K. Hayhoe, L.B. Johnson, J.J. Magnuson, S. Polasky, S.K. Robinson, B.J. Shuter, M.M. Wander, D.J. Wuebbles, D.R. Zak, R.L. Lindroth, S.C. Moser, and M.L. Wilson. 2003. Confronting climate change in the Great Lakes region: Impacts on our communities and ecosystems. Cambridge, Mass.: Union of Concerned Scientists/Washington, D.C.: Ecological Society of America.

Knapp, R.A., and K.R. Matthews. 2000. Non-native fish introductions and the decline of the mountain yellow-legged frog from within protected areas. Conserv. Biol. 14:428–438.

Knowles, L.L., and B.C. Carstons. 2007. Estimating a geographically explicit model of population divergence for statistical phylogeography. Evolution 61:477–493.

Knowles, L.L., and W.P. Maddison. 2002. Statistical phylogeography. Mol. Ecol. 11:2623–2635.

Kohler, W. 1925. The mentality of apes. London: Routledge & Kegan Paul.

Kolbe, J.J., and F.J. Janzen. 2002. Impact of nest-site selection on nest success and nest temperature in natural and disturbed habitats. Ecology 83:269–281.

Kolbe, J.J., R.E. Glor, L.R. Schettino, A.D. Lara, and A. Larson. 2004. Genetic variation increases during biological invasion by a Cuban lizard. Nature 431:177–181.

Kotler, B.P., J.S. Brown, R.H. Slotow, W.L. Goodfriend, and M. Strauss. 1993. The influence of snakes on the foraging behavior of gerbils. Oikos 67:309–316.

Kozak, K.H., R.A. Blaine, and A. Larson. 2006. Gene lineages and eastern North American palaeodrainage basins: Phylogeography and speciation in salamanders of the *Eurycea bislineata* species complex. Mol. Ecol. 15:191–207.

Krajik, K. 2004. All downhill from here? Science 303:1600–1602.

Kuch, U., J.S. Keogh, J. Weigel, L.A. Smith, and M. Dietrich. 2005. Phylogeography of Australia's King Brown Snake (*Pseudechis australis*) reveals Pliocene divergence and Pleistocene dispersal of a top predator. Naturwissenschaften 92:121–127.

Kuhner, M.K., J. Yamato, and J. Felsenstein. 1995. Estimating effective population size and mutation rate from sequence data using Metropolis-Hastings sampling. Genetics 140:1421–1430.

Kumazawa, Y. 2004. Mitochondrial DNA sequences of five squamates: Phylogenetic affiliation of snakes. DNA Res. 11:137–144.

Kumazawa, Y., H. Ota, M. Nishida, and T. Ozawa. 1998. The complete nucleotide sequence of a snake (*Dinodon semicarinatus*) mitochondrial genome with two identical control regions. Genetics 150:313–329.

Kunda, Z. 1999. Social cognition: Making sense of people. Cambridge, Mass.: MIT Press.

Kunz, T.H. 2001. Seeing in the dark: Recent technological advances for the study of free-ranging bats. Bat Res. News 42:91.

Lackey, R.T. 2001. Values, policy, and ecosystem health. BioScience 51:437–443.

Lacki, M.J., J.W. Hummer, and H.J. Webster. 1992. Mine-drainage treatment wetland as habitat for herpetofaunal wildlife. Environ. Manage. 16:513–520.

Lacki, M.J., J.W. Hummer, and J.L. Fitzgerald. 2005. Population patterns of Copperbelly Water Snakes (*Nerodia erythrogaster neglecta*) in a riparian corridor impacted by mining and reclamation. Am. Midl. Nat. 153:357–369.

Lader, M.H., and A.M. Matthews. 1968. A physiological model of phobic anxiety and desensitization. Behav. Res. Ther. 6:411–421.

Lanave, C., G. Preparata, C. Saccone, and G. Serio. 1984. A new method for calculating evolutionary substitution rates. J. Mol. Evol. 20:86–93.

Landres, P.B., J. Verner, and J.W. Thomas. 1988. Ecological uses of vertebrate indicator species: A critique. Conserv. Biol. 2:316–328.

Lang, P.J. 1969. The mechanics of desensitization and the laboratory study of human fear. *In* C.M. Franks, ed., Behavior therapy: Appraisal and status, pp. 160–191. New York: McGraw-Hill.

Lang, P.J., and A.D. Lazovik. 1963. Experimental desensitization of a phobia. J. Abnorm. Soc. Psychol. 66:519–525.

Langley, W.M., H.W. Lipps, and J.F. Theis. 1989. Responses of Kansas motorists to snake models on rural highways. Trans. Kansas Acad. Sci. 92:43–48.

Langton, T.E.S., ed. 1989. Amphibians and roads. ACO Polymer Products, Shefford, U.K.

Lantuéjoul, C. 2002. Geostatistical simulation: Models and algorithms. New York: Springer.

Larget, B. 2006. Introduction to Markov chain Monte Carlo methods in molecular evolution. *In* R. Nielson, ed., Statistical methods in molecular evolution, pp. 45–62. New York: Springer.

Larkin, J.L., J.J. Cox, M.W. Wichrowski, M.R. Dzialak, and D.S. Maehr. 2004. Influences on release-site fidelity of translocated elk. Restor. Ecol. 12:97–105.

Larsen, K.W., and P.T. Gregory. 1989. Population size and survivorship of the Common Garter Snake, *Thamnophis sirtalis,* near the northern limit of its distribution. Holarctic Ecol. 12:81–86.

Lawson, R. 1987. Molecular studies of thamnophiine snakes. 1. The phylogeny of the genus *Nerodia*. J. Herpetol. 21:140–157.

Lawson, R., and H.C. Dessauer. 1979. Biochemical genetics and systematics of garter snakes in the *Thamnophis elegans-couchii-ordinoides* complex. Occ. Pap. Mus. Zool. La. State Univ. 56:1–24.

Lawson, R., and R.B. King. 1996. Gene flow and melanism in Lake Erie garter snake populations. Biol. J. Linn. Soc. 59:1–19.

Lawson, R., and C.S. Leib. 1990. Variation and hybridization in *Elaphe bairdi* (Serpentes: Colubridae). J. Herpetol. 24:280–292.

Lawson, R., A.J. Meier, P.G. Frank, and P.E. Moler. 1991. Allozyme variation and systematics of the *Nerodia fasciata-Nerodia clarkii* complex of water snakes (Serpentes: Colubridae). Copeia 1991:638–659.

Lawson, R., J.B. Slowinski, B.I. Crother, and F.T. Burbrink. 2005. Phylogeny of the Colubroidea (Serpentes): New evidence from mitochonrial and nuclear genes. Mol. Phylogenet. Evol. 37:581–601.

Leaché, A.D., and J.A. McGuire. 2006. Phylogenetic relationships of horned lizards (*Phrynosoma*) based on nuclear and mitochondrial data: Evidence for a misleading mitochondrial gene tree. Mol. Phylogenet. Evol. 39:628–644.

Leaché, A.D., and T.W. Reeder. 2002. Molecular systematics of the eastern fence lizard (*Sceloporus undulatus*): A comparison of parsimony, likelihood, and Bayesian approaches. Syst. Biol. 51:44–68.

Leberg, P.L., and G.D. Firmin. 2008. Role of inbreeding depression and purging in captive breeding and restoration programmes. Mol. Ecol. 17:334–343.

Lebreton, J.D., G. Hemery, J. Clobert, and H. Coquillart. 1992. Modeling survival and testing biological hypotheses using marked animals: A unified approach with case studies. Ecol. Monogr. 62:67–118.

Lee, J.R., and C.R. Peterson. 2003. Herpetological inventory of Craters of the Moon National Monument 1999–2001. Final Report to Craters of the Moon National Monument and Preserve, National Park Service, Pocatello, Idaho.

Lefsky, M.A., W.B. Cohen, G.G. Parker, and D.J. Harding. 2002. Lidar remote sensing for ecosystem studies. BioScience 52: 19–30.

LeGalliard, J., P.S. Fitze, R. Ferriere, and L. Clobert. 2005. Sex ratio bias, male aggression, and population collapse in lizards. Proc. Nat. Acad. Sci. U.S.A. 102:18231–18236.

Legge, J.T. 1996. Final report on the status and distribution of the Eastern Massasauga Rattlesnake (*Sistrurus catenatus catenatus*) in Michigan. Unpublished report to U.S. Fish and Wildlife Service, Region 3 Endangered Species Office, Fort Snelling, Minn.

Lehtinen, R.M., J.-B. Ramanamanjato, and J.G. Raveloarison. 2003. Edge effects and extinction proneness in a herpetofauna from Madagascar. Biodiv. Conserv. 12:1357–1370.

Leinonen, T., R.B. O'Hara, J.M. Cano, and J. Merilä. 2008. Comparative studies of quantitative trait and neutral marker divergence: A meta-analysis. J. Evol. Biol. 21:1–17.

Lemmon, E.M., A.R. Lemmon, J.T. Collins, J.A. Lee-Yaw, and D.C. Cannatella. 2007. Phylogeny-based delimitation of species boundaries and contact zones in the trilling chorus frogs. Mol. Phylogenet. Evol. 44:1068–1082.

Lenz, S., and A.D. Schmidt. 2002. *Bericht zum Stand des Erprobungs- und Entwicklungsvorhabens "Würfelnatter" der DGHT*, Pt. 3: *Erprobungsstandort Mosel*. Elaphe 10:53–59.

Leopold, A. 1949. A Sand County almanac. Oxford, U.K.: Oxford University Press.

Lerner, J., J. Mackey, and F. Casey. 2007. What's in Noah's wallet? Land conservation spending in the United States. BioScience 57:419–423.

Levell, J. 1997. A field guide to reptiles and the law. 2nd ed. Lanesboro, Minn.: Serpent's Tale Books.

Levin, P.S., E.E. Holmes, K.R. Piner, and C.J. Harvey. 2006. Shifts in a Pacific ocean fish assemblage: The potential influence of exploitation. Conserv. Biol. 20:1181–1190.

Levin, S.A. 1992. The problem of pattern and scale in ecology. Ecology 73:1943–1967.

Levis, D.J. 1969. The phobic test apparatus: An objective measure of human avoidance behavior to small objects. Behav. Res. Ther. 7:309–315.

Leynaud, G. C., and E.H. Bucher. 2005. Restoration of degraded Chaco woodlands: Effects on reptile assemblages. For. Ecol. Manage. 213:384–390.

Li, H., and J.F. Reynolds. 1995. On definition and quantification of heterogeneity. Oikos 73:280–284.

Li, J.-L. 1995. China Snake Island. Dalian: Liaoning Science and Technology Press.

Li, M., B.G. Fry, and R.M. Kini. 2005. Putting the brakes on snake venom evolution: The unique molecular evolutionary patters of *Aipysurus eydouxii* (marbled sea snake) phospholipase A_2 toxins. Mol. Biol. Evol. 22:934–941.

Lifson, N., and R. McClintock. 1966. Theory of use of the turnover rates of body water for measuring energy and material balance. J. Theor. Biol. 12:46–74.

Lillywhite, H.B. 1987. Temperature, energetics and physiological ecology. *In* R.A. Seigel, J.T. Collins, and S.S. Novak, eds., Snakes: Ecology and evolutionary biology, pp. 442–477. New York: McGraw-Hill.

Lin, Z.X., X. Ji, L.G. Luo, and X.M. Ma. 2005. Incubation temperature affects hatching success, embryonic expenditure of energy and hatchling phenotypes of a prolonged egg-retaining snake, *Deinagkistodon actus* (Viperidae). J. Therm. Biol. 30:289–297.

Lincoln, F.C. 1930. Calculating waterfowl abundance on the basis of banding returns. USDA Circular 118:1–4.

Lind, A.J., H.H. Welsh Jr., and D.A. Tallmon. 2005. Garter snake population dynamics from a 16-year study: Considerations for ecological monitoring. Ecol. Appl. 15:294–303.

Lindell, L.E. 1997. Annual variation in growth rate and body condition of Adders, *Vipera berus:* Effects of food availability and weather. Can. J. Zool. 75:261–270.

Lindenmayer, D.B., C.R. Margules, and D.B. Botkin. 2000. Indicators of biodiversity for ecologically sustainable forest management. Conserv. Biol. 14:941–950.

Lindstedt, D.M. 2005. Renewable resources at stake: Barataria-Terrebonne estuarine system in southeast Louisiana. J. Coastal Res. 44:162–175.

Linnaeus, C. 1758. *Systema naturae per regina tria naturae, secundum classes, ordines, genera, species, cum characteribus, differentiis, synonymis, locis,* 10th ed. Stockholm, Sweden.

Lips, K.R., F. Brem, R. Brenes, J.D. Reeve, R.A. Alford, J. Voyles, C. Carey, L. Livo, A.P. Pessier, and J.P. Collins. 2006. Emerging infectious disease and the loss of biodiversity in a neotropical amphibian community. Proc. Nat. Acad. Sci. U.S.A. 103:3165–3170.

Litt, A.R., L. Provencher, G.W. Tanner, and R. Franz. 2001. Herpetofaunal responses to restoration treatments of longleaf pine sandhills in Florida. Restor. Ecol. 9:462–474.

Litvaitis, J.A. 1993. Response of early successional vertebrates to historic changes in land use. Conserv. Biol. 7:866–873.

Liu, L., and D.K. Pearl. 2007. Species trees from gene trees: Reconstructing Bayesian posterior distributions of a species phylogeny using estimated gene tree distributions. Syst. Biol. 56:504–514.

Lloyd, B.D., and R.G. Powlesland. 1994. The decline of Kakapo Strigops habroptilus and attempts at conservation by translocation. Biol. Conserv. 69:75–85.

LoBue, V., and J.S. DeLouche. 2008. Detecting the snake in the grass. Psychol. Sci. 19:284–289.

Lockwood, M.A., C.P. Griffin, M.E. Morrow, C.J. Randel, and N.J. Silvy. 2005. Survival, movments, and reproduction of released captive-reared Attwater's prairie-chicken. J. Wildl. Manage. 69:1251–1258.

Lodge, D.M. 1993. Biological invasions: Lessons for ecology. Trends Ecol. Evol. 8:133–137.

Loehle, C., T.B. Wigley, P.A. Shipman, S.F. Fox, S. Rutzmoser, R.E. Thill, and M.A. Melchiors. 2005. Herpetofaunal species richness responses to forest landscape structure in Arkansas. For. Ecol. Manage. 209:293–308.

Lohoefener, R., and L. Lohmeier. 1986. Experiments with gopher tortoise (Gopherus polyphemus) relocation in southern Mississippi. Herpetol. Rev. 17:37–40.

Lougheed, S.C., H.L. Gibbs, K.A. Prior, and P.J. Weatherhead. 1999. Hierarchical patterns of genetic population structure in Black Rat Snakes (Elaphe obsoleta obsoleta) as revealed by microsatellite DNA analysis. Evolution 53:1995–2001.

Lougheed, S.C., H.L. Gibbs, K.A. Prior, and P.J. Weatherhead. 2000. A comparison of RAPD versus microsatellite DNA markers in population studies of the Massasauga Rattlesnake. J. Hered. 91:458–463.

Lourdais, O., X. Bonnet, D. DeNardo, and G. Naulleau. 2002. Do sex divergences in reproductive ecophysiology translate into dimorphic demographic patterns? Popul. Ecol. 44:241–249.

Lourdais, O., R. Shine, X. Bonnet, M. Guillon, and G. Naulleau. 2004. Climate effects embryonic development in a viviparous snake, Vipera aspis. Oikos 104:551–560.

Lowe, A., S. Harris, and P. Ashton. 2004. Ecological genetics: Design, analysis, and application. Malden, Mass.: Blackwell.

Luikart, G., P.R. England, D. Tallmon, S. Jordon, and P. Taberlet. 2003. The power and promise of population genomics: From genotyping to genome typing. Nat. Rev. Genet. 4:981–994.

Luiselli, L. 2006. Site occupancy and density of sympatric Gabon viper (Bitis gabonica) and nose-horned viper (Bitis nasicornis). J. Trop. Ecol. 22:555–564.

Luiselli, L., and D. Capizzi. 1997. Influences of area, isolation and habitat features on distribution of snakes in Mediterranean fragmented woodlands. Biodiv. Conserv. 6:1339–1351.

Luiselli, L., M. Capula, and R. Shine. 1997. Food habits, growth rates, and reproductive biology of Grass Snakes, Natrix natrix (Colubridae) in the Italian Alps. J. Zool. (London) 241:371–380.

Lukoschek, V., M. Waycott, and G. Dunshea. 2005. Isolation and characterization of microsatellite loci from the Australasian Sea Snake, Aipysurus laevis. Mol. Ecol. Notes 5:879–881.

Lukoschek, V., M. Waycott, and H. Marsh. 2007. Phylogeography of the Olive Sea Snake, Aipysurus laevis (Hydrophiinae) indicates Pleistocene range expansion around northern Australia but low contemporary gene flow. Mol. Ecol. 16:3406–3422.

Lutterschmidt W.I., and H.K. Reinert. 1990. The effect of ingested transmitters upon the temperature preference of the Northern Water Snake, Nerodia s. sipedon. Herpetologica 46:39–42.

Lynch, M., and T.J. Crease. 1990. The analysis of population survey data on DNA sequence variation. Mol. Biol. Evol. 7:377–394.

Lynch, M., J. Conery, and R. Büger. 1995. Mutation accumulation and the extinction of small populations. Am. Nat. 146:489–518.

MacArthur, R.H., and E.O. Wilson. 1967. The theory of island biogeography. Princeton: Princeton University Press.

MacInnes, C.D., E.H. Dunn, D.H. Rusch, F. Cooke, and F.G. Cooch. 1990. Advancement of goose nesting dates in the Hudson Bay region 1951–86. Can. Field-Nat. 104:295–297.

MacKenzie, D.I., J.D. Nichols, G.B. Lachman, S. Droege, J.A. Royle, and C.A. Langtimm. 2002. Estimating site occupancy rates when detection probabilities are less than one. Ecology 83:2248–2255.

MacKenzie, D.I., J.D. Nichols, J.A. Royle, K.H. Pollock, L.L. Bailey, J.E. Hines. 2006. Occupancy estimation and modeling: Inferring patterns and dynamics of species occurrence. Burlington, Mass.: Academic Press.

Mackessy, S.P. 2005. Desert Massasauga Rattlesnake (*Sistrurus catenatus edwardsii*): A technical conservation assessment. [online]. Denver, Colo.: U.S. Forest Service, Rocky Mountain Region, Species Conservation Project.

Mackessy, S.P., K. Williams, and K.G. Ashton. 2003. Ontogenetic variation in venom composition and diet of *Crotalus oreganus concolor*: A case of venom paedomorphosis? Copeia 2003:769–782

Macmillan, S. 1995. Restoration of an exptirpated Red-sided Garter Snake *Thamnophis sirtalis parietalis* population in the interlake region of Manitoba, Canada. Biol. Conserv. 72:13–16

Maddison, W.P. 1997. Gene trees in species trees. Syst. Biol. 46:523–536.

Mader, H.J. 1984. Animal habitat isolation by roads and agricultural fields. Biol. Conserv. 29:81–96.

Madsen, T. 1984. Movements, home range size and habitat use of radio-tracked Grass Snakes (*Natrix natrix*) in southern Sweden. Copeia 1984:707–713.

Madsen, T., and M. Osterkamp. 1982. Notes on the biology of the fish-eating snake *Lycodonomorphus bicolor* in Lake Tanganyika. J. Herpetol. 17:186–189.

Madsen, T., and R. Shine. 1992. A rapid, sexually selected shift in mean body size in a population of snake. Evolution 46:1220–1224.

Madsen, T., and R. Shine. 1993a. Costs of reproduction in a population of European Adders. Oecologia 94:488–495.

Madsen, T., and R. Shine. 1993b. Phenotypic plasticitiy in body sizes and sexual size dimorphism in European Grass Snakes. Evolution 47:321–325.

Madsen, T., and R. Shine. 1994. Components of lifetime reproductive success in Adders (*Vipera berus*). J. Anim. Ecol. 63:561–568.

Madsen, T., and R. Shine. 1998. Spatial subdivision within a population of tropical pythons (*Liasis fuscus*) in a superficially homogeneous habitat. Aust. J. Ecol. 23:340–348.

Madsen, T., and R. Shine. 1999a. The adjustment of reproductive threshold to prey abundance in a capital breeder. J. Anim. Ecol. 68:571–580.

Madsen, T., and R. Shine. 1999b. Life history consequences of nest-site variation in tropical pythons (*Liasis fuscus*). Ecology 80:989–997.

Madsen, T., and R. Shine. 2000a. Rain, fish and snakes: Climatically driven population dynamics of Arafura Filesnakes in tropical Australia. Oecologia 124:208–215.

Madsen, T., and R. Shine. 2000b. Silver spoons and snake body sizes: Prey availability early in life influences long-term growth rates of free ranging pythons. J. Anim. Ecol. 69:952–958.

Madsen, T., and B. Stille. 1988. The effect of size dependent mortality on color morphs in male Adders, *Vipera berus*. Oikos 52:73–78.

Madsen, T., and B. Újvári. 2006. MHC class I variation associates with parasite resistance and longevity in tropical pythons. J. Evol. Biol. 19:1973–1978.

Madsen, T., R. Shine, J. Loman, and T. Håkansson. 1992. Why do female Adders copulate so frequently? Nature 335:440–441.

Madsen, T., R. Shine, J. Loman, and T. Håkansson. 1993. Determinants of mating success in male Adders, *Vipera berus*. Anim. Behav. 45:491–499.

Madsen, T., B. Stille, and R. Shine. 1995. Inbreeding depression in an isolated population of Adders *Vipera berus*. Biol. Conserv. 75:113–118.

Madsen, T., R. Shine, M. Olsson, and H. Wittzell. 1999. Restoration of an inbred Adder population. Nature 402:34–35.

Madsen, T., M. Olsson, H. Wittzell, B. Stille, A. Gullberg, R. Shine, S. Andersson, and H. Tegelström. 2000. Population size and genetic diversity in sand lizards (*Lacerta agilis*) and Adders (*Vipera berus*). Biol. Conserv. 94:257–262.

Madsen, T., B. Újvári, and M. Olsson. 2004. Novel genes continue to enhance population growth in Adders (*Vipera berus*). Biol. Conserv. 120:145–147.

Madsen, T., B. Újvári, M. Olsson, and R. Shine. 2005. Paternal alleles enhance female reproductive success in tropical pythons. Mol. Ecol. 14:1783–1787.

Madsen, T., B. Újvári, R. Shine, and M. Olsson. 2006. Rain, rats and pythons: Climate-driven population dynamics of predators and prey in tropical Australia. Austr. Ecol. 31:30–37.

Mahalanobis, P.C. 1936. On the generalized distance in statistics. Proc. Natl. Inst. Sci. India 12:49–55.

Malhotra, A., and R.S. Thorpe. 2004. Maximizing information in systematic revisions: A combined molecular and morphological analysis of a cryptic pitviper complex (*Trimeresurus stejnegeri*). Biol. J. Linn. Soc. 82:219–235.

Manel, S., M.K. Schwartz, G. Luikart, and P. Taberlet. 2003. Landscape genetics: Combining landscape ecology and population genetics. Trends Ecol. Evol. 18:189–197.

Manel, S., O.E. Gaggiotti, and R.S. Waples. 2005. Assignment methods: Matching biological questions with appropriate techniques. Trends Ecol. Evol. 20:136–142.

Manier, M.K. 2005. Population genetics, ecology and evolution of a vertebrate metacommunity. PhD dissertation, Oregon State University, Corvalis.

Manier, M.K., and S.J. Arnold. 2005. Population genetic analysis identifies source-sink dynamics for two sympatric garter snake species (*Thamnophis elegans* and *Thamnophis sirtalis*). Mol. Ecol. 14:3965–3976.

Manier, M.K., and S.J. Arnold. 2006. Ecological correlates of population genetic structure: A comparative approach using a vertebrate metacommunity. Proc. R. Soc. Biol. Sci. Ser. B 273:3001–3009.

Manier, M.K., C.M. Seyler, and S.J. Arnold. 2007. Adaptive divergence between ecotypes of the Terrestrial Garter Snake, *Thamnophis elegans*, assessed with F_{ST}-Q_{ST} comparisons. J. Evol. Biol. 20:1705–1719.

Mannan, R.W., M.L. Morrison, and E.C. Meslow. 1984. Comment: The use of guilds in forest bird management. Wildl. Soc. Bull. 12:426–430.

Manni, F., E. Guérard, and E. Heyer. 2004. Geographic patterns of (genetic, morphologic, linguistic) variation: How barriers can be detected by using Monmonier's algorithm. Hum. Biol. 76:173–190.

Mao, J.J., Y.C. Lai, and G. Norval. 2006. An improved technique for scale-clipping of small snakes. Herpetol. Rev. 37:426–427.

Marcellini, D.L., and T.A. Jenssen. 1988. Visitor behavior in the National Zoo's Reptile House. Zoo Biol. 7:329–338.

Marcellini, D.L., and J.B. Murphy. 1998. Education in a zoological park or aquarium: An ontogeny of learning opportunities. Herpetologica 54(Suppl.):S12–S16.

Marjoram, P., and P. Donnelly. 1994. Pairwise comparisons of mitochondrial DNA sequences in subdivided populations and implications for early human evolution. Genetics 136:673–683.

Markland, F.S., K. Shieh, Q. Zhou, V. Gloubkov, R.P. Sherwin, V. Richters, and R. Sposto. 2001. A novel snake venom disintegrin that inhibits human ovarian cancer dissemination and angiogenesis in an orthotopic nude mouse model. Haemeostasis 31:183–191.

Marsh, D.M., and P.C. Trenham. 2001. Metapopulation dynamics and amphibian conservation. Conserv. Biol. 15:40–49.

Marsh, D.M., G.S. Milam, N.P. Gorham, and N.G. Beckman. 2005. Forest roads as partial barriers to terrestrial salamander movement. Conserv. Biol. 19:2004–2008.

Marshall, J.C., J.V. Manning, and B.A. Kingsbury. 2006. Movement and macrohabitat selection of the eastern massasauga in a fen habitat. Herpetologica 62:141–150.

Martin, W.H. 2002. Life history constraints on the Timber Rattlesnake (*Crotalus horridus*) at its climatic limits. *In* G.W. Schuett, M. Höggren, M.E. Douglas, and H.W. Greene, eds., Biology of the vipers, pp. 285–306. Eagle Mountain, Utah: Eagle Mountain Publishing.

Mason, R.T. 1993. Chemical ecology of the Red-sided Garter Snake, *Thamnophis sirtalis parietalis*. Brain Behav. Evol. 41:261–268.

Mason, R.T. 1994. The potential of pheromonal control of reptilian populations. *In* P. Sorensen, ed., Luring lampreys: Assessing the feasibility of using odorants to control sea lamprey in the Great Lakes, pp. 24–34. Ann Arbor: Great Lakes Fishery Commission.

Mason, R.T. 1999. Integrated pest management: The case for pheromonal control of habu (*Trimeresurus flavoviridis*) and Brown Treesnakes (*Boiga irregularis*). *In* G.H. Rodda, Y. Sawai, D. Chiszar, and H. Tanaka, eds., Problem snake management: The habu and the Brown Treesnake, pp. 196–205. Ithaca: Cornell University Press.

Mason-Gamer, R.J., and E.A. Kellogg. 1996. Testing for phylogenetic conflict among molecular data sets in the Tribe Triticeae (Gramineae). Syst. Biol. 45:524–545.

Masters, P. 1996. The effects of fire-driven succession on reptiles in *Spinifex* grasslands at Uluru National Park, Northern Territory. Wildl. Res. 23:39–48.

Mathews, A.E. 1989. Conflict, controversy, and compromise: The Concho Water Snake (*Nerodia harteri paucimaculata*) versus the Stacy Dam and Reservoir. Environ. Manage. 13:297–307.

Matthews, K.R., R.A. Knapp, and K.L. Pope. 2002. Garter snake distributions in high-elevation aquatic ecosystems: Is there a link with declining amphibian populations and nonnative trout introductions? J. Herpetol. 36:16–22.

Mayr, E. 1963. Animal species and evolution. Cambridge, Mass.: Harvard University Press.

Mazerolle, M.J. 2006. Improving data analysis in herpetology: Using Akaike's information criterion (AIC) to assess the strength of biological hypotheses. Amphibia-Reptilia 27:169–180.

McCallum, M.L., and J.L. McCallum. 2006. Publication trends in natural history and field studies in herpetology. Herpetol. Conserv. Biol. 1:62–67.

McCarty, J.P. 2001. Ecological consequences of recent climate change. Conserv. Biol. 15:320–331.

McCauley, D.J., F. Keesing, T.P. Young, B.F. Allan, and R.M. Pringle. 2006. Indirect effects of large herbivores on snakes in an African savanna. Ecology 87:2657–2663.

McCracken, G.F., G.M. Burghardt, and S.E. Houts. 1999. Microsatellite markers and multiple paternity in the garter snake *Thamnophis sirtalis*. Mol. Ecol. 8:1475–1480.

McGarigal, K., and B.J. Marks. 1995. FRAGSTATS: Spatial pattern analysis program for quantifying landscape structure. General Technical Report PNW-351. Portland, Ore.: U.S. Forest Service, Pacific Northwest Research Station.

McKay, J.K., and R.G. Latta. 2002. Adaptive population divergence: Markers, QTL and traits. Trends Ecol. Evol. 17:285–291.

McKinstry, M.C., and S.H. Anderson. 2002. Survival, fates, and success of transplanted beavers, *Castor canadensis*, in Wyoming. Can. Field-Nat. 116:60–68.

McLellan, B.B., and D.M. Shackleton. 1988. Grizzly bears and resource extraction industries: Effects of roads on behaviour, habitat use and demography. J. Appl. Ecol. 25:451–460.

McLeod, R.F., and J.E. Gates. 1998. Response of herpetofaunal communities to forest cutting and burning at Chesapeake farms, Maryland. Am. Midl. Nat. 139:164–177.

McPhee, M.E. 2003. Generations in captivity increases behavioral variance: Considerations for captive breeding and reintroduction programs. Biol. Conserv. 115:71–77.

Means, D.B., and H.W. Campbell. 1980. Effects of prescribed burning on amphibians and reptiles. *In* G.W. Wood, ed., Prescribed fire and wildlife in southern forests, pp. 89–97. Georgetown, S.C.: Baruch Forest Science Institute, Clemson University.

Meffe, G.K., and C.R. Carroll. 1997. Principles of conservation biology. 2nd ed. Sunderland, Mass.: Sinauer.

Mendelson, J.R., and W.B. Jennings. 1992. Shifts in the relative abundance of snakes in a desert grassland. J. Herpetol. 26:38–45.

Mendez, M., T. Waller, P.A. Micucci, E. Alvarenga, and J.C. Morales. 2007. Genetic population structure of the yellow anaconda (*Eunectes notaeus*) in northern Argentina: Management implications. *In* R.W. Henderson and R. Powell, eds., Biology of boas and pythons, pp. 405–414. Eagle Mountain, Utah: Eagle Mountain Publishing.

Merckelbach, H.L.G.J. 1989. Preparedness and classical conditioning of fear: A critical inquiry. PhD dissertation, Rijksuniversiteit, Utrecht.

Merilä, J., and P. Crnokrak. 2001. Comparison of genetic differentiation at marker loci and quantitative traits. J. Evol. Biol. 14:892–903.

Merilä, J., A. Forsman, and L.E. Lindell. 1992. High frequency of ventral scale anomalies in *Vipera berus* populations. Copeia 1992:1127–1130.

Merkle, D.A. 1985. Genetic variation in the eastern cottonmouth, *Agkistrodon piscivorus piscivorus* (Lacépède) (Reptilia: Crotalidae) at the northern edge of its range. Brimleyana 11:55–61.

Merriam, C.H. 1894. Laws of temperature control of the geographic distribution of terrestrial animals and plants. Nat. Geogr. 6:229–238.

Meshaka, W.E., Jr., B.P. Butterfield, and J.B. Hauge. 2004. Exotic amphibians and reptiles of Florida. Malabar, Fla.: Krieger.

Metropolis, N., A.W. Rosenbluth, M.N. Rosenbluth, A.H. Teller, and E. Teller. 1953. Equations of state calculations by fast computing machines. J. Chem. Phys. 21:1087–1091.

Meylan, A.B., and D. Ehrenfeld. 2000. Conservation of marine turtles. *In* M.W. Klemens, ed., Turtle conservation, pp. 96–125. Washington, D.C.: Smithsonian Institution Press.

Miller, G.L. 2000. Nature's fading chorus: Classic and contemporary writings on amphibians. Washington, D.C.: Island Press.

Mills, L.S., and F.W. Allendorf. 1996. The one-migrant-per-generation rule in conservation and management. Conserv. Biol. 6:1509–1518.

Millspaugh, J.J., and J.M. Marzluff. 2001. Radio tracking and animal populations. San Diego: Academic Press.

Milne, T., and C.M. Bull. 2000. Burrow choice by individuals of different sizes in the endangered pygmy blue tongue lizard, *Tiliqua adelaidensis*. Biol. Conserv. 95:295–301.

Mineka, S., R. Kier, and V. Price. 1980. Fear of snakes in wild- and laboratory-reared rhesus monkeys (*Macaca mulatta*). Anim. Learn. Behav. 8:653–663.

Mineka, S., M. Davidson, M. Cook, and R. Kier. 1984. Observational conditioning of snake fear in rhesus monkeys. J. Abnorm. Psychol. 93:355–372.

Mitchell, P.C. 1922. Monkeys and the fear of snakes. Proc. Zool. Soc. Lond. 1922:347–348.

Mitchell, P.C., and R.I. Pocock. 1907. On the feeding of reptiles in captivity with observations on the fear of snakes by other vertebrates. Proc. Zool. Soc. Lond. 1907:785–794.

Mitro, M.G. 2003. Demography and viability analysis of a diamondback terrapin population. Can. J. Zool. 81:716–726.

Mittermeier, R.A., J.L. Carr, I.R. Swingland, T.B. Werner, and R.B. Mast. 1992. Conservation of amphibians and reptiles. *In* K. Adler, ed., Herpetology: Current research on the biology of amphibians and reptiles, pp. 59–80. Oxford, Ohio: Society for the Study of Amphibians and Reptiles.

Mittermeier, R.A., P.R. Gil, M. Hoffman, J. Pilgrim, T. Brooks, C.G. Mittermeier, J. Lamoreux, and G. Fonseca. 2004. Hotspots revisited: Earth's biologically richest and most endangered terrestrial ecosystems. Mexico City: Cemex.

Moehrenschlager, A., and D.W. Macdonald. 2003. Movement and survival parameters of translocated and resident swift foxes *Vulpes velox*. Anim. Conserv. 6:199–206.

Mollard, A., and A. Torre. 2004. Proximity, territory and sustainable management at the local level: An introduction. Int. J. Sustainable Dev. 7:221–236.

Montgomery, S. 2001. The snake scientist. New York: Houghton Mifflin.

Moore, J.A., and J.C. Gillingham. 2006. Spatial ecology and multi-scale habitat selection by a threatened rattlesnake: The eastern massasauga (*Sistrurus catenatus catenatus*). Copeia 2006:742–750.

Moore, J.L., A. Balmford, T. Brooks, N.D. Burgess, L.A. Hansen, C. Rahbek, and P.H. Williams. 2003. Performance of sub-saharan vertebrates as indicator groups for identifying priority areas for conservation. Conserv. Biol. 17:207–218.

Moore, R.D., R.A. Griffiths, C.M. O'Brien, A. Murphy, and D. Jay. 2004. Induced defenses in an endangered amphibian in response to an introduced snake predator. Oecologia 141:139–147.

Moore, W.S. 1995. Inferring phylogenies from mtDNA variation: Mitochondrial-gene trees versus nuclear-gene trees. Evolution 49:718–726.

Morando, M., L.J. Avila, and J.W. Sites Jr. 2003. Sampling strategies for delimiting species: genes, individuals, and populations in the *Liolaemus elongatus-kriegi* complex (Squamata; Liolaemidae) in Andean—Patagonian South America. Syst. Biol. 52:159–185.

Moriarty, E.C., and D.C. Cannatella. 2004. Phylogenetic relationships of North American chorus frogs (genus *Pseudacris*), from 12S and 16S mtDNA. Mol. Phylogenet. Evol. 30:409–420.

Moriarty, J.J., and M. Linck. 1997. Reintroduction of Bull Snakes into a recreated area. *In* J.J. Moriarty and D. Jones, eds., Minnesota's amphibians and reptiles, their conservation and status, pp. 43–54. Lanesboro, Minn.: Serpent's Tale Natural History Book Distributors.

Morin, P.A., G. Luikart, R.K. Wayne, and the SNP Workshop Group. 2004. SNPs in ecology, evolution and conservation. Trends Ecol. Evol. 19:208–216.

Morin, P.J. 1981. Predatory salamanders reverse the outcome of competition among three species of anuran tadpoles. Science 212:1284–1286.

Moritz, C., and D.P. Faith. 1998. Comparative phylogeography and the identification of genetically divergent areas for conservation. Mol. Ecol. 7:419–429.

Morris, R., and D. Morris. 1965. Men and snakes. New York: McGraw-Hill.

Morrison, M.L. 1986. Bird populations as indicators of environmental change. *In* R.F. Johnston, ed., Current ornithology, pp. 429–451. New York: Plenum Press.

Morrison, M.L., and L.S. Hall. 2002. Standard terminology: Toward a common language to advance ecological understanding and application. *In* J.M. Scott, P.J. Heglund, M.L. Morrison, J.B. Haufler, M.G. Raphael, W.A. Wall, and F.B. Samson, eds., Predicting species occurrences: Issues of accuracy and scale. Washington, D.C.: Island Press.

Motzkin, G., and D.R. Foster. 2002. Grasslands, heathlands and shrublands in coastal New England: Historical interpretations and approaches to conservation. J. Biogeogr. 29:1569–1590.

Moulis, R.A. 1997. Predation by the imported fire ant (*Solenopsis invicta*) on loggerhead sea turtle (*Caretta caretta*) nests on Wassaw National Wildlife Refuge, Georgia. Chelonian Conserv. Biol. 2:433–436.

Mountain, J.L., A. Knight, M. Jobin, C. Gignoux, A. Miller, A.A. Lin, and P.A. Underhill. 2002. SNPSTRs: Empirically derived, rapidly typed, autosomal haplotypes for inference of population history and mutational processes. Genome Res. 12:1766–1772.

Mueller, R.L. 2006. Evolutionary rates, divergence times, and the performance of mitochondrial genes in Bayesian phylogenetic analysis. Syst. Biol. 55:289–300.

Mullen, E.B., and P. Ross. 1997. Survival of relocated tortoises: Feasibility of relocating tortoises as a successful mitigation tool. *In* J. Van Abbema, ed., Proceedings: Conservation, restoration, and management of tortoises and turtles—an international conference, pp.140–146. New York: New York Turtle and Tortoise Society.

Mullin, S.J., R.J. Cooper, and W.H.N. Gutzke. 1998. The foraging ecology of the Gray Rat Snake (*Elaphe obsoleta spiloides*). III. Searching for different prey types in structurally varied habitats. Can. J. Zool. 76:548–555.

Murcia, C. 1995. Edge effects in fragmented forests: Implications for conservation. Trends Ecol. Evol. 10:58–62.

Murphy, J.B., and D. Chiszar. 1989. Herpetological masterplanning for the 1990s. Int. Zoo Yearb. 28:1–7.

Murphy, J.B., and D.E. Jacques. 2006. Death from snakebite: The entwined histories of Grace Olive Wiley and Wesley H. Dickinson. Bull. Chicago Herpetol. Soc. (Special Suppl.):1–20.

Murphy, J.C., and R.W. Henderson. 1997. Tales of giant snakes: A historical natural history of anacondas and pythons. Malabar, Fla.: Krieger.

Murphy, R.W., V. Kovac, O. Haddrath, G.S. Allen, A. Fishbein, and N.E. Mandrak. 1995. mtDNA gene sequence, allozyme, and morphological uniformity among Red Diamond Rattlesnakes, *Crotalus ruber* and *Crotalus exsul*. Can. J. Zool. 73:270–281.

Murphy, R.W., J.W. Sites Jr., D.G. But, and C.H. Haufler. 1996. Proteins: Isozyme electrophoresis. *In* D.M. Hillis, C. Moritz, and B.K. Maple, eds., Molecular systematics, 2nd ed., pp. 51–120. Sunderland, Mass.: Sinauer.

Murray, E.J., and F. Foote. 1979. The origin of fear of snakes. Behav. Res. Ther. 17:489–493.

Mushinsky, H.R. 1987. Foraging ecology. *In* R.A. Seigel, J.T. Collins, and S.S. Novak, eds., Snakes: Ecology and evolutionary biology, pp. 302–334. New York: McGraw-Hill.

Mushinsky, H.R., and J.J. Hebrard. 1977. Food partitioning by five species of water snakes in Louisiana. Herpetologica 33:162–166.

Mushinsky, H.R., J.J. Hebrard, and D.S. Vodopich. 1982. Ontogeny of water snake foraging ecology. Ecology 63:1624–1629.

Nagy, K.A. 1980. CO_2 production in animals: Analysis of potential errors in the doubly labeled water method. Am. J. Physiol. 238:R466–R473.

Nagy, K.A., and D.P. Costa. 1980. Water flux in animals: Analysis of potential errors in the tritiated water method. Am. J. Physiol. 238:R454–R465.

Nagy, Z.T., R. Lawson, U. Joger, and M. Wink. 2004. Molecular systematics of racers, whipsnakes and relatives (Reptilia: Colubridae) using mitochondrial and nuclear markers. J. Zool. Syst. Evol. Res. 42:223–233.

Naulleau, G. 1984. Les serpents de France. Revue Française d'Aquariologie 11:1–56.

Nee, S., E.C. Holmes, A. Rambaut, and P.H. Harvey. 1995. Inferring population history from molecular phylogenies. Philos. Trans. R. Soc. Lond. B 349:25–31.

Nei, M. 1982. Evolution of human races at the gene level. *In* B. Bonne-Tamir, T. Cohen, and R.M. Goodman, eds., Human genetics, Pt. a: The unfolding genome, pp. 167–181. New York: Alan R. Liss.

Nei, M., and F. Tajima. 1981. Genetic drift and estimation of effective population size. Genetics 98:625–640.

Nelson, R.D., and H. Salwasser. 1982. The forest service wildlife and fish habitat relationships program. Trans. N. Am. Wildl. Nat. Resour. Conf. 47:174–183.

Nichols, J.D., and B.K. Williams. 2006. Monitoring for conservation. Trends Ecol. Evol. 21:668–673.

Nichols, R. 2001. Gene trees and species trees are not the same. Trends Ecol. Evol. 16:358–364.

Nicols, A. 1883. On snakes. *In* A. Nicols, ed., Zoological notes, pp. 1–58. London: L. Upcott Gill.

Nielsen, R. 1997. A likelihood approach to populations samples of microsatellite alleles. Genetics 146:711–716.

Nielsen, R. 2006. Statistical methods in molecular evolution. New York: Springer.

Nilson, G., C. Andren, and B. Flardh. 1990. *Vipera albizona,* a new mountain viper from central Turkey, with comments on isolating effects of the Anatolian diagonal. Amphib-Reptilia 20:355–375.

Nilson, G., C. Andren, Y. Ioannidis, and M. Dimaki. 1999. Ecology and conservation of the Milos viper, *Macrovipera schweizeri* (Werner, 1935). Amphib-Reptilia 20:355–375.

Nishimura, M. 1999. Structure and application of the slanting nylon-net fence to prevent dispersal of habu (*Trimeresurus flavoviridis*). *In* G.H. Rodda, Y. Sawai, D. Chiszar, and H. Tanaka, eds., Problem snake management: The habu and the Brown Treesnake, pp. 313–318. Ithaca: Cornell University Press.

Norris, J.L., and K.H. Pollock. 1996. Nonparametric MLE under two closed capture recapture models with heterogeneity. Biometrics 52:639–649.

Norris, S. 2007. Ghosts in our midst: Coming to terms with amphibian extinctions. BioScience 57:311–316.

Nowak, E.M., T. Hare, and J. McNally. 2002. Management of "nuisance" vipers: Effects of translocation on Western Diamond-backed Rattlesnakes (*Crotalus atrox*). *In* G.W. Schuett, M. Hoggren, M.E. Douglas, and H.W. Greene, eds., Biology of the vipers, pp. 533–560. Eagle Mountain, Utah: Eagle Mountain Publishing.

[NRC] National Resources Council 1986. Ecological knowledge and environmental problem-solving: Concepts and case studies. Washington, D.C.: National Academy Press.

Nunney, L. 2000. The limits to knowledge in conservation genetics: The value of effective population size. Evol. Biol. 32:179–194.

Nunney, L., and D.R. Elam. 1994. Estimating the effective population size of conserved populations. Conserv. Biol. 8:175–184.

Nylander, J.A.A. 2004. MrModeltest v2. Program distributed by the author, Evolutionary Biology Centre, Uppsala University.

Nylander, J.A.A., F. Ronquist, J.P. Huelsenbeck, and J.L. Nieves-Aldrey. 2004. Bayesian phylogenetic analysis of combined data. Syst. Biol. 53:47–67.

O'Brien, S., B. Robert, and H. Tiandray. 2005. Hatch size, somatic growth rate and size-dependent survival in the endangered ploughshare tortoise. Biol. Conserv. 126:141–145.

O'Connor, M.P., A.E. Sieg, and A.E. Dunham. 2006. Linking physiological effects on activity and resource use to population level phenomena. Integr. Comp. Biol. 46:1093–1109.

O'Donnell, R.P., and S.J. Arnold. 2005. Evidence for selection on thermoregulation: Effects of temperature on embryo mortality in the garter snake *Thamnophis elegans*. Copeia 2005:930–934.

Odum, A., and M.J. Goode. 1994. The species survival plan for *Crotalus durissus unicolor*: A multifaceted approach to conservation of an insular rattlesnake. *In* J.B. Murphy, K. Adler, and J.T. Collins, eds., Captive management and conservation of amphibians and reptiles, pp. 363–368. Ithaca: Society for the Study of Amphibians and Reptiles.

Odum, E.P. 1971. Fundamentals of ecology. Philadelphia: W.B. Saunders.

Oglesby, R.T., and C.R. Smith. 1995. Climate changes in the northeast. *In* E.T. LaRoe, G.S. Farris, C.E. Puckett, P.D. Doran, and M.J. Mac, eds., Our living resources: A report to the nation on the distribution, abundance, and health of U.S. plants, animals, and ecosystems, pp. 390–391. Washington, D.C.: U.S. Dept of Interior, National Biological Service.

Öhman, A., and S. Mineka. 2001. Fears, phobias, and preparedness: Toward an evolved module of fear and fear learning. Psychol. Rev. 108:483–522.

Öhman, A., A. Eriksson, and C. Olofsson. 1975. One-trial learning and superior resistance to extinction of autonomic responses conditioned to potentially phobic stimuli. J. Comp. Physiol. Psych. 88:619–627.

Öhman, A., D. Lundqvist, and F. Esteves. 2001. The face in the crowd revisited: A threat advantage with schematic stimuli. J. Pers. Soc. Psychol. 80:381–396.

Oldak, P.D. 1976. Comparison of the scent glands secretion lipids of twenty-five snakes: Implications for biochemical systematics. Copeia 1976:320–326.

Oldham, R.S., and R.N. Humphries. 2000. Evaluating the success of great crested newt (*Triturus cristatus*) translocations. Herpetol. J. 10:183–190.

Olmo, E. 1986. Reptilia. Berlin: Gebruder Borntraeger.

Olson, M.A., and R.V. Kendrick. 2008. Origins of attitudes. *In* W.D. Crano and R. Prislin, eds., Attitude and attitude change, pp. 111–130. New York: Psychology Press.

Olsson, M., and T. Madsen. 2001. Promiscuity in sand lizards (*Lacerta agilis*) and Adder snakes (*Vipera berus*): Causes and consequences. J. Hered. 92:190–197.

Orrock, J.L., and J.F. Pagels. 2003. Tree communities, microhabitat characteristics, and small mammals associated with the endangered rock vole, *Microtus chrotorrhinus*, in Virginia. Southeast. Nat. 2:547–558.

Ota, H. 2000. Current status of the threatened amphibians and reptiles of Japan. Popul. Ecol. 42:5–9.

Ota, H., S. Iwanaga, K. Itoman, M. Nishimura, and A. Mori. 1991. Reproductive mode of a natricine snake, *Amphiesma pryeri* (Colubridae: Squamata), from the Ryukyu Archipelago, with special reference to the viviparity of *A. p. ishigakiensis*. Biol. Mag. Okinawa 29:37–43.

Otis, D.L., K.P. Burnham, G.C. White, and D.R. Anderson. 1978. Statistical inference from capture data on closed animal populations. Wildl. Monogr. 62:7–135.

Owen, C. 1742. An essay towards a natural history of serpents. London: John Gray.

Oyler-McCance, S.J., J. St. John, J.M. Parker, and S.H. Anderson. 2005. Characterization of microsatellite loci isolated in Midget Faded Rattlesnake (*Crotalus viridis concolor*). Mol. Ecol. Notes 5:452–453.

Ozaki, K., M. Isono, T. Kawahara, S. Iida, T. Kudo, and K. Fukuyama. 2006. A mechanistic approach to evaluation of umbrella species as conservation surrogates. Conserv. Biol. 20:1507–1515.

Paetkau, D., R. Slade, M. Burdens, and A. Estoup. 2004. Genetic assignment methods for the direct, real-time estimation of migration rate: A simulation-based exploration of accuracy and power. Mol. Ecol. 13:55–65.

Page, R.D.M, and E.C. Holmes. 1998. Molecular evolution: A phylogenetic approach. Oxford, U.K.: Blackwell.

Paik, N.K., and S.Y. Yang. 1987. Genetic variation in natural populations of the Cat-Snake (*Elaphe dione*). Korean J. Zool. 30:211–218.

Palumbi, S.R., F. Cipriano, and M.P. Hare. 2001. Predicting nuclear gene coalescence from mitochondrial data: The three-times rule. Evolution 55:859–868.

Panchal, M. 2007. The automation of nested clade phylogeographic analysis. Bioinformatics 23:509–510.

Panksepp, J. 1998. Affective neuroscience. New York: Oxford University Press.

Paquin, M.M., G.D. Wylie, and E.J. Routman. 2006. Population structure of the Giant Garter Snake, *Thamnophis gigas*. Conserv. Genet. 7:25–36.

Parent, C.E., and P.J. Weatherhead. 2000. Behavioral and life-history responses of Eastern Massasauga Rattlesnakes (*Sistrurus catenatus catenatus*) to human disturbance. Oecologia 125:170–178.

Parker, S., and K.A. Prior. 1999. Population monitoring of the Massasauga Rattlesnake (*Sistrurus catenatus catenatus*) in Bruce Peninsula National Park, Ontario, Canada. *In* B. Johnson and M. Wright, eds., Second international symposium and workshop on the conservation of the Eastern Massasauga Rattlesnake, *Sistrurus catenatus catenatus*: Population and habitat management issues in urban, bog, prairie and forested ecosystems, pp. 63–66. Toronto: Toronto Zoo.

Parker, W.S., and W.S. Brown. 1973. Species composition and population changes in two complexes of snake hibernacula in northern Utah. Herpetologica 29:319–326.

Parker, W.S., and M.V. Plummer. 1987. Population ecology. *In* R.A. Seigel, J.T. Collins, and S.S. Novak, eds., Snakes: Ecology and evolutionary biology, pp. 253–301. New York: McGraw-Hill.

Parkinson, C.L., K.R. Zamudio, and H.W. Greene. 2000. Phylogeography of the pitviper clade *Agkistrodon*: Historical ecology, species status, and conservation of cantils. Mol. Ecol. 9:411–420.

Parmesan, C., and G. Yohe. 2003. A globally coherent fingerprint of climate change impacts across natural systems. Nature 421:37–42.

Parmesan, C., N. Ryrholm, C. Stefanescu, J.K. Hill, C.D. Thomas, H. Descimon, B. Huntley, L. Kaila, J. Kullberg, T. Tammaru, W.J. Tennent, J.A. Thomas, and M. Warren. 1999. Poleward shifts in geographical ranges of butterfly species associated with regional warming. Nature 399:579–583.

Patten, M.A., and D.T. Bolger. 2003. Variation in top-down control of avian reproductive success across a fragmentation gradient. Oikos 101:479–488.

Paul, G.L. 1969a. Outcome of systematic desensitization I: Background, procedures, and uncontrolled reports of individual treatment. In C.M. Franks, ed., Behavior therapy: Appraisal and status, pp. 63–104. New York: McGraw-Hill.

Paul, G.L. 1969b. Outcome of systematic desensitization II: Controlled investigations of individual treatment, technique variations, and current status. In C.M. Franks, ed., Behavior therapy: Appraisal and status, pp. 105–159. New York: McGraw-Hill.

Pearson, D., and R. Shine. 2002. Expulsion of intraperitoneally-implanted radiotransmitters by Australian pythons. Herpetol. Rev. 33:261–263.

Pearson, D., R. Shine, and A. Williams. 2002. Geographic variation in sexual size dimorphism within a single snake species (Morelia spilota, Pythonidae). Oecologia 131:418–426.

Pearson, D., R. Shine, and A. Williams. 2005. Spatial ecology of a threatened python (Morelia spilota imbricata) and the effects of anthropogenic habitat change. Austr. Ecol. 30:261–274.

Pedrono, M., and A. Sarovy. 2000. Trial release of the world's rarest tortoise Geochelone yniphora in Madagascar. Biol. Conserv. 95:333–342.

Peel, D., J.R. Ovenden, and L.L. Peel. 2004. NeEstimator: Software for estimating effective population size (ver. 1.3). Brisbane, Australia: Queensland Government, Department of Primary Industries and Fisheries.

Pellegrino, K.C.M., M.T. Rodrigues, A.N. Waite, M. Morando, Y. Yonenaga-Yassuda, and J.W. Sites Jr. 2005. Phylogeography and species limits in the Gymnodactylus darwinii complex (Gekkonidae, Squamata): Genetic structure coincides with river systems in the Brazilian Atlantic forest. Biol. J. Linn. Soc. 85:13–26.

Perison, D., J.P. Phelps, C. Pavel, and R. Kellison. 1997. The effects of timber harvest in a South Carolina blackwater bottomland. For. Ecol. Manage. 90:171–185.

Perry, M.C., C.B. Sibrel, and G.A. Gough. 1996. Wetlands mitigation: Partnership between an electric power company and a federal wildlife refuge. Environ. Manage. 20:933–939.

Pesole, G., C. Gissi, A. de Chirico, and C. Saccone. 1999. Nucleotide substitution rate of mammalian mitochondrial genomes. J. Mol. Evol. 48:427–434.

Peterson, B.L., B.E. Kus, and D.H. Deutschman. 2004. Determining nest predators of the Least Bell's Vireo through point counts, tracking stations, and video photography. J. Field Ornithol. 75:89–95.

Peterson, C.R. 1982. Body temperature variation in free-living garter snakes (Thamnophis elegans vagrans). PhD dissertation, Washington State University, Pullman.

Peterson, C.R., and M.E. Dorcas. 1992. The use of automated data acquisition techniques in monitoring amphibian and reptile populations. In D.R. McCullough and R.H. Barrett, eds., Wildlife 2001: Populations, pp. 369–378. London: Elsevier Applied Science.

Peterson, C.R., and M.E. Dorcas. 1994. Automated data acquisition. In R.W. McDiarmid, W.R. Heyer, M. Donnelly, and L. Hayek, eds., Measuring and monitoring biological diversity—standard methods for amphibians, pp. 47–57. Washington, D.C.: Smithsonian Institution Press.

Peterson, C.R., A.R. Gibson, and M.E. Dorcas. 1993. Snake thermal ecology: The causes and consequences of body-temperature variation. In R.A. Seigel and J.T. Collins, eds., Snakes: Ecology and behavior, pp. 241–314. New York: McGraw-Hill.

Peterson, C.R., J.O. Cossel Jr., D. Pilliod, and B.M. Bean. 2002. The occurrence, distribution, relative abundance, and habitat relationships of amphibians and reptiles on the Idaho Army National Guard Orchard Training Area, Ada County, Idaho. Final Report to the Idaho Army National Guard, Boise, Idaho.

Phillips, B.L., and R. Shine. 2004. Adapting to an invasive species: Toxic cane toads induce morphological change in Australian snakes. Proc. Nat. Acad. Sci. U.S.A. 101:17150–17155.

Phillips, B.L., and R. Shine. 2006. An invasive species induces rapid adaptive change in a native predator: Cane toads and Black Snakes in Australia. Proc. R. Soc. Lond. B Biol. Sci. 273:1545–1550.

Phillips, B.L., G.P. Brown, and R. Shine. 2003. Assessing the potential impact of cane toads on Australian snakes. Conserv. Biol. 17:1738–1747.

Phillpot, P. 1996. Visitor viewing behaviour in the Gaherty Reptile Breeding Centre, Jersey Wildlife Preservation Trust: A preliminary study. Dodo: J. Wildl. Preserv. Trust 32:193–202.

Piccione, P.A. 1990. Mehen, mysteries, and resurrection from the coiled serpent. J. Amer. Res. Center Egypt 27:43–52.

Pilgrim, M.A. 2001. Offspring size variation in five species of snakes (*Sistrurus miliarius, Sistrurus catenatus, Thamnophis sirtalis, Thamnophis proximus* and *Nerodia rhombifer*): Implications for optimal offspring size theory. MS thesis, Southeastern Louisiana University, Hammond.

Pilgrim, M.A. 2005. Linking microgeographic variation in Pigmy Rattlesnake (*Sistrurus miliarius*) life history and demography with diet composition: A stable isotope approach. PhD dissertation, University of Arkansas, Fayetteville.

Pilgrim, M.A. 2007. Expression of maternal isotopes in offspring: Implications for interpreting ontogenetic shifts in isotopic composition of consumer tissues. Isotopes Environ. Health Stud. 43:155–163.

Placyk, J.S., Jr., G.M. Burghardt, R.L. Small, R.B. King, G.S. Casper, and J.W. Robinson. 2007. Post-glacial recolonization of the Great Lakes region by the Common Gartersnake (*Thamnophis sirtalis*) inferred from mtDNA sequences. Mol. Phylogenet. Evol. 43:452–467.

Pledger, S. 2000. Unified maximum likelihood estimates for closed capture-recapture models using mixtures. Biometrics 56:434–442.

Plummer, M.V. 1997. Population ecology of Green Snakes (*Opheodrys aestivus*) revisited. Herpetol. Monogr. 11:102–123.

Plummer, M.V., and N.E. Mills. 2000. Spatial ecology and survivorship of resident and translocated Hognose Snakes (*Heterodon platirhinos*). J. Herpetol. 34:565–575.

Polis, G.A., and S.J. McCormick. 1987. Intraguild predation and competition among desert scorpions. Ecology 68:332–343.

Pollock, K.H. 1982. A capture-recapture design robust to unequal probabilities of capture. J. Wildl. Manage. 46:757–760.

Pollock, K.H. 2002. The use of auxiliary variables in capture-recapture modelling: An overview. J. Appl. Stat. 29:85–102.

Pollock, K.H., J.D. Nichols, C. Brownie, and J.E. Hines. 1990. Statistical inference for capture-recapture experiments. Wildl. Monogr. 107:1–97.

Pook, C.E., W. Wüster, and R.S. Thorpe. 2000. Historical biogeography of the Western Rattlesnake (Serpentes: Viperidae: *Crotalus viridis*), inferred from mitochondrial DNA sequence information. Mol. Phylogenet. Evol. 15:269–282.

Posada, D., and T.R. Buckley. 2004. Model selection and model averaging in phylogenetics: Advantages of Akaike information criterion and Bayesian approaches over likelihood ratio tests. Syst. Biol. 53:793–808.

Posada, D., and K.A. Crandall. 1998. Modeltest: Testing the model of DNA substitution. Bioinformatics 14:817–818.

Posada, D., and K.A. Crandall. 2001. Selecting the best-fit model of nucleotide substitution. Syst. Biol. 50:580–601.

Posada, D., K.A. Crandall, and A.R. Templeton. 2000. GeoDis: A program for the cladistic nested analysis of the geographical distribution of genetic haplotypes. Mol. Ecol. 9:487–488.

Pounds, J.A., M.P.L. Fogden, and J.H. Campbell. 1999. Biological response to climate change on a tropical mountain. Nature 398:611–615.

Prentice, E.F., T.A. Flagg, C.S. McCutcheon, D.F. Brastow, and D.C. Cross. 1990. Equiptment, methods, and an automated data-entry station for PIT tagging. Am. Fish. Soc. Symp. 7:335–340.

Price, R.M. 1982. Dorsal snake scale microdermatoglyphics: Ecological indicator or taxonomic tool? J. Herpetol. 16:294–306.

Pringle, R.M., J.K. Webb, and R. Shine. 2003. Canopy structure, microclimate, and habitat selection by a nocturnal snake, *Hoplocephalus bungaroides*. Ecology 84:2668–2679.

Prior, K.A., and P.J. Weatherhead. 1994. Response of free-ranging Eastern Massasauga Rattlesnakes to human disturbance. J. Herpetol. 28:255–257.

Prior, K.A., and P.J. Weatherhead. 1996. Habitat features of Black Rat Snake hibernacula in Ontario. J. Herpetol. 30:211–218.

Prior, K.A., and P.J. Weatherhead. 1998. Status of the Black Rat Snake, *Elaphe obsoleta obsoleta*, in Canada. Ottawa: Committee on the Status of Endangered Wildlife in Canada.

Prior, K.A., H.L. Gibbs, and P.J. Weatherhead. 1997. Population genetic structure in the Black Rat Snake: Implications for management. Conserv. Biol. 11:1147–1158.

Prior, K.A., G. Blouin-Demers, and P.J. Weatherhead. 2001. Sampling biases in demographic analyses of Black Rat Snakes (*Elaphe obsoleta*). Herpetologica 57:460–469.

Pritchard, J.K., M. Stephens, and P. Donnelly. 2000. Inference of population structure using multilocus genotype data. Genetics 155:945–959.

Prival, D.B., M.J. Goode, D.E. Swann, C.R. Schwalbe, and M.J. Schroff. 2002. Natural history of a northern population of Twin-spotted Rattlesnakes, *Crotalus pricei*. J. Herpetol. 36:598–607.

Proctor, M.F., B.N. McLellan, C. Strobeck, and R.M.R. Barclay. 2004. Gender-specific dispersal distances of grizzly bears estimated by genetic analysis. Can. J. Zool. 82:1108–1118.

Prosser, M.R. 1999. Sexual selection in Northern Water Snakes, *Nerodia sipedon sipedon*. PhD dissertation, McMaster University, Hamilton.

Prosser, M.R., H.L. Gibbs, and P.J. Weatherhead. 1999. Microgeographic population genetic structure in the Northern Water Snake, *Nerodia sipedon sipedon* detected using microsatellite DNA loci. Mol. Ecol. 8:329–333.

Prosser, M.R., P.J. Weatherhead, H.L. Gibbs, and G.P. Brown. 2002. Genetic analysis of the mating system and opportunity for sexual selection in Northern Water Snakes (*Nerodia sipedon*). Behav. Ecol. 13:800–807.

Prugnolle, F., and T. de Meeus. 2002. Inferring sex-biased dispersal from population genetic tools: A review. Heredity 88:161–165.

Puorto, G., M.G. Salomão, R.D.G. Theakston, R.S. Thorpe, D.A. Warrell, and W. Wüster. 2001. Combining mitochondrial DNA sequences and morphological data to infer species boundaries: Phylogeography of lanceheaded pitvipers in the Brazilian Atlantic forest, and the status of *Bothrops pradoi* (Squamata: Serpentes: Viperidae). J. Evol. Biol. 14:527–538.

Purkis, H.M., and O.V. Lipp. 2007. Autonomic attention does not equal automatic fear: Preferential attention without implicit valence. Emotion 7:314–323.

Pybus O.G., A. Rambaut, and P.H. Harvey. 2000. An integrated framework for the inference of viral population history from reconstructed genealogies. Genetics 155:1429–1437.

Pybus, O.G., A.J. Drummond, T. Nakano, B.H. Robertson, and A. Rambaut. 2003. The epidemiology and iatrogenic transmission of hepatitis C virus in Egypt: A Bayesian coalescent approach. Mol. Biol. Evol. 20:381–387.

Quammen, D. 2000. The boilerplate rhino: Nature in the eye of the beholder. New York: Scribner.

Quick, J.S., H.K. Reinert, E.R. DeCuba, and R.A. Odum. 2005. Recent occurrence and dietary habits of *Boa constrictor* on Aruba, Dutch West Indies. J. Herpetol. 39:304–307.

Rachman, S. 1967. Systematic desensitization. Psychol. Bull. 67:93–103.

Rachman, S. 2002. Fears born and bred: Non-associative fear acquisition? Behav. Res. Ther. 40:121–126.

Ramakrishnan, U., R.G. Coss, J. Schank, A. Dharawat, and S. Kim. 2005. Snake species discrimination by wild bonnet macaques (*Macaca radiata*). Ethology 111:337–356.

Ramos-Onsins, S.E., and J. Rozas. 2002. Statistical properties of new neutrality tests against population growth. Mol. Biol. Evol. 19:2092–2100.

Ramp, D., V.K. Wilson, and D.B. Croft. 2006. Assessing the impacts of roads in peri-urban reserves: Road-based fatalities and road usage by wildlife in the Royal National Park, New South Wales, Australia. Biol. Conserv. 129:348–359.

Rand, D.M., and L.M. Kann. 1996. Excess amino acid polymorphism in mitochondrial DNA: Contrasts among genes from Drosophila, mice, and humans. Mol. Biol. Evol. 13:735–748.

Rannala, B. 2002. Identifiability of parameters in MCMC Bayesian inference of phylogeny. Syst. Biol. 51:754–760.

Rapport, D.J., G. Bohm, D. Buckingham, J.J. Cairns, R. Costanza, J.R. Karr, H.A.M. de Kriuijf, R. Levins, A.J. McMichael, N.O. Nielsen, and W.G. Whitford. 1999. Ecosystem health: The concept, the ISEH, and the important tasks ahead. Ecosyst. Health 5:82–90.

Rawlings, L.H., and S.C. Donnellan. 2003. Phylogeographic analysis of the green python, *Morelia viridis*, reveals cryptic diversity. Mol. Phylogenet. Evol. 27:36–44.

Ray, D.A. 2007. SINEs of progress: Mobile element applications for molecular ecology. Mol. Ecol. 16:19–33.

Ray, J.M., and R.B. King. 2006. The temporal and spatial scale of microevolution: Fine scale color pattern variation in the Lake Erie Watersnake, *Nerodia sipedon insularum*. Evol. Ecol. Res. 8:915–925.

Ray, N., M. Currat, and L. Excoffier. 2003. Intra-deme molecular diversity in spatially expanding populations. Mol. Biol. Evol. 20.76–86.

Reagan, S.R., J.M. Ertel, and V.L. Wright. 2000. David and Goliath retold: Fire ants and alligators. J. Herpetol. 34:475–478.

Redi, F. 1675. *Observationes de viperis*. Amsterdam: Sumptibus Andreae Frisii.

Reed, D.H., and R. Frankham. 2001. How closely correlated are molecular and quantitative measures of genetic diversity: A meta-analysis. Evolution 55:1095–1103.

Reed, J.M., L.S. Mills, J.B. Dunning, E.S. Menges, K.S. McKelvey, R. Frye, S.R. Beissinger, M.C. Anstett, and P. Miller. 2002. Emerging issues in population viability analysis. Conserv. Biol. 16:7–19.

Reed, R.N. 2005. An ecological risk assessment of nonnative boas and pythons as potentially invasive species in the United States. Risk Anal. 25:753–766.

Reed, R.N., and R. Shine. 2002. Lying in wait for extinction: Ecological correlates of conservation status among Australian elapid snakes. Conserv. Biol. 16:451–461.

Reijnen, R., R. Foppen, C. Terbraak, and J. Thissen. 1995. The effect of car traffic on breeding bird populations in woodland. III. Reduction in relation to the proximity of main roads. J. Appl. Ecol. 32:187–202.

Reinert, H.K. 1984. Habitat variation within sympatric snake populations. Ecology 65:1673–1682.

Reinert, H.K. 1991. Translocation as a conservation strategy for amphibians and reptiles: Some comments, concerns, and observations. Herpetologica 47:357–363.

Reinert, H.K. 1992. Radiotelemetric field studies of pitvipers: Data acquisition and analysis. In J.A. Campbell and E.D. Brodie Jr., eds., Biology of the pitvipers, pp. 185–197. Tyler, Tex.: Selva.

Reinert, H.K. 1993. Habitat selection in snakes. In R.A. Seigel and J.T. Collins, eds., Snakes: Ecology and behavior, pp. 201–240. New York: McGraw Hill.

Reinert, H.K., and D. Cundall. 1982. An improved surgical implantation method for radio-tracking snakes. Copeia 1982:703–705.

Reinert, H.K., and R.R. Rupert. 1999. Impacts of translocation on behavior and survival of Timber Rattlesnakes, *Crotalus horridus*. J. Herpetol. 33:45–61.

Reinert, H.K., and R.T. Zappalorti. 1988. Timber Rattlesnakes (*Crotalus horridus*) of the pine barrens: Their movement patterns and habitat preference. Copiea 1988:964–978.

Relyea, R.A. 2005. The impact of insecticides and herbicides on the biodiversity and productivity of aquatic communities. Ecol. Appl. 15:618–627.

Renfrew, R.B., and C.A. Ribic. 2003. Grassland passerine nest predators near pasture edges identified on videotape. Auk 120:371–383.

Renken, R.B., W.K. Gram, D.K. Fantz, S.C. Richter, T.J. Miller, K.B. Ricke, B. Russell, and X. Wang. 2004. Effects of forest management on amphibians and reptiles in Missouri Ozark forests. Conserv. Biol. 18:174–188.

Reynolds, R.P., and N.J. Scott Jr. 1987. Use of a mammalian resource by a Chihuahuan snake community. *In* N.J. Scott Jr., ed., Herpetological communities: A symposium of the Society for the Study of Amphibians and Reptiles and the Herpetologists' League, August 1977, p. 99. Washington, D.C.: U.S. Fish and Wildlife Service.

Reznick, D., H. Rodd, and L. Nunney. 2004. Empirical evidence of rapid evolution. *In* R. Ferrière, U. Dieckmann, and D. Couvet, eds., Evolutionary conservation biology, pp. 101–118. New York: Cambridge University Press.

Richardson, M.L., P.J. Weatherhead, and J.D. Brawn. 2006. Habitat use and activity of Prairie Kingsnakes (*Lampropeltis calligaster calligaster*) in Illinois. J. Herpetol. 40:424–428.

Ridenhour, B.J. 2004. The coevolutionary process: The effects of population structure on a predator-prey system. PhD dissertation, Indiana University, Bloomington.

Ridenhour, B.J., E.D. Brodie Jr., and E.D. Brodie III. 2006. Patterns of genetic differentiation in *Thamnophis* and *Taricha* from the Pacific Northwest. J. Biogeogr. 34:724–735.

Ridgeway, R.L., R.M. Silverstein, and M.N. Inscoe, eds. 1990. Behavior modifying chemicals for insect management. New York: Marcel Dekker.

Ringold, P.L., J. Alegria, R.L. Czaplewski, B.S. Mulder, T. Tolle, and K. Burnett. 1996. Adaptive monitoring design for ecosystem management. Ecol. Appl. 6:745–747.

Rivas, J.A., and G.M. Burghardt. 2005. Snake mating systems, behavior, and evolution: The revisionary implications of recent findings. J. Comp. Psychol. 119:447–454.

Rivera, P.C., M. Chiaraviglio, G. Perez, and C.N. Gardenal. 2005. Protein polymorphism in populations of *Boa constrictor occidentalis* (Boidae) from Codoba province, Argentina. Amphib-Reptilia 26:175–181.

Rivera, P.C., C.N. Gardenal, and M. Chiaraviglio. 2006. Sex-biased dispersal and high levels of gene flow among local populations in the argentine boa constrictor, *Boa constrictor constrictor*. Austr. Ecol. 31:948–955.

Roark, A.W., and M.E. Dorcas. 2000. Regional body temperature variation in Corn Snakes measured using temperature-sensitive passive integrated transponders. J. Herpetol. 34:481–485.

Robinson J.W. 2005. Post-Pleistocene glacial retreat, colonization routes and geographic variation of the Northern Watersnake, *Nerodia sipedon*, in the Great Lakes region. MS thesis, Northern Illinois University, DeKalb.

Rodda, G.H., and E.W. Campbell. 2002. Distance sampling of forest snakes and lizards. Herpetol. Rev. 33:271–274.

Rodda, G.H., and T.H. Fritts. 1992a. The impact of the introduction of the colubrid snake, *Boiga irregularis* on Guam's lizards. J. Herpetol. 26:166–174.

Rodda, G.H., and T.H. Fritts. 1992b. Sampling techniques for an arboreal snake, *Boiga irregularis*. Micronesica 25:23–40.

Rodda, G.H., T.H. Fritts, and P.J. Conry. 1992. Origin and population growth of the Brown Tree Snake, *Boiga irregularis*, on Guam. Pacific Sci. 46:46–57.

Rodda, G.H., T.H. Fritts, C.S. Clark, S.W. Gotte, and D. Chiszar. 1999a. A state-of-the-art trap for the Brown Treesnake. *In* G.H. Rodda, Y. Sawai, D. Chiszar, and H. Tanaka, eds., Problem snake management: The habu and the Brown Treesnake, pp. 285–305. Ithaca: Cornell University Press.

Rodda, G.H., T.H. Fritts, M.J. McCoid, and E.W. Campell. 1999b. An overview of the biology of the Brown Treesnake (*Boiga irregularis*), a costly introduced pest on Pacific Islands. *In* G.H. Rodda, Y. Sawai, D. Chiszar, and H. Tanaka, eds., Problem snake management: The habu and the Brown Treesnake, pp. 44–80. Ithaca: Cornell University Press.

Rodda, G.H., T.H. Fritts, M.J. McCoid, and E.W. Campell. 1999c. Population trends and limiting factors in *Boiga irregularis. In* G.H. Rodda, Y. Sawai, D. Chiszar, and H. Tanaka, eds., Problem snake management: The habu and the Brown Treesnake, pp. 236–253. Ithaca: Cornell University Press.

Rodda, G.H., Y. Sawai, D. Chiszar, and H. Tanaka. 1999d. Problem snake management: The habu and the Brown Treesnake. Ithaca: Cornell University Press.

Rodda, G.H., E.W.I. Campbell, T.H. Fritts, and C.S. Clark. 2005. The predictive power of visual searching. Herpetol. Rev. 36:259–264.

Rodda, G.H., J.L. Farley, R. Bischof, and R.N. Reed. 2007a. New developments in snake barrier technology: fly-ash covered wall offers a feasible alternative for permanent barriers to brown treesnakes (*Boiga irregularis*). Herpetol. Conserv. Biol. 2:157–163.

Rodda, G.H., J.A. Savidge, C.L. Tyrrell, M.T. Christy, and A.R. Ellingson. 2007b. Size bias in visual searches and trapping of Brown Treesnakes on Guam. J. Wildl. Manage. 71:656–661.

Rodriguez, A., G. Crema, and M. Delibes. 1996. Use of non-wildlife passages across a high speed railway by terrestrial vertebrates. J. Appl. Ecol. 33:1527–1540.

Rodríguez, F., J.F. Oliver, A. Marín, and J.R. Medina. 1990. The general stochastic model of nucleotide substitution. J. Theor. Biol. 142:485–501.

Rodriquez-Robles, J.A. 1998. Alternative perspectives on the diet of gopher snakes (*Pituophis catenifer*, Colubridae): literature records versus stomach contents of wild and museum speciemens. Copeia 1998:463–466.

Rodríguez-Robles, J.A., and J.M. de Jesús-Escobar. 1999. Molecular systematics of New World lampropeltinine snakes (Colubridae): Implications for biogeography and evolution of food habits. Biol. J. Linn. Soc. 68:355–385.

Rodríguez-Robles, J.A., G.R. Stewart, and T.J. Papenfuss. 2001. Mitochondrial DNA-based phylogeography of North American rubber boas, *Charina bottae* (Serpentes: Boidae). Mol. Phylogenet. Evol. 18:227–237.

Rodríguez-Robles, J.A., T. Jezkova, and M.A. García. 2007. Evolutionary relationships and historical biogeography of *Anolis desechensis* and *A. monensis,* two lizards endemic to small islands in the eastern Caribbean Sea. J. Biogeogr. 34:1546–1558.

Roe, J.H., and A. Georges. 2007. Heterogeneous wetland complexes, buffer zones, and travel corridors: Landscape management for freshwater reptiles. Biol. Conserv. 135:67–76.

Roe, J.H., B.A. Kingsbury, and N.R. Herbert. 2003. Wetland and upland use patterns in semi-aquatic snakes: Implications for wetland conservation. Wetlands 23:1003–1014.

Roe, J.H., B.A. Kingsbury, and N.R. Herbert. 2004. Comparative water snake ecology: Conservation of mobile animals that use temporally dynamic resources. Biol. Conserv. 118:79–89.

Roe, J.H., J. Gibson, and B.A. Kingsbury. 2006. Beyond the wetland border: Estimating the impact of roads for two species of water snakes. Biol. Conserv. 130:161–168.

Rogers, A.R. 2001. Recent telemetry technology. *In* J.J. Millspaugh and J.M. Marzluff, eds., Radio tracking and animal populations, pp. 79–121. San Diego: Academic Press.

Rogers, A.R., and H. Harpending. 1992. Population growth makes waves in the distribution of pairwise genetic differences. Mol. Biol. Evol. 9:552–569.

Rogers, A.R., A.E. Fraley, M.J. Bamshad, W.S. Watkins, and L.B. Jorde. 1996. Mitochondrial mismatch analysis is insensitive to the mutational process. Mol. Biol. Evol. 13:895–902.

Ronquist, F., and J.P. Huelsenbeck. 2003. MrBayes 3: Bayesian phylogenetic inference under mixed models. Bioinformatics 19:1572–1574.

Root, T.L., J.T. Price, K.R. Hall, S.H. Schneider, C. Rosenzweig, and J.A. Pounds. 2003. Fingerprints of global warming on wild animals and plants. Nature 421:57–60.

Rose, F.L., and K.W. Selcer. 1989. Genetic divergence of the allopatric populations of *Nerodia harteri*. J. Herpetol. 23:261–267.

Rosen, P.C., and C.H. Lowe. 1994. Highway mortality of snakes in the Sonoran Desert of southern Arizona. Biol. Conserv. 68:143–148.

Rosenberg, N.A., and M. Nordborg. 2002. Genealogical trees, coalescent theory, and the analysis of genetic polymorphisms. Nat. Rev. Genet. 3:380–390.

Rosenblum, E.B., H.E. Hoekstra, and M.W. Nachman. 2004. Adaptive reptile color variation and the evolution of the Mc1r gene. Evolution 58:1794–1808.

Ross, B., T.S. Fredericksen, E. Ross, W. Hoffman, M.L. Morrison, J. Beyea, M.B. Lester, B.N. Johnson, and N.J. Fredericksen. 2000. Relative abundance and species richness of herpetofauna in forest stands in Pennsylvania. For. Sci. 46:139–146.

Rossman, D.E., N.B. Ford, and R.A. Seigel. 1996. The garter snakes: Evolution and ecology. Norman: University of Oklahoma Press.

Row, J.R., and G. Blouin-Demers. 2006. Thermal quality influences habitat selection at multiple spatial scales in Milksnakes. Ecoscience 13: 443–450.

Row, J.R., G. Blouin-Demers, and P.J. Weatherhead. 2007. Demographic effects of road mortality in Black Ratsnakes (*Elaphe obsoleta*). Biol. Conserv. 137:117–124.

Rowe, K.C., E.J. Heske, and K.N. Paige. 2006. Comparative phylogeography of eastern chipmunks and white-footed mice in relation to the individualistic nature of species. Mol. Ecol. 15:4003–4020.

Rozas, J., J.C. Sánchez-DelBarrio, X. Messeguer, and R. Rozas. 2006. DnaSP (ver. 4.10.8). Universitat de Barcelona, Barcelona.

Russell, F. 1980. Snake venom poisoning. Great Neck, N.Y.: Scholium International.

Russell, K.R., D.H. Van Lear, and D.C. Guynn Jr. 1999. Prescribed fire effects on herpetofauna: Review and management implications. Wildl. Soc. Bull. 27:374–384.

Russell, K.R., T.B. Wigley, W.M. Baughman, H.G. Hanlin, and W.M. Ford. 2004. Responses of southeastern amphibians and reptiles to forest management: A review. *In* H.M. Rauscher and K. Johnsen, eds. Southern forest science: Past, present, and future. General Technical Report SRS-75, pp. 319–334. Asheville, N.C.: U.S. Forest Service, Southern Research Station.

Ruston, S.P., S.J. Ormerod, and G. Kerby. 2004. New paradigms for modeling species distributions? J. Appl. Ecol. 41:193–200.

Rye, L.A. 2000. Analysis of area of intergradation between described subspecies of the Common Garter Snake, *Thamnophis sirtalis*, in Canada. PhD dissertation, University of Guelph, Guelph.

Safina, C., and J. Burger. 1983. Effects of a human disturbance on reproductive success in the black skimmer. Condor 85:164–171.

Sage, J.R. 2005. Spatial ecology, habitat utilization, and hibernation ecology of the eastern massasauga (*Sistrurus catenatus catenatus*) in a disturbed landscape. MS thesis, Purdue University, Fort Wayne.

Saint Girons, H., and E. Kramer. 1963. Le cycle sexuel chez Vipera berus *(L.) en montagne*. Rev. Suisse Zool. 70:191–221.

Saint Girons, H., and P. Pfeffer. 1971. *Le cycle sexual des serpentes du Cambodge*. Ann. Sci. Nat. Zool. (Paris), 12th ser. 13:543–572.

Sajdak, R.A., and R.W. Henderson. 1991. Status of West Indian racers in the Lesser Antilles. Oryx 25:33–38.

Sakai, A.K., F.W. Allendorf, J.S. Holt, D.M. Lodge, J. Molofsky, K.A. With, S. Baughman, R.J. Cabin, J.E. Cohen, N.C. Ellstrand, D.E. McCauley, P. O'Neil, I.M. Parker, J.N. Thompson, and S.G. Weller. 2001. The population biology of invasive species. Annu. Rev. Ecol. Syst. 32:305–332.

Sambrook, J., E.F. Fritsch, and T. Maniatis. 2001. Molecular cloning: A laboratory manual. 3rd ed. New York: Cold Spring Harbor Laboratory Press.

Sanderson, E.W., K.H. Redford, A. Vedder, P.B. Coppolillo, and S.E. Ward. 2002. A conceptual model for conservation planning based on landscape species requirements. Landsc. Urban Plan. 58:41–56.

Sanderson, M.J. 2002. Estimating absolute rates of molecular evolution and divergence times: A penalized likelihood approach. Mol. Biol. Evol. 19:101–109.

Sanderson, M.J. 2003. r8s: Inferring absolute rates of evolution and divergence times in the absence of a molecular clock. Bioinformatics 19:301–302.

Sanz, L., H.L. Gibbs, S.P. Mackessy, and J.J. Calvete. 2006. Venom proteomes of closely related *Sistrurus* rattlesnakes with divergent diets. J. Proteome Res. 5:2098–2112.

Sarrazin, F., and S. Legendre. 2000. Demographic approach to releasing adults versus young in reintroductions. Conserv. Biol. 14:488–500.

Sasa, M. 1997. *Cerrophidion godmani* in Costa Rica: A case of extremely low allozyme variation. J. Herpetol. 31:569–572.

Sasa, M., and R. Barrantes. 1998. Allozyme variation in populations of *Bothrops asper* (Serpentes: Viperidae) in Costa Rica. Herpetologica 54:462–469.

Sattler, P., and S. Guttman. 1976. An electrophoretic analysis of *Thamnophis sirtalis* from western Ohio. Copeia 1976:352–356.

Savidge, J.A. 1987. Extinction of an island forest avifauna by an introduced predator. Ecology 68:660–668.

Savidge, J.A. 1991. Population characteristics of the introduced Brown Tree Snake (*Boiga irregularis*) on Guam. Biotropica 23:294–300.

Schaafsma, W., and G.N. van Vark. 1979. Classification and discrimination problems with applications, Pt. IIa. Stat. Neerlandica 33:91–126.

Schaeffer, D.J., E.E. Herricks, and H.W. Kerster. 1988. Ecosystem health: I. Measuring ecosystem health. Environ. Manage. 12:445–455.

Schindler, D.E., and S.C. Lubetkin. 2004. Using stable isotopes to quantify material transport in food webs. *In* G.A. Polis, M.E. Power, and G.R. Huxel, eds., Food webs at the landscape level, pp. 25–42. Chicago: University of Chicago Press.

Schlaepfer, M.A., M.C. Runge, and P.W. Sherman. 2002. Ecological and evolutionary traps. Trends Ecol. Evol. 17:474–480.

Schlaepfer, M.A., C. Hoover, and C.K. Dodd Jr. 2005. Challenges in evaluating the impact of the trade in amphibians and reptiles on wild populations. BioScience 55:256–264.

Schmidt, A.D., and S. Lenz. 2001. *Bericht zum Stand der erprobungs- und entwicklungsvorhaben "Würfelnatter"der DGHT*, Pt. 1: *Erprobungsstandort Elbe*. Elaphe 3:60–66.

Schmitz, O.J., P.A. Hambäck, and A.P. Beckerman. 2000. Trophic cascades in terrestrial systems: A review of the effects of carnivore removals on plants. Am. Nat. 155:141–153.

Schneider, C.J., M. Cunningham, and C. Moritz. 1998. The comparative phylogeography and the history of endemic vertebrates in the Wet Tropics rainforests of Australia. Mol. Ecol. 7:487–498.

Schwaner, T.D. 1990. Geographic variation in scale and skeletal anomalies of Tiger Snakes (Elapidae: *Notechis scutatus-ater* complex) in Southern Australia. Copeia 1990:1168–1173.

Schwartz, M.K., G. Luikart, and R.S. Waples. 2007. Genetic monitoring as a promising tool for conservation and management. Trends Ecol. Evol. 22:25–33.

Schwarzkopf, L., and R. Shine. 1992. Costs of reproduction in lizards: Escape tactics and susceptibility to predation. Behav. Ecol. Sociobiol. 31:17–25.

Scott, I.A.W., C.M. Hayes, J.S. Keogh, and J.K. Webb. 2001. Isolation and characterization of novel microsatellite markers from the Australian Tiger Snakes (Elapidae: *Notechis*) and amplification in the closely related genus *Hoplocephalus*. Mol. Ecol. Notes 1:117–119.

Scott, J.M., P.J. Heglund, M.L. Morrison, J.B. Haufler, M.G. Raphael, W.A. Wall, and F.B. Samson, eds. 2002a. Predicting species occurrences: Issues of accuracy and scale. Washington, D.C.: Island Press.

Scott, J.M., C.R. Peterson, J.W. Karl, E. Strand, L.K. Svancara, and N.M. Wright. 2002b. A gap analysis of Idaho: Final report. Idaho Cooperative Fish and Wildlife Research Unit, Moscow, Idaho.

Scott, N.J., Jr., and R.A. Seigel. 1992. The management of amphibian and reptile populations: Species priorities and methodological and theoretical constraints. *In* D.R. McCullough and R.H. Barrett, eds., Wildlife 2001: Populations, pp. 343–368. London: Elsevier Science.

Sealy, J. 1997. Short-distance translocations of Timber Rattlesnakes in a North Carolina state park: A successful conservation and management program. Sonoran Herpetol. 10:94–99.

Seba, A. 1734–1735. *Locupletissimi rerum naturalium thesauri,* Vol. 1 (1734), Vol. 2 (1735). Amsterdam: Janssonio-Waesbergios, J. Wetstenium, and Gul Smith.

Seber, G.A.F. 1982. The estimation of animal abundance and related parameters. 2nd ed. New York: Macmillan.

Secor, S.M., and J. Diamond. 1998. A vertebrate model of extreme physiological regulation. Nature 395:659–662.

Secor, S.M., and K.A. Nagy. 1994. Bioenergetic correlates of foraging mode for the snakes *Crotalus cerastes* and *Masticophis flagellum.* Ecology 75:1600–1614.

Seebacher, F. 2005. A review of thermoregulation and physiological performance in reptiles: What is the role of phenotypic flexibility? J. Comp. Physiol. B Biochem. Syst. Environ. Physiol. 175:453–461.

Seigel, R.A. 1993. Summary: Future research on snakes, or how to combat "lizard envy." *In* R.A. Seigel and J.T. Collins, eds., Snakes: Ecology and behavior, pp. 395–402. New York: McGraw-Hill.

Seigel, R.A., and J.T. Collins, eds. 1993. Snakes: Ecology and behavior. New York: McGraw-Hill.

Seigel, R.A., and C.K. Dodd Jr. 2000. Manipulation of turtle populations for conservation: Halfway technologies or viable options? *In* M.W. Klemens, ed., Turtle conservation, pp. 218–238. Washington, D.C.: Smithsonian Institution Press.

Seigel, R.A., and H.S. Fitch. 1985. Annual variation in reproduction in snakes in a fluctuating environment. J. Anim. Ecol. 54:497–505.

Seigel, R.A., and N.B. Ford. 1987. Reproductive ecology. *In* R.A. Seigel, J.T. Collins, and S.S. Novak, eds., Snakes: Ecology and evolutionary biology, pp. 210–252. New York: McGraw-Hill.

Seigel, R.A., and N.B. Ford. 1991. Phenotypic plasticity in the reproductive characteristics of an oviparous snake, *Elaphe guttata:* Implications for life history studies. Herpetologica 47:301–307.

Seigel, R.A., and C.A. Sheil. 1999. Population viability analysis: Applications for the conservation of massasaugas. *In* B. Johnson and M. Wright, eds., Second international symposium and workshop on the conservation of the Eastern Massasauga Rattlesnake, *Sistrurus catenatus catenatus:* Population and habitat management issues in urban, bog, prairie, and forested ecosystems, pp. 17–22. Toronto: Toronto Zoo.

Seigel, R.A., M.M. Huggins, and N.B. Ford. 1987. Reduction in locomotor ability as a cost of reproduction in gravid snakes. Oecologia 73:481–485.

Seigel, R.A., J.W. Gibbons, and T.K. Lynch. 1995. Temporal changes in reptile populations: Effects of a severe drought on aquatic snakes. Herpetologica 51:424–434.

Seigel, R.A., C.A. Sheil, and J.S. Doody. 1998. Changes in a population of an endangered rattlesnake *Sistrurus catenatus* following a severe flood. Biol. Conserv. 83:127–131.

Seigel, R.A., R.B. Smith, and N.A. Seigel. 2003. Swine flu or 1918 pandemic? Upper respiratory tract disease and the sudden mortality of gopher tortoises (*Gopherus polyphemus*) on a protected habitat in Florida. J. Herpetol. 37:137–144.

Semlitsch, R.D. 1987a. Density-dependent growth and fecundity in the paedomorphic salamander *Ambystoma talpoideum.* Ecology 68:1003–1008.

Semlitsch, R.D. 1987b. Paedomorphosis in *Ambystoma talpoideum:* Effects of density, food, and pond drying. Ecology 68:994–1002.

Semlitsch, R.D., and J.W. Gibbons. 1985. Phenotypic variation in metamorphosis and paedomorphosis in the salamander *Ambystoma talpoideum*. Ecology 66:1123–1130.

Setser, K., and J.F. Cavitt. 2003. Effects of burning on snakes in Kansas, USA, tallgrass prairie. Nat. Areas J. 23:315–319.

Severini, M.A. 1651. *Vipera pythia*. Patavii: Typis Pauli Frambotti Bibliop.

Seyfarth, R.M., and D.L. Cheney. 1980. The ontogeny of vervet monkey alarm-calling behavior: A preliminary report. Z. Tierpsychol. 54:37–56.

Seyfarth, R.M., D.L. Cheney, and P. Marler. 1980. Monkey responses to three different alarm calls: Evidence for predator classification and semantic communication. Science 210:801–803.

Seymour, M.M., and R.B. King. 2003. Lake Erie Water Snake (*Nerodia sipedon insularum*) recovery plan. Fort Snelling, Minn.: U.S. Fish and Wildlife Service, Region 3.

Shaffer, M.L. 1994. Population viability analysis: Determining nature's share. *In* G.K. Meffe and C.R. Carroll, eds., Principles of conservation biology, pp. 195–196. Sunderland, Mass.: Sinauer.

Shedlock, A.M., K. Takahashi, and N. Okada. 2004. SINEs of speciation: Tracking lineages with retroposons. Trends Ecol. Evol. 19:545–553.

Sheppard, S.K., and J.D. Harwood. 2005. Advances in molecular ecology: Tracking trophic links through predator-prey food-webs. Funct. Ecol. 19:751–762.

Sherbrooke, W.C. 2006. Habitat use and activity patterns of two Green Ratsnakes (*Senticolis triaspis*) in the Chiricahua Mountains, Arizona. Herpetol. Rev. 37:34–37.

Shetty, S., and R. Shine. 2002. Philopatry and homing behavior of sea snakes (*Laticauda colubrina*) from two adjacent islands in Fiji. Conserv. Biol. 16:1422–1426.

Shine, R. 1979. Activity patterns in Australian elapid snakes (Squamata: Serpentes: Elapidae). Herpetologica 35:1–11.

Shine, R. 1980. "Costs" of reproduction in reptiles. Oecologia 46:92–100.

Shine, R. 1981. Venomous snakes in cold climates: Ecology of the Australian genus *Drysdalia* (Serpentes: Elapidae). Copeia 1981:14–25.

Shine, R. 1986. Predation upon Filesnakes (*Acrochordus arafurae*) by aboriginal hunters: Selectivity with respect to size, sex and reproductive condition. Copeia 1986:238–239.

Shine, R. 1989. Constraints, allometry, and adaptation: Food habits and reproductive biology of Australian Brownsnakes (*Pseudonaja*: Elapidae). Herpetologica 45:195–207.

Shine, R. 1991. Australian snakes: A natural history. Ithaca: Cornell University Press.

Shine, R. 1994. Allometric patterns in the ecology of Australian snakes. Copeia 1994:851–867.

Shine, R. 2003. Reproductive strategies in snakes. Proc. R. Soc. Lond. B Biol. Sci. 270:995–1004.

Shine, R., and X. Bonnet. 2000. Snakes: A new "model organism" in ecological research? Trends Ecol. Evol. 15:221–222.

Shine, R., and M. Fitzgerald. 1995. Variation in mating systems and sexual size dimorphism between populations of the Australian python *Morelia spilota* (Serpentes: Pythonidae). Oecologia 103:490–498.

Shine, R., and M. Fitzgerald. 1996. Large snakes in a mosaic rural landscape: The ecology of carpet pythons, *Morelia spilota* (Serpentes: Pythonidae) in coastal eastern Australia. Biol. Conserv. 76:113–122.

Shine, R., and T. Madsen. 1996. Is thermoregulation unimportant for most reptiles?: An example using water pythons (*Liasis fuscus*) in tropical Australia. Physiol. Zool. 69:252–269.

Shine, R., and T. Madsen. 1997. Prey abundance and predator reproduction: Rats and pythons on a tropical Australian floodplain. Ecology 78:1078–1086.

Shine, R., and R.T. Mason. 2001. Serpentine cross-dressers (behaviour of Manitoban garter snake). Nat. Hist. 110:56–58.

Shine, R., and R.T. Mason. 2004. Patterns of mortality in a cold-climate population of garter snakes (*Thamnophis sirtalis parietalis*). Biol. Conserv. 120:201–210.

Shine, R., and M. Wall. 2007. Why is intraspecific niche partitioning more common in snakes than in lizards? *In* S.M. Reilly, L.D. McBrayer, and D. Miles, eds., Lizard ecology, pp. 173–208. Cambridge, U.K.: Cambridge University Press.

Shine, R., P. Harlow, J.S. Keogh, and Boeadi. 1995. Biology and commercial utilization of acrochordid snakes, with special reference to karung (*Acrochordus javanicus*). J. Herpetol. 29:352–360.

Shine, R., T.R.L. Madsen, M.J. Elphick, and P.S. Harlow. 1996. The influence of nest temperatures and maternal brooding on hatchling phenotypes in water pythons. Ecology 78:1713–1721.

Shine, R., P.S. Harlow, J.S. Keogh, and Boeadi. 1998a. The influence of sex and body size on food habits of a giant tropical snake, *Python reticulatus*. Funct. Ecol. 12:248–258.

Shine, R., J.K. Webb, M. Fitzgerald, and J. Summer. 1998b. The impact of bush-rock removal on an endangered snake species, *Hoplocephalus bungaroides* (Serpentes: Elapidae). Wildl. Res. 25:285–295.

Shine, R., Ambariyanto, P.S. Harlow, and Mumpuni. 1999. Reticulated pythons in Sumatra: Biology, harvesting and sustainability. Biol. Conserv. 87:349–357.

Shine, R., M.M. Olsson, M.P. LeMaster, I.T. Moore, and R.T. Mason. 2000. Effects of sex, body size, temperature and location on the antipredator tactics of free-ranging gartersnakes (*Thamnophis sirtalis*, Colubridae). Behav. Ecol. 11:239–245.

Shine, R., M.P. LeMaster, I.T. Moore, M.M. Olsson, and R.T. Mason. 2001. Bumpus in the snake den: Effects of sex, size, and body condition on mortality of Red-sided Garter Snakes. Evolution 55:598–604.

Shine, R., E.G. Barrott, and M.J. Elphick. 2002a. Some like it hot: Effects of forest clearing on nest temperatures of montane reptiles. Ecology 83:2808–2815.

Shine, R., R.N. Reed, S. Shetty, M. LeMaster, and R.T. Mason. 2002b. Reproductive isolating mechanisms between two sympatric sibling species of sea-snakes. Evolution 56:1655–1662.

Shine, R., L. Sun, E. Zhao, and X. Bonnet. 2002c. A review of 30 years of ecological research on the Shedao pit-viper. Herpetol. Nat. Hist. 9:1–14.

Shine, R., M.P. LeMaster, M. Wall, T. Langkilde, and R. Mason. 2004a. Why did the snake cross the road? Effects of roads on movement and mate-location by garter snakes (*Thamnophis sirtalis parietalis*). Ecol. Soc. 9(1):9. Available at: http://www.ecology andsociety.org/vol9/iss1/art.

Shine, R., B. Phillips, T. Langkilde, D.I. Lutterschmidt, H. Waye, and R.T. Mason. 2004b. Mechanisms and consequences of sexual conflict in garter snakes (*Thamnophis sirtalis*, Colubridae). Behav. Ecol. 15:654–660.

Shine, R., B. Phillips, H. Waye, M. LeMaster, and R.T. Mason. 2004c. Species isolating mechanisms in a mating system with male mate choice (garter snakes, *Thamnophis*). Can. J. Zool. 82:1091–1099.

Shine, R., T. Langkilde, M. Wall, and R.T. Mason. 2005. The fitness correlates of scalation asymmetry in garter snakes (*Thamnophis sirtalis parietalis*). Funct. Ecol. 19:306–314.

Shipman, P.A., S.F. Fox, R.E. Thill, J.P. Phelps, and D.M. Leslie Jr. 2004. Reptile communities under diverse forest management in the Ouachita Mountains, Arkansas. *In* J.M. Guldin, ed., Ouachita and Ozark Mountains Symposium: Ecosystem management research, October 26–28, 1999, Hot Springs, Arkansas. General Technical Report SRS-74, pp. 174–182. Asheville, N.C.: U.S. Forest Service, Southern Research Station.

Shoemaker, K.T. 2007. Habitat manipulation as a viable strategy for the conservation of the Massasauga Rattlesnake in New York State. MS thesis, State University of New York, Syracuse.

Siddall, M.E. 1998. Success of parsimony in the four-taxon case: Long-branch repulsion by likelihood in the Farris Zone. Cladistics 14:209–220.

Sideleva, O., N. Ananjeva, S. Ryabov, and N. Orlov. 2003. The comparison of morphological and molecular characters inheritance in family groups of rat snakes of *Elaphe* genus (Serpentes: Colubridae). Russ. J. Herpetol. 10:149–156.

Simmons, J. 2002. Herpetological collecting and collections management, rev. ed. Herpetological Ciruclars. Tyler, Tex.: Society for the Study of Reptiles and Amphibians.

Simonsen, K.L., G.A. Churchill, and C.F. Aquadro. 1995. Properties of statistical tests of neutrality for DNA polymorphism data. Genetics 141:413–429.

Simossis, V.A., and J. Heringa. 2005. PRALINE: A multiple sequence alignment toolbox that integrates homology-extended and secondary structure information. Nucleic Acids Res. 33 (Online): W289–W294.

Sinclair, E.A., R.L. Bezy, J.L.R. Camarillo, K. Bolles, K.A. Crandall, and J.W. Sites Jr. 2004. Testing species boundaries in an ancient species complex with deep phylogeographic structure: Genus *Xantusia* (Squamata: Xantusiidae). Am. Nat. 163:396–414.

Sites, J.W., Jr., and J.C. Marshall. 2004. Operational criteria for delimiting species. Annu. Rev. Ecol. Syst. 35:199–229.

Slatkin, M. 1995. A measure of population subdivision based on microsatellite allele frequencies. Genetics 139:457–462.

Slatkin, M., and R.R. Hudson. 1991. Pairwise comparisons of mitochondrial DNA sequences in stable and exponentially growing populations. Genetics 129:555–562.

Smallwood, K.S., J. Beyea, and M.L. Morrison. 1999. Using the best scientific data for endangered species conservation. Environ. Manage. 24:421–435.

Smith, B.E., and N.T. Stephens. 2003. Conservation assessment of the Pale Milk Snake in the Black Hills National Forest, South Dakota and Wyoming. Denver, Colo.: U.S. Forest Service, Rocky Mountain Region.

Smith, K.G., and J.D. Clark. 1994. Black bears in Arkansas: Characteristics of a successful translocation. J. Mammal. 75:309–320.

Smith, L.L., and C.K. Dodd Jr. 2003. Wildlife mortality on U.S. Highway 441 across Paynes Prairie, Alachua County, Florida. Fla. Sci. 66:128–140.

Smith, M.A., and D.M. Green. 2005. Dispersal and the metapopulation paradigm in amphibian ecology and conservation: Are all amphibian populations metapopulations? Ecography 28:110–128.

Smith, W.J. 1977. The behavior of communicating: An ethological approach. Cambridge, Mass.: Harvard University Press.

Snow, R.W., K.L. Krysko, K.M. Enge, L. Oberhofer, A. Warren-Bradley, and L. Wilkins. 2007. Introduced populations of *Boa constrictor* (Boidae) and *Python molurus bivittatus* (Pythonidae) in southern Florida. *In:* R. Henderson, R. Powell, G. Schuett, and M. Douglass, eds., Biology of the Boas and Pythons, pp. 416–438. Eagle Mountain, Utah: Eagle Mountain Publishing.

Sokal, R.R., and S.J. Rohlf. 1995. Biometry: The principles and practice of statistics in biological research. 3rd ed. New York: W.H. Freeman.

Soltis, D.E., A.B. Morris, J.S. McLachlan, P.S. Manos, and P.S. Soltis. 2006. Comparative phylogeography of unglaciated eastern North America. Mol. Ecol. 15:4261–4293.

Soulé, M.E. 1991. Conservation: Tactics for a constant crisis. Science 253:744–750.

Souter, N.J., C.M. Bull, and M.N. Hutchinson. 2004. Adding burrows to enhance a population of the endangered pygmy blue tongue lizard, *Tiliqua adelaidensis*. Biol. Conserv. 116:403–408.

Sparks, T.H., and T.J. Yates. 1997. The effect of spring temperature on the appearance dates of British butterflies 1883–1993. Ecography 20:368–374.

Speakman, J.R. 1997. Doubly labelled water: Theory and practice. London: Chapman & Hall.

Spear, S.F., C.R. Peterson, M.D. Matocq, and A. Storfer. 2005. Landscape genetics of the blotched tiger salamander *Ambystoma tigrinum melanostictum*. Mol. Ecol. 14:2553–2564.

Spellerberg, I.F. 1975. Conservation and management of Britain's reptiles based on their ecological and behavioural requirements: A progress report. Biol. Conserv. 7:289–300.

Spellerberg, I.F. 1988. Ecology and management of reptile populations in forests. Q. J. For. 82:99–109.

Spetich, M.A. 2002. Upland oak ecology symposium: A synthesis. In M.A. Spetich, ed., Upland oak ecology symposium: History, current conditions, and sustainability, pp. 3–9. Fayetteville, Ark.: U.S. Forest Service, Southern Research Station.

Stafford, D.P., F.W. Plapp Jr., and R.R. Fleet. 1977. Snakes as indicators of environmental contamination: relation of detoxifying enzymes and pesticide residues to species occurrence in three aquatic ecosystems. Arch. Environ. Contam. Toxicol. 5:15–27.

Stake, M.M., and D.A. Cimprich. 2003. Using video to monitor predation at black-capped vireo nests. Condor 105:348–357.

Stake, M.M., J. Faaborg, and F.R. Thompson. 2004. Video identification of predators at golden-cheeked warbler nests. J. Field Ornithol. 74:337–344.

Standora, M.M. 2002. Landscape level GIS modeling of Eastern Massasauga Rattlesnake (Sistrurus catenatus catenatus) habitat in Michigan. MS thesis, Purdue University, Fort Wayne.

Standora, M.M., and B. Kingsbury. 2002. Using GIS to model habitat for the Eastern Massasauga Rattlesnake in Michigan. Final report to the Michigan Department of Natural Resources, Lansing, Mich.

Stanford, K.M., and R.B. King. 2004. Growth, survival and reproduction in a northern Illinois population of the Plains Gartersnake, Thamnophis radix. Copeia 2004:465–478.

Stanley Price, M.R. 1989. Animal re-introductions: The Arabian oryx in Oman. Cambridge, U.K.: Cambridge University Press.

Stapley, J., C.M. Hayes, J.K. Webb, and J.S. Keogh. 2005. Novel microsatellite loci identified from the Australian Eastern Small-eyed Snake (Elapidae: Rhinocephalus nigrescens) and cross species amplification in the related genus Suta. Mol. Ecol. Notes 5:54–56.

Steele, C.A., and A. Storfer. 2006. Coalescent-based hypothesis testing supports multiple Pleistocene refugia in the Pacific Northwest for the Pacific Giant Salamander (Dicamptodon tenebrosus). Mol. Ecol. 15:2477–2487.

Steen, D.A., M.J. Aresco, S.G. Beilke, B.W. Compton, E.P. Condon, C.K. Dodd Jr., H. Forrester, J.W. Gibbons, J.L. Greene, G. Johnson, T.A. Langen, M.J. Oldham, D.N. Oxier, R.A. Saumure, F.W. Schueler, J.M. Sleeman, L.L. Smith, J.K. Tucker, and J.P. Gibbs. 2006. Relative vulnerability of female turtles to road mortality. Anim. Conserv. 9:269–273.

Stevens, R.A. 1973. A report on the lowland viper, Atheris superciliaris (Peters), from the Lake Chilwa floodplain of Malawi. Arnoldia 22:1–22.

Stevenson, R.D. 1985. Body size and limits to the daily range of body temperature in terrestrial ectotherms. Am. Nat. 125:102–117.

Stevenson, R.D., and W.A. Woods Jr. 2006. Condition indices for conservation: New uses for evolving tools. Integrat. Comp. Biol. 46:1169–1190.

Stockwell, C.A., and M.V. Ashley. 2004. Rapid adaptation and conservation. Conserv. Biol. 18:272–273.

Stockwell, C.A., A.P. Hendry, and M.T. Kinnison. 2003. Contemporary evolution meets conservation biology. Trends Ecol. Evol. 18:94–101.

Struhsaker, T.T. 1967. Auditory communication among vervet monkeys (Cercopithecus aethiops). In S.A. Altmann, ed., Social communication among primates, pp. 281–324. Chicago: University of Chicago Press.

Stuart, B.L., J. Smith, K. Davey, P. Din, and S.G. Platt. 2000. The harvest of and trade from Tonlap Sap, Cambodia. Traffic Bull. 18(3):115–124.

Sullivan, B.K. 1981. Distribution and relative abundance of snakes along a transect in California. J. Herpetol. 15:247–248.

Sullivan, B.K. 2000. Long-term shifts in a snake population: A California site revisited. Biol. Conserv. 94:321–325.

Sullivan, B.K., M.A. Kwiatkowski, and G.W. Schuett. 2004. Translocation of urban gila monsters: A problematic conservation tool. Biol. Conserv. 117:235–242.

Sun, L.X., R. Shine, Z. Debi, and T. Zhengren. 2001. Biotic and abiotic influences on activity patterns of insular pit-vipers (*Gloydius shedaoensis*, Viperidae) from northeastern China. Biol. Conserv. 97:387–398.

Šurinová, M. 1971. An analysis of the popularity of animals. Int. Zoo Yearb. 11:165–167.

Sutherland, W.J., A.S. Pullin, P.M. Dolman, and T.M. Knight. 2004. The need for evidence-based conservation. Trends Ecol. Evol. 19:305–308.

Swaisgood, R.R., and D.J. Shepherdson. 2005. Scientific approaches to enrichment and stereotypes in zoo animals: What's been done and where should we go next. Zoo Biol. 24:499–518.

Swofford D.L. 2000. PAUP*: Phylogenetic analysis using parsimony (*and other methods). 4th ed. Sunderland, Mass.: Sinauer.

Swofford, D.L., P.J. Waddell, J.P. Huelsenbeck, P.G. Foster, P.O. Lewis, and J.S. Rogers. 2001. Bias in phylogenetic estimation and its relevance to the choice between parsimony and likelihood methods. Syst. Biol. 50:525–539.

Szabo, P.S. 2007. Noah at the ballot box: Status and challenges. BioScience 57:424–427.

Szaro, R.C., S.J. Belfit, J.K. Aitkin, and J.N. Rinne. 1985. Impact of grazing on a riparian garter snake. *In* R.R. Johnson, C.D. Ziebell, D.R. Patton, P.F. Folliott, and E.H. Hamre, eds., Riparian ecosystems and their management: Reconciling conflicting uses. General Technical Report RM-120, pp. 359–363. Fort Collins, Colo.: U.S. Forest Service, Rocky Mountain Forest and Range Experiment Station.

Tajima, F. 1989. Statistical method for testing the neutral mutation hypothesis. Genetics 123:585–595.

Tajima, F. 1996. The amount of DNA phylogeography: Methods of evaluating and minimizing inference errors. Genetics 143:1457–1465.

Tallmon, D.A., G. Luikart, and M.A. Beaumont. 2004a. Comparative evaluation of a new effective population size estimator based on approximate Bayesian computation. Genetics 167:977–988.

Tallmon, D.A., G. Luikart, and R.S. Waples. 2004b. The alluring simplicity and complex reality of genetic rescue. Trends Ecol. Evol. 19:489–496.

Tanaka, H., Y. Hayashi, and A. Nakamura. 1999. Factors affecting annual incidence of habu bites, and how residents develop and transfer cognition of high risk sites. *In* G.H. Rodda, Y. Sawai, D. Chiszar, and H. Tanaka, eds., Problem snake management: The habu and Brown Treesnake, pp. 130–146. Ithaca: Cornell University Press.

Tavaré, S. 1986. Some probabilistic and statistical problems on the analysis of DNA sequences. *In* R.M. Miura, ed., Some mathematical questions in biology—DNA sequence analysis, pp. 57–86. Providence, R.I.: American Math Society.

Taylor, E.N., M.A. Malawy, D.M. Browning, S.V. Lemar, and D.F. DeNardo. 2005. Effects of food supplementation on the physiological ecology of female Western Diamond-Backed Rattlesnakes (*Crotalus atrox*). Oecologia 144:206–213.

Taylor, S.S., I.G. Jamieson, and D.P. Armstrong. 2005. Successful island reintroductions of New Zealand robins and saddlebacks with small numbers of founders. Anim. Conserv. 8:415–420.

Templeton, A.R. 1998. Nested clade analysis of phylogeographic data: Testing hypotheses about gene flow and population history. Mol. Ecol. 7:381–397.

Templeton, A.R. 2004. Statistical phylogeography: Methods of evaluating and minimizing inference errors. Mol. Ecol. 13:789–809.

Templeton, A.R., E. Routman, and C.A. Phillips. 1995. Separating population structure from population history: A cladistic analysis of the geographical distribution of mitochondrial DNA haplotypes in the tiger salamander, *Ambystoma tigrinum*. Genetics 140:767–782.

Terborgh, J., L. Lopez, V.P. Nunez, M. Rao, G. Shahabuddin, G. Orihuela, M. Riveros, R. Ascanio, G.H. Adler, T.D. Lambert, and L. Balbas. 2001. Ecological meltdown in predator-free forest fragments. Science 294:1923–1926.

Thomas, J.W. 1982. Needs for and approaches to wildlife habitat assessment. Trans. N. Am. Wildl. Nat. Res. Conf. 51:203–214.

Thompson, C.M., and K. McGarigal. 2002. The influence of research scale on bald eagle habitat selection along the lower Hudson River, New York. Landsc. Ecol. 17: 569–586.

Thompson, F.R., III, and D.E. Burhans. 2004. Differences in predators of artificial and real songbird nests: Evidence of bias in artificial nest studies. Conserv. Biol. 18:373–380.

Thompson, J.S., and B.I. Crother. 1998. Allozyme variation among disjunct populations of the Florida Green Watersnake (*Nerodia floridana*). Copeia 1998:715–719.

Thompson, L.M., F.T. van Manen, and T.L. King. 2005. Geostatistical analysis of allele presence patterns among American black bears in eastern North Carolina. Ursus 16:59–69.

Thorne, J.L., and H. Kishino. 2002. Divergence time and evolutionary rate estimation with multilocus data. Syst. Biol. 51:689–702.

Thorne, J.L., and H. Kishino. 2005. Estimation of divergence times from molecular sequence data. *In* R. Nielsen, ed., Statistical methods in molecular evolution, pp. 235–256. New York: Springer Verlag.

Toda, M., M. Nishida, M.-C. Tu, T. Hikida, and H. Ota. 1999. Genetic variation, phylogeny and biogeography of the pit vipers of the Genus *Trimeresurus sensu lato* (Reptilia: Viperidae) in the subtropical East Asian islands. *In* H. Ota, ed., Tropical island herpetofauna: Origin, current diversity, and conservation, pp. 249–270. Amsterdam: Elsevier.

Todd, B.D., and B.B. Rothermel. 2006. Assessing quality of clearcut habitats for amphibians: Effects on abundances versus vital rates in the southern toad (*Bufo terrestris*). Biol. Conserv. 133:178–185.

Todd, B.D., C.T. Winne, J.D. Willson, and J.W. Gibbons. 2007. Getting the drift: Effects of timing, trap type, and taxon on herpetofaunal drift fence surveys. Am. Midl. Nat. 158:292–305.

Topsell, E. 1658. The history of four-footed beasts and serpents and insects, vol. 2: The history of serpents. London: G. Sawbridge.

Tracy, C.R., and K.A. Christian. 1986. Ecological relations among space, time and thermal niche axes. Ecology 67:609–615.

Trani, M.K. 2002a. The influence of spatial scale on landscape pattern description and wildlife habitat assessment. *In* J.M. Scott, P.J. Heglund, M.L. Morrison, J.B. Haufler, M.G. Raphael, W.A. Wall, and F.B. Samson, eds., Predicting species occurrences: Issues of accuracy and scale. Washington, D.C.: Island Press.

Trani, M.K. 2002b. TERRA-5: Maintaining species in the south. Southern forest resource assessment draft report. Asheville, N.C.: U.S. Forest Service, Southern Research Station.

Travis, E.K., F.H. Vargas, J. Merkel, N. Gottdenker, R.E. Miller, and P.G. Parker. 2006. Hematology, plasma chemistry, and serology of the flightless cormorant (*Phalacrocorax harrisi*) in the Galápagos Islands, Ecuador. J. Wildl. Dis. 42:133–141.

Truett, J.C., J.L.D. Dullum, M.R. Matchett, E. Owens, and D. Seery. 2001. Translocating prairie dogs: A review. Wildl. Soc. Bull. 29:863–872.

Tscharntke, T., A.M. Klein, A. Kruess, I. Steffan-Dewenter, and C. Thies. 2005. Landscape perspectives on agricultural intensification and biodiversity—ecosystem service management. Ecol. Lett. 8:857–874.

Tuberville, T.D., J.R. Bodie, J.B. Jensen, L. Laclaire, and G. J. Whitfield. 2000. Apparent decline of the Southern Hog-nosed Snake, *Heterodon simus*. J. Elisha Mitchell Sci. Soc. 116:19–40.

Tuberville, T.D., E.E. Clark, K.A. Buhlmann, and J.W. Gibbons. 2005. Translocation as a conservation tool: Site fidelity and movement of repatriated gopher tortoises (*Gopherus polyphemus*). Anim. Conserv. 8:349–358.

Turner, F.B. 1977. The dynamics of populations of squamates, crocodilians and rhynchocephalians. *In* C. Gans and D.W. Tinkle, eds., Biology of the Reptilia, pp. 157–264. New York: Academic Press.

Turner, M.G. 2005. Landscape ecology in North America: Past, present, and future. Ecology 86:1967–1974.

Turner, M.G., R. Gardner, and R.V. O'Neil. 2001. Landscape ecology in theory and practice: Pattern and process. New York: Springer-Verlag.

Tzika, A.C., S. Koenig, R. Miller, G. Garcia, C. Remy, and M.C. Milinkovitch. 2008a. Population structure of an endemic vulnerable species, the Jamaican Boa (*Epicrates subflavus*). Mol. Ecol. 17:533–544.

Tzika, A.C., C. Remy, R. Gibson, and M.C. Milinkovitch. 2009. Molecular genetic analysis of a captive-breeding program: The vulnerable endemic Jamaican Yellow Boa. Conserv. Genet. 10:69–77.

Újvári, B., and Z. Korsós. 2000. Use of radiotelemetry on snakes: A review. Acta Zool. Acad. Sci. Hungaricae 46:115–146.

Újvári, B., Z. Korsos, and T. Pechy. 2000. Life history, population characteristics and conservation of the Hungarian meadow viper (*Vipera ursinii rakosiensis*). Amphib-Reptilia 21:267–278.

Újvári, B., T. Madsen, T. Kotenko, M. Olsson, R. Shine, and H. Wittzell. 2002. Low genetic diversity threatens imminent extinction for the Hungarian meadow viper (*Vipera ursinii rakosiensis*). Biol. Conserv. 105:127–130.

Újvári, B., T. Madsen, and M. Olsson. 2005. Discrepancy in mitochondrial and nuclear polymorphism in meadow vipers (*Vipera ursinii*) questions the unambiguous use of mtDNA in conservation studies. Amphib-Reptilia 26:287–292.

Urban, D.L., R.V. O'Neill, and H.H. Shugart Jr. 1987. Landscape ecology: A hierarchical perspective can help scientists understand spatial patterns. BioScience 37:199–127.

Urbina-Cardona, J.N., M. Olivares-Pérez, and V.H. Reynoso. 2006. Herpetofauna diversity and microenvironment correlates across a pasture-edge-interior ecotone in tropical rainforest fragments in the Los Tuxtlas Biospere Reserve of Veracruz, Mexico. Biol. Conserv. 132:61–75.

Ursenbacher, S., M. Carlsson, V. Helfer, H. Tegelstrom, and L. Fumagalli. 2006. Phylogeography and Pleistocene refugia of the Adder (*Vipera berus*) as inferred from mitochondrial DNA sequence data. Mol. Ecol. 15:3425–3437.

Ursenbacher, S., J-C. Monney, and L. Fumagalli. 2008. Limited genetic diversity and high differentiation among the remnant Adder (*Vipera berus*) populations in the Swiss and French Jura Mountains. Conserv. Genet. 9:in press.

U.S. Fish and Wildlife Service. 2003. Lake Erie Watersnake (*Nerodia sipedon insularum*) recovery plan. U.S. Fish and Wildlife Service, Fort Snelling, Minn. Available at: http://midwest.fws.gov/reynoldsburg/endangered/lews.html or http://ecos.fws.gov/tess_public/servlet/gov.doi.tess_public.servlets.EntryPage.

Utiger, U., N. Helfenberger, B. Schätti, C. Schmidt, M. Ruf, and V. Ziswiler. 2002. Molecular systematics and phylogeny of Old and New World ratsnakes, *Elaphe* auct., and related genera (Reptilia, Squamata, Colubridae). Russ. J. Herpetol. 9:105–124.

Van Mierop, L.H.S., and S.M. Barnard. 1978. Further observations on thermoregulation in the brooding female *Python molurus bivittatus* (Serpentes: Boidae). Copeia 1978:615–621.

van Veller, M.G.P., M. Zandee, and D.J. Kornet. 1999. Two requirements for obtaining valid common patterns under different assumptions in vicariance biogeography. Cladistics 15:393–406.

Verner, J., M.L. Morrison, and C.J. Ralph. 1986. Introduction. In J. Verner, M.L. Morrison, and C.J. Ralph, eds., Wildlife 2000: Modeling habitat relationships of terrestrial vertebrates, pp. xi–xv. Madison: University of Wisconsin Press.

Vickery, P.D., J.R. Herkert, F.L. Knopf, J. Ruth, and C.E. Keller. 1999. Grassland birds: An overview of threats and recommended management strategies. In Cornell Lab of Ornithology, Strategies for bird conservation: The partners in flight planning process. Available at:,. http://birds.cornell.edu/pifcapemay/ (accessed March 5, 2007).

Vidal, N., and S.B. Hedges. 2002. Higher-level relationships of caenophidian snakes inferred from four nuclear and mitochondrial genes. C.R. Biologies 325:987–995.

Vidal, N., and S.B. Hedges. 2004. Molecular evidence for a terrestrial origin of snakes. Proc. R. Soc. Lond. B (Suppl.) 271:S226–S229.

Villarreal, X., J. Bricker, H.K. Reinert, L. Gelbert, and L.M. Bushar. 1996. Isolation and characterization of microsatellite loci for use in population genetic analysis in the Timber Rattlesnake, *Crotalus horridus*. J. Hered. 87:152–155.

Virginian-Pilot. 1994. Habitat study: A snake in the grass? September 8, 2004. Hampton Roads, Virginia: Landmark Communications, Inc.

Vitousek, P.M., C.M. D'Antonio, and L.L. Loope. 1996. Biological invasions as a global environmental change. Am. Sci. 84:218–228.

Vitt, L.J., T.C.S. Avila-Pires, J.P. Caldwell, and V.R.L. Oliveira. 1998. The impact of individual tree harvesting on thermal environments of lizards in Amazonian rain forest. Conserv. Biol. 12:654–664.

Vogt, V. 2006. *Har funnet verdens eldste rituelle handling: Tilba Pytonslangen for 70 00 år siden* [World's oldest ritual discovered: Worshipped the python 70,000 years ago]. Apollon 2006:27–31.

Voris, H.K., and H.H. Voris. 1995. Commuting on the tropical tides: The life of the yellow-lipped sea krait. Ocean Realm April:57–61.

Vucetich, J.A., and T.A. Waite. 2000. Is one migrant per generation sufficient for the genetic management of fluctuating populations? Anim. Conserv. 3:261–266.

Vucetich, J.A., and T.A. Waite. 2001. Migration and inbreeding: The importance of recipient population size for genetic management. Conserv. Genet. 2:167–171.

Wakely, J. 2007. Coalescent theory: An introduction. Greenwood Village, Colo.: Roberts and Co.

Waldron, J.L., J.D. Lanham, and S.H. Bennett. 2006. Using behaviorally-based seasons to investigate Canebrake Rattlesnake (*Crotalus horridus*) movement patterns and habitat selection. Herpetologica 62:389–398.

Walk, R.D. 1956. Self ratings of fear in a fear-invoking situation. J. Abnorm. Soc. Psychol. 52:171–178.

Walther, E., and T. Langer. 2008. Attitude formation and change through association: An evaluative conditioning account. *In* W.D. Crano and R. Prislin, eds., Attitudes and attitude change, pp. 87–110. New York: Psychology Press.

Walther, G.-R., E. Post, P. Convey, A. Menzel, C. Parmesan, T.J.C. Beebee, J.-M. Fromentin, O. Hoegh-Guldberg, and F. Bairlein. 2002. Ecological responses to recent climate change. Nature 419:389–395.

Wang, J. 2004. Application of the one-migrant-per-generation rule to conservation and management. Conserv. Biol. 18:332–343.

Waples, R.S. 1989. A generalized approach for estimating effective population size from temporal changes in allele frequency. Genetics 121:379–391.

Waser, P., and C. Strobeck. 1998. Genetic signatures of interpopulation dispersal. Trends Ecol. Evol. 13:43–44.

Weary, G.C. 1969. An improved method of marking snakes. Copeia 1969:854–855.

Weatherhead, P.J., and G. Blouin-Demers. 2004a. Long-term effects of radiotelemetry on Black Ratsnakes. Wildl. Soc. Bull. 32:900–906.

Weatherhead, P.J., and G. Blouin-Demers. 2004b. Understanding avian nest predation: Why ornithologists should study snakes. J. Avian Biol. 35:185–190.

Weatherhead, P.J., and M.B. Charland. 1985. Habitat selection in an Ontario population of the snake, *Elaphe obsoleta*. J. Herpetol. 19:12–19.

Weatherhead, P.J., and I.C. Robertson. 1990. Homing to food by Black Rat Snakes (*Elaphe obsoleta*). Copeia 1990:1164–1165.

Weatherhead, P.J., F.E. Barry, G.P. Brown, and M.R.L. Forbes. 1995. Sex ratios, mating behavior and sexual size dimorphism of the Northern Water Snake, *Nerodia sipedon*. Behav. Ecol. Sociobiol. 36:301–311.

Weatherhead, P.J., G.P. Brown, M.R. Prosser, and K.J. Kissner. 1998. Variation in offspring sex ratios in the Northern Water Snake (*Nerodia sipedon*). Can. J. Zool. 76:2200–2206.

Weatherhead, P.J., G.P. Brown, M.R. Prosser, and K.J. Kissner. 1999. Factors affecting neonate size variation in Northern Water Snakes, *Nerodia sipedon*. J. Herpetol. 33:577–589.

Weatherhead, P.J., G. Blouin-Demers, and K.A. Prior. 2002. Synchronous variation and long-term trends in two populations of Black Rat Snakes. Conserv. Biol. 16:1602–1608.

Webb, J.K., and R. Shine. 1997a. A field study of spatial ecology and movements of a threatened snake species, *Hoplocephalus bungaroides*. Biol. Conserv. 82:203–217.

Webb, J.K., and R. Shine. 1997b. Out on a limb: Conservation implications of tree-hollow use by a threatened snake species (*Hoplocephalus bungaroides*: Serpentes, Elapidae). Biol. Conserv. 81:21–33.

Webb, J.K., and R. Shine. 1998a. Ecological characteristics of a threatened snake species, *Hoplocephalus bungaroides* (Serpentes, Elapidae). Anim. Conserv. 1:185–193.

Webb, J.K., and R. Shine. 1998b. Thermoregulation by a nocturnal elapid snake (*Hoplocephalus bungaroides*) in south-eastern Australia. Physiol. Zool. 71:680–692.

Webb, J.K., and R. Shine. 2000. Paving the way for habitat restoration: Can artificial rocks restore degraded habitats of endangered reptiles? Biol. Conserv. 92:93–99.

Webb, J.K., and M.J. Whiting. 2005. Why don't small snakes bask? Juvenile Broad-headed Snakes trade thermal benefits for safety. Oikos 110:515–522.

Webb, J.K., G.P. Brown, and R. Shine. 2001. Body size, locomotor speed and antipredator behaviour in a tropical snake (*Tropidonophis mairii*, Colubridae): The influence of incubation environments and genetic factors. Funct. Ecol. 15:561–568.

Webb, J.K., B.W. Brook, and R. Shine. 2002a. Collectors endanger Australia's most threatened snake, the Broad-headed Snake *Hoplocephalus bungaroides*. Oryx 36:170–181.

Webb, J.K., B.W. Brook, and R. Shine. 2002b. What makes a species vulnerable to extinction? Comparative life-history traits of two sympatric snakes. Ecol. Res. 17:59–67.

Webb, J.K., B.W. Brook, and R. Shine. 2003. Does foraging mode influence life history traits? A comparative study of growth, maturation and survival of two species of sympatric snakes from south-eastern Australia. Austr. Ecol. 28:601–610.

Webb, J.K., R.M. Pringle, and R. Shine. 2004. How do nocturnal snakes select diurnal retreat sites? Copeia 2004:919–925.

Webb, J.K., R. Shine, and K.A. Christian. 2005a. Does intraspecfic niche partitioning in native predators influence its response to an invasion by a toxic prey? Austr. Ecol. 30:201–209.

Webb, J.K., R. Shine, and R.M. Pringle. 2005b. Canopy removal restores habitat quality for an endangered snake in a fire suppressed landscape. Copeia 2005:894–900.

Webb, S.D. 1990. Historical biogeography. *In* R.L. Myers and J.J. Ewel, eds., Ecosystems of Florida, pp. 70–102. Orlando: University of Central Florida Press.

Wedekind, C. 2002. Manipulating sex ratios for conservation: Short-term risks and long-term benefits. Anim. Conserv. 5:13–20.

Weir, B.S., and C.C. Cockerham. 1984. Estimating F-statistics for the analysis of population structure. Evolution 38:1358–1370.

Weir, J. 1990. The Sweetwater rattlesnake round-up: A case study in environmental ethics. Conserv. Biol. 6:116–127.

Welsh, H.H., Jr., and S. Droege. 2001. A case for using plethodontid salamanders for monitoring biodiversity and ecosystem integrity of North American forests. Conserv. Biol. 15:558–568.

Werman, S.D. 1992. Phylogenetic relationships of central and south American pitvipers of the genus *Bothrops* (*sensu lato*): Cladistic analysis of biochemical and anatomical characters. *In* J.A. Campbell and E.D. Brodie Jr., eds., Biology of the pitvipers, pp. 21–40. Tyler, Tex.: Selva.

Whiles, M.R., and J.W. Grubaugh. 1993. Biodiversity and coarse woody debris in southern forests. *In* J.W. McMinn and D.A. Crossley Jr., eds., Biodiversity and coarse woody debris in southern forests. General Technical Report SE-94, pp. 94–100. Athens, Ga.: U.S. Forest Service, Southeastern Forest Experiment Station.

Whitaker, P.B., and R. Shine. 2000. Sources of mortality of large elapid snakes in an agricultural landscape. J. Herpetol. 34:121–128.

Whitaker, P.B., and R. Shine. 2003. A radiotelemetric study of movements and shelter-site selection by free-ranging brownsnakes (*Pseudonaja textilis*, Elapidae). Herpetol. Monogr. 17:130–144.

Whitaker, Z. 1989. Snakeman. Bombay: India Magazine Books.

White, G.C., and K.P. Burnham. 1999. Program MARK: Survival estimation from populations of marked animals. Bird Stud. 46:120–139.

White G.C., and R.A. Garrott. 1990. Analysis of wildlife radio-tracking data. San Diego: Academic Press.

White, J., and S. Barry. 1984. Families, frogs, and fun: Developing a family learning lab in a zoo, HERPlab: A case study. Washington, D.C.: Office of Education, National Zoological Park, Smithsonian Institution.

White, J., and D.L. Marcellini. 1986. HERPlab: A family learning centre at National Zoological Park. Int. Zoo Yearb. 24–25:340–343.

Whiteman, N.K., K.D. Matson, J.L. Bollmer, and P.G. Parker. 2006. Disease ecology in the Galapagos Hawk (*Buteo galapagoensis*): Host genetic diversity, parasite load and natural antibodies. Proc. R. Soc. Lond. B Biol. Sci. 273:797–804.

Whiting, M.J., J.R. Dixon, and B.D. Greene. 1997. Spatial ecology of the Concho Water Snake (*Nerodia harteri paucimaculata*) in a large lake system. J. Herpetol. 31:327–335.

Whitley, A.R., P. Spruell, and F.W. Allendorf. 2006. Can common species provide valuable information for conservation? Mol. Ecol. 15:2767–2786.

Whitlock, M.C., and D.E. McCauley. 1999. Indirect measures of gene flow and migration: $F_{ST} \neq 1/(4Nm + 1)$. Heredity 82:117–125.

Whittaker, R.H. 1970. Communities and ecosystems. New York: Macmillan.

Whittall, J.B., A. Medina-Marino, E.A. Zimmer, and S.A. Hodges. 2006. Generating single-copy nuclear gene data in a recent adaptive radiation. Mol. Phylogenet. Evol. 39:124–134.

Wicklum, D., and R.W. Davies. 1995. Ecosystem health and integrity? Can. J. Bot. 73:997–1000.

Wiens, J.A. 1989. Spatial scaling in ecology. Funct. Ecol. 3:385–397.

Wiens, J.J., and C.H. Graham. 2005. Niche conservatism: Integrating evolution, ecology, and conservation biology. Annu. Rev. Ecol. Evol. Syst. 36:519–539.

Wigley, T.B., K.V. Miller, D.S. deCalesta, and M.W. Thomas. 2000. Herbicides as an alternative to prescribed burning for achieving wildlife management objectives. *In* W.M. Ford, K.R. Russell, and C.E. Moorman, eds., The role of fire in nongame wildlife management and community restoration: Traditional uses and new directions. Proceedings of a special workshop, September 15, 2000, Nashville, Tenn. General Technical Report NE-288, pp. 124–138. Newton Square, Penn.: U.S. Forest Service, Northeastern Research Station.

Wikelski M., R.W. Kays, J. Kasdin, K. Thorup, J.A. Smith, W.W. Cochran, and G.W. Swenson Jr. 2007. Going wild—what a global small-animal tracking system could do for experimental biologists. J. Exp. Biol. 210:181–186.

Wilbur, H.M. 1976. Density-dependent aspects of metamorphosis in *Ambystoma* and *Rana sylvatica*. Ecology 57:1289–1296.

Wilbur, H.M., and J.P. Collins. 1973. Ecological aspects of amphibian metamorphosis. Science 182:1305–1314.

Wilcove, D.S., and T. Eisner. 2000. Whatever happened to natural history? Chron. Higher Educ., Sept. 14, B24.

Wilcove, D.S., D. Rothstein, J. Dubow, A. Phillips, and E. Losos. 1998. Quantifying threats to imperiled species in the United States. BioScience 48:607–615.

Wiles, G.J., C.F. Aguon, G.W. Davis, and D.J. Grout. 1995. The status and distribution of endangered animals and plants in northern Guam. Micronesia 28:31–49.

Wiles, G.J., J. Bart, R. Beck Jr., and C.F. Agoun. 2003. Impacts of the Brown Tree Snake: Patterns of decline and species persistence in Guam's avifauna. Conserv. Biol. 17:1350–1360.

Wilgers, D.J., and E.A. Horne. 2006. Effects of different burn regimes on tallgrass prairie herpetofaunal species diversity and community composition in the Flint Hills, Kansas. J. Herpetol. 40:73–84.

Wilkinson, J.A., J.L. Glenn, R.C. Straight, and J.W. Sites Jr. 1991. Distribution and genetic variation in venom A and B populations of the Mojave Rattlesnake (*Crotalus scutulatus scutulatus*) in Arizona. Herpetologica 47:54–68.

Williamson-Natesan, E.G. 2005. Comparison of methods for detecting bottlenecks from microsatellite loci. Conserv. Genet. 6:551–562.

Wills, C.A., and S.J. Beaupre. 2000. An application of randomization for detecting evidence of thermoregulation in Timber Rattlesnakes (*Crotalus horridus*) from northwest Arkansas. Physiol. Biochem. Zool. 73:325–334.

Willson, J.D., and M.E. Dorcas. 2003. Quantitative sampling of stream salamanders: Comparison of dipnetting and funnel trapping techniques. Herpetol. Rev. 34:128–130.

Willson, J.D., C.T. Winne, and L.A. Fedewa. 2005. Unveiling escape and capture rates of aquatic snakes and salamanders (*Siren* spp. and *Amphiuma means*) in commercial funnel traps. J. Freshwater Ecol. 20:397–403.

Willson, J.D., C.T. Winne, M.E. Dorcas, and J.W. Gibbons. 2006. Post-drought responses of semi-aquatic snakes inhabiting an isolated wetland: Insights on different strategies for persistence in a dynamic habitat. Wetlands 26:1071–1078.

Willson, J.D., C.T. Winne, and M.B. Keck. 2008. Empirical tests of biased body size distributions in aquatic snake captures. Copeia 2008:401–408.

Wilson, E.O. 1984. Biophilia: The human bond with other species. Cambridge, Mass.: Harvard University Press.

Wilson, E.O. 1992. The diversity of life. New York: W.W. Norton.

Wilson, E.O. 1994. Naturalist. Washington, D.C.: Island Press.

Wilson, E.O. 2006. The creation: An appeal to save life on earth. New York: W.W. Norton.

Wilson, G.A., and B. Rannala. 2003. Bayesian inference of recent migration rates using multilocus genotypes. Genetics 163:1177–1191.

Wilson, L.D., and J.R. McCranie. 2003. Herpetofaunal indicator species as measures of environmental stability in Honduras. Caribb. J. Sci. 39:50–67.

Wilson, T.P., and D. Mauger. 1999. Home range and habitat use of *Sistrurus catenatus catenatus* in eastern Will County, Illinois. *In* B. Johnson and M. Wright, eds., Second international symposium and workshop on the conservation of the Eastern Massasauga Rattlesnake, *Sistrurus catenatus catenatus:* Population and habitat management issues in urban, bog, prairie and forested ecosystems, pp. 125–134. Toronto: Toronto Zoo.

Winkle, W., and H. Hudde. 1997. Long-term trends in reproductive traits of tits (*Parus major, P. caeruleus*) and pied flycatchers (*Ficedula hypoleuca*). J. Avian Biol. 28:187–190.

Winne, C.T. 2005. Increases in capture rates of an aquatic snake (*Seminatrix pygaea*) using naturally baited minnow traps: Evidence for aquatic funnel trapping as a measure of foraging activity. Herpetol. Rev. 36:411–413.

Winne, C.T., M.E. Dorcas, and S.M. Poppy. 2005. Population structure, body size, and seasonal activity of Black Swamp Snakes (*Seminatrix pygaea*). Southeast. Nat. 4:1–14.

Winne, C.T., J.D. Willson, K.M. Andrews, and R.N. Reed. 2006a. Efficacy of marking snakes with disposable medical cautery units. Herpetol. Rev. 37:52–54.

Winne, C.T., J.D. Willson, and J.W. Gibbons. 2006b. Income breeding allows an aquatic snake (*Seminatrix pygaea*) to reproduce normally following prolonged drought-induced aestivation. J. Anim. Ecol. 75:1352–1360.

Winne, C.T., J.D. Willson, B.D. Todd, K.M. Andrews, and J.W. Gibbons. 2007. Enigmatic decline of a protected population of Eastern Kingsnakes, *Lampropeltis getula*. Copeia 2007:507–519.

Wisconsin Department of Natural Resources. 2007. Proposed incidental take authorization for the proposed development of the Target store development, city of Oak Creek,

Milwaukee County, Wisconsin. Available at: http://www.dnr.wi.gov/org/land/er/take/ target.htm (accessed June 18, 2007).

Wisdom, M.J., R.S. Holthausen, B.C. Wales, C.D. Hargis, V.A. Saab, D.C. Lee, W.J. Hann, T.D. Rich, M.M. Rowland, W.J. Murphy, and M.R. Eames. 2000. Source habitats for terrestrial vertebrates of focus in the interior Columbia Basin: Broad scale trends and management implications. General Technical Bulletin PNW-GTR-485. Portland, Ore.: U.S. Forest Service, Pacific Northwest Research Station.

Wisely, S.M., J.J. Ososky, and S.W. Buskirk. 2002. Morphological changes to black-footed ferrets (Mustela nigripes) resulting from captivity. Can. J. Zool. 80:1562–1568.

Wisely, S.M., R.M. Santymire, T.M. Livieri, P.E. Marinai, J.S. Kreeger, D.E. Wildt, and J. Howard. 2005. Environment influences morphology and development for in situ and ex situ population of the black-footed ferret (Mustela nigripes). Anim. Conserv. 8:321–328.

Wolf, C.M., B. Griffith, C. Reed, and S.A. Temple. 1996. Avian and mammalian translocations: Update and reanalysis of 1987 survey data. Conserv. Biol. 10:1142–1154.

Wolf, C.M., T. Garland, and B. Griffith. 1998. Predictors of avian and mammalian translocation success: Reanalysis with phylogentically independent contrasts. Biol. Conserv. 86:243–255.

Wolff, J.O. 1996. Population fluctuations of mast-eating rodents are correlated with production of acorns. J. Mammal. 77:850–856.

Wolin, L.R., J.M. Ordy, and A. Dillman. 1963. Monkey's fear of snakes: A study of its basis and generality. J. Genet. Psychol. 103:207–226.

Wolpe, J. 1958. Psychotherapy by reciprocal inhibition. Palo Alto, Calif.: Stanford University Press.

Wolpin, M., and J. Raines. 1966. Visual imagery, expected roles and extinction as possible factors in reducing fear and avoidance behavior. Behav. Res. Ther. 4:25–37.

Woodhouse, N., J. Rouse, and R. Black. 2002. Monitoring pre-traffic conditions on Highway 69 extension as part of the Eastern Massasauga Rattlesnake and Eastern Hog-nosed Snake–Highway 69 extension impact study. Report for the Ontario Ministry of Transportation, Toronto.

Wright, J.D. 2006. Traffic mortality of reptiles. In C. Seburn and C. Bishop, eds., Ecology, conservation and status of reptiles in Canada. Special issue. Herpetol. Conserv. 2:169–182.

Wright, J.D., and A. Didiuk. 1998. Status of the Plains Hognose Snake (Heterodon nasicus nasicus) in Alberta. Alberta Wildlife Status Report no. 15. Alberta: Alberta Environmental Protection.

Wright, S. 1931. Evolution in Mendelian populations. Genetics 16:97–256.

Wüster, W., and R.S. Thorpe. 1994. Naja siamensis, a cryptic species of venomous snake revealed by mtDNA sequencing. Experientia 50:75–79.

Wüster, W., M. da Graca Salomão, J.A. Quijada-Mascareñas, R.S. Thorpe, and Butantan-British Bothrops Systematics Project. 2002. Origin and evolution of the South American pitviper fauna: Evidence from mitochondrial DNA sequence data. In G.W. Schuett, M. Höggren, M.E. Douglas, and H.W. Greene, eds., Biology of the vipers, pp. 111–128. Salt Lake City, Utah: Eagle Mountain Publishing.

Wüster, W., J.E. Ferguson, J.A. Quijada-Mascareñas, C.E. Pook, M.G. Salomão, and R.S. Thorpe. 2005a. No rattlesnakes in the rainforests: Reply to Gosling and Bush. Mol. Ecol. 14:3619–3621.

Wüster, W., J.E. Ferguson, J.A. Quijada-Mascareñas, C.E. Pook, M.G. Salomão, and R.S. Thorpe. 2005b. Tracing an invasion: Landbridges, refugia, and the phylogeography of the neotropical rattlesnake (Serpentes: Viperidae: Crotalus durissus). Mol. Ecol. 14:1095–1108.

Wüster, W., S. Crookes, I. Inrich, Y. Mane, C.E. Pook, J.-F. Trape, and D.G. Broadley. 2007. The phylogeny of cobras inferred from mitochondrial DNA sequences: Evolution of venom spitting and the phylogeography of the African spitting cobras (Serpentes: Elapidae: Naja nigricollis complex). Mol. Phylogenet. Evol. 45:437–453.

Wylie, G.D., M.L. Casazza, L.L. Martin, and M. Carpenter. 2002. Monitoring Giant Garter Snakes at Colusa National Wildlife Refuge: 2002 progress report. Sacramento, Calif.: U.S. Forest Service, Western Ecological Research Center.

Yahner, R.H. 2004. Wildlife response to more than 50 years of vegetation maintenance on a Pennsylvania, U.S., right-of-way. J. Arboric. 30:123–126.

Yahner, R.H., W.C. Bramble, and W.R. Byrnes. 2001a. Effect of vegetation maintenance of an electric transmission right-of-way on reptile and amphibian populations. J. Arboric. 27:24–29.

Yahner, R.H., W.C. Bramble, and W.R. Byrnes. 2001b. Response of amphibian and reptile populations to vegetation maintenance of an electric transmission line right-of-way. J. Arboric. 27:215–221.

Yanes, M., J.M. Velasco, and F. Suárez. 1995. Permeability of roads and railways to vertebrates: The importance of culverts. Biol. Conserv. 71:217–222.

Yang, Z., 1996. Among-site rate variation and its impact on phylogenetic analysis. Trends Ecol. Evol. 11:367–372.

Yerkes, R.M., and A.W. Yerkes. 1936. Nature and condition of avoidance (fear) responses in chimpanzees. J. Comp. Psychol. 25:507–528.

Zaidan, F.I., and S.J. Beaupre. 2003. Effects of body mass, meal size, fast length, and temperature on specific dynamic action in the Timber Rattlesnake (*Crotalus horridus*). Physiol. Biochem. Zool. 76:447–458.

Zamudio, K.R., and H.W. Greene. 1997. Phylogeography of the bushmaster (*Lachesis muta*: Viperidae): Implications for neotropical biogeography, systematics, and conservation. Biol. J. Linn. Soc. 62:421–442.

Zamudio, K.R., K.B. Jones, and R.H. Ward. 1997. Molecular systematics of short-horned lizards: Biogeography and taxonomy of a widespread species complex. Syst. Biol. 46:284–305.

Zappalorti, R.T., and H.K. Reinert. 1994. Artificial refugia as a habitat-improvement strategy for snake conservation. *In* J.B. Murphy, K. Adler, and J.T. Collins, eds., Captive management and conservation of amphibians and reptiles, pp. 369–375. Ithaca: Society for the Study of Amphibians and Reptiles.

Zhang, D.X., and G.M. Hewitt. 1996. Nuclear integrations: Challenges for mitochondrial DNA markers. Trends Ecol. Evol. 11:247–251.

Zhang, Y., M.C. Westfall, K.C. Hermes, and M.E. Dorcas. 2008. Physiological and behavioral control of heating and cooling rates in rubber boats, *Charina bottae*. J. Therm. Biol. 33:7–11.

Zhao, E., and K. Adler. 1993. Herpetology of China. Oxford, Ohio: Society for the Study of Amphibians and Reptiles.

Zhivotovsky, L.A. 2001. Estimating divergence time and the use of microsatellite genetic distances: Impacts of population growth and gene flow. Mol. Biol. Evol. 18:700–709.

Zhou, Z., and Z. Jiang. 2004. International trade status and crisis for snake species in China. Conserv. Biol. 18:1386–1394.

Zhou, Z., and Z. Jiang. 2005. Identifying snake species threatened by economic exploitation and international trade in China. Biodiv. Conserv. 14:3525–3536.

Zink, R.M. 2002. Methods in comparative phylogeography, and their application to studying evolution in the North American aridlands. Integ. Comp. Biol. 42:953–959.

Zink, R.M., and G.F. Barrowclough. 2008. Mitochondrial DNA under siege in avian phylogeography. Mol. Ecol. 17:2107–2121.

Zuffi, M.A.L., F. Giudici, and P. Ioale. 1999. Frequency and effort of reproduction in female *Vipera aspis* from a southern population. Acta Oecol. 20:633–638.

Zweifel, R.G. 1981. Genetics of color pattern polymorphism in the California Kingsnake. J. Hered. 72:238–244.

Zweig, M.H., and G. Campbell. 1993. Receiver-operating characteristic (ROC) plots: A fundamental evaluation tool in clinical medicine. Clin. Chem. 39:561–577.

Taxonomic Index

Page numbers that are italicized indicate the occurrence of a taxon in either a table or a figure.

Subject Index

9 780801 445651